Vladimir A. Milashev

Explosion Pipes

Translated by R. E. Sorkina

With 51 Figures

Springer-Verlag
Berlin Heidelberg New York
London Paris Tokyo

Professor Dr. V. A. MILASHEV
All-Union Research Institute
for Geology and Mineral Resources
of the World Ocean
(VNIIOkeangeologia)
190121, Leningrad, USSR

Title of the original Russian edition: Trubki vzryva
© by Nedra, Leningrad 1984

ISBN 3-540-18649-2 Springer-Verlag Berlin Heidelberg New York
ISBN 0-387-18649-2 Springer-Verlag New York Berlin Heidelberg

Library of Congress Cataloging-in-Publication Data. Milashev, V. A. (Vladimir Arkad'evich).
[Trubki vzryva. English] Explosion pipes / Vladimir A. Milashev ; translated by R. E. Sorkina.
Translation of: Trubki vzryva. Includes index. ISBN 0-387-18649-2 (U.S.) 1. Kimberlite.
2. Necks (Geology) I. Title. QE462.K5M54413 1988 551.8'8–dc 19 87-36916

© Springer-Verlag Berlin Heidelberg 1988
Printed in Germany

Printing: Druckhaus Beltz, Hemsbach/Bergstr.
Bookbinding: J. Schäffer GmbH & Co. KG., Grünstadt
2132/3130-543210

Preface

Explosion pipes are very common among geological formations. They have received widespread attention owing to their peculiar shape, intrusive contacts with enclosing rocks, diverse infillings, and associated economic minerals. In the second half of the last century these small, almost vertical, channels, usually subisometric in plan near the surface and narrowing and flattening downward, received a name reflecting the most probable mechanism of their origin.

Early workers studying volcanic pipe bodies attributed the formation of such channels without question to strong endogenic explosions. A better knowledge of the mineralogy and structural and textural features of rocks filling pipe bodies and of the whole diatreme inner structure has been the subject more complicated, and by the middle of our century the need for reappraisal of once obvious ideas became apparent. This reappraisal has revealed a theoretical weakness in the explosion concept and its inconsistency with important original data. Many hypotheses have been advanced, some quite fantastic. From time to time critical reviews about the origin of diatremes have been published, dealing with tubular bodies composed of rocks of similar composition (kimberlites, basalts).

Volcanic rock types ranging from ultramatic and alkali-ultrabasic to acidic and carbonatites occur among magmatic rocks in pipes. Rocks composing these bodies are marked by diverse structural and textural features, so that in addition to massive and amygdaloidal volcanic rocks we find breccia and tufficite breccia as well as a distinctive breccia composed almost entirely of fragments of sedimentary or metamorphic rocks which were intruded by the pipes.

Explosion pipes have attracted and continue to attract attention not only because of their peculiar structure and diverse composition, but mainly because of the abundance of associated economic minerals. Among these, diamonds are the most important; in addition to industrial concentrations of iron, uranium, REE and the like have been found in the pipes.

The fact that explosion pipes contain a wide spectrum of ore and nonmetalliferous minerals explains why bodies similar in appearance and inner structure have been studied by different groups of geologists interested in one or another mineral deposit. Therefore, sampling and examination techniques, as well as the interpretation of results, depend on the particular economic minerals and their conditions of formation and distribution.

At present, numerous publications are devoted to the geology, mineralogy, and environment of origin of particular groups of pipes. These geological bodies have, however, not been described as a whole.

This monograph will help to fill the gap in the literature. The reader will find a synthesis on pipe bodies of different composition, a comparative analysis of results, along with a discussion of the physical state of the magmatic material, dynamics, energetics, and mechanism of their formation.

The section of the book on economic minerals associated with explosion pipes and conditions of formation and distribution of the pipes follows the discussion of diatreme origin. These subjects will be the subject of further study.

Leningrad, Spring 1988   V.A. Milashev

Contents

Part III   Mechanism and Conditions of Pipe Formation

Part 1 Kimberlite and Lamproite Pipes

CHAPTER 1

Morphology of Kimberlite Bodies

Kimberlite Bodies: Major Types and Models

Kimberlite rocks occur mainly as diverse tubular bodies: the so-called explosion pipes or diatremes. Dykes are also fairly common, and in some regions we find kimberlite sills and complex irregular-shaped bodies apparently related to karst caves filled with kimberlite melts.

As a whole, explosion pipes are cone-shaped bodies with the tapering downward. In plan, at the level of the present exposure surface they range in size from 0.01 to 140 ha with the cross-section area decreasing with depth. Individual pipes (Crown Diamond Mine, Udachnaya, Moskvichka) have small "blows" near the present exposure surface; below it, however, the bodies are conical in shape, like typical diatremes.

At the lower and middle horizons of a diatreme succession, the lithology of the enclosing rocks does not greatly affect their shape, although it is reflected in details of the outline. For example, one of the well-known kimberlite pipes in Yakutia does not vary in shape at places where it intrudes very weak marl and fairly strong limestones (Poturoev 1976). The mechanical strength of rocks intruded by diatremes affects their shape only when weak rocks occur at the apical parts of slightly eroded pipes where wide funnels are formed.

There is a functional relationship between the "rate" with which the diatreme cross-section area decreases with depth and the slope of the "walls" of a pipe vent. The change in gradient from pipe to pipe is locally quite important. While the cross-section area of the Kimberley Pipe decreases by 8% per 100 m and that of the De Beers and Mir Pipes by 13 and 14%, respectively, it is 56% and 68% in the Roberts Victor and Saint Augustine Pipes, respectively. However, irrespective of some differences, kimberlite pipes, to a first approximation, have a funnel shape with an apical angle (between generators) of $10°$ to $15°$. Hence the size of an exposed kimberlite diatreme allows its volume to be estimated approximately. To simplify the estimation we constructed a nomograph helping first to discuss an area observed on the surface and then the anticipated volume of a pipe approximated by the shape of a cone, the angle between its generators being $12°$ (Milashev 1974c). Similar results were obtained for the upper and middle horizons of 11 African diatremes (Fig. 1, Table 1) (Hawthorne 1975). A somewhat smaller value ($82°$) as compared to our data ($84°$) for an average slope of diatreme rim can be attributed to the fact that in his estimates Hawthorne (1975) had not used the lower and some middle horizons of explosion pipes.

The cross-section area of kimberlite pipes not only decreases with depth, but they also become flattened, and as a result gradually degenerate into dykes. These

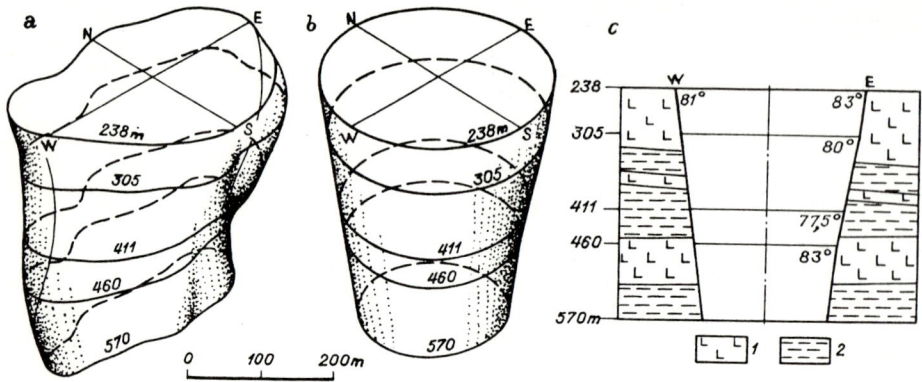

Fig. 1. Isometric projections and vertical section of the Jagersfontein kimberlite pipes. (After Hawthorne 1975; Cornelissen and Verwoerd 1975). Isometric projections of: **a** various horizons of the pipe; **b** circles equal in area to pipe sections at appropriate levels; **c** vertical section of an ideal pipe. *1* Karroo dolerite; *2* Karroo shale

Table 1. Average slope for walls of kimberlite pipes. (Hawthorne 1975)

Pipe	Depth, m	Number of estimated horizons	Average slope (degrees)
Premier	170–538	5	84
Finsch	88–348	2	82
Dutoitspan	0–411	4	79
Koffyfontein	46–244	3	85
Wesselton	250–470	3	81
Jagersfontein	238–570	5	81
Bultfontein	296–670	3	83
De Beers	105–300	3	84
Kimberley	91–294	4	80
West End	0–165	2	81
Kao	0– 90	2	85
Average for all pipes			82

changes in various pipes take place at different depths. For example, the Kimberley, Saint Augustine, and De Beers Pipes turn into dykes at depths of 1073 m, 244 m, and 732 m, respectively.

The kimberlite pipes at the present exposure surface exhibit great differences in shape. Some of them are almost regularly rounded or oval with even and rugged margins. Elongate and strongly extended lenticular flattened tubular bodies passing into dykes across the strike are very common. There are also diatremes of amoeboid, mushroom-like, and other odd shapes. Some workers have attempted to classify diatremes according to their shape in plan and recognized about ten varieties (rounded, oval, pyriform, quadrangular, dumb-bell-shaped, irregular isometric, and the like) (Trofimov 1967). However, this classification, being very complicated and theoreti-

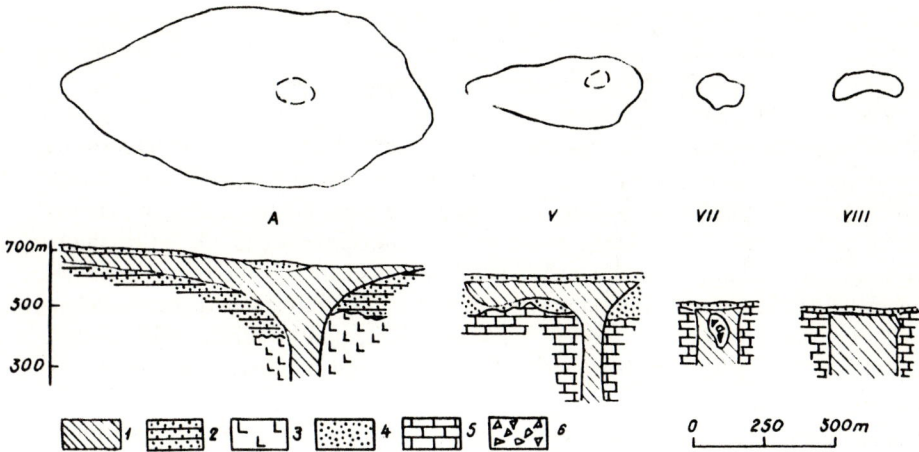

Fig. 2. Shape in plan and in cross-section of some kimberlite pipes of the Bakwanga field. (Meyer de Stadelhofen 1963). *1* Kimberlite breccia; *2* Mesozoic sandstone; *3* dolerite; *4* deluvial deposits; *5* carbonate rocks; *6* large xenoliths

cally not substantiated, cannot be used in practice and has therefore not been widely accepted.

Knowing that the cross-section of an explosion pipe becomes less isometric with depth and that root parts change into dykes, we subdivide them into two morphological groups linked by gradual transitions, viz., explosion pipes proper (isometric and oval) and flattened-tubular bodies (strongly elongate and lenticular in plan). The short to long axis ratio is 1:1 to 1:3 and 1:3 to 1:10 of the former and the latter, respectively. Kimberlite pipes, isometric and oval in shape, filled with massive large porphyritic rocks, are assigned to stocks by some investigators, who rule out their relation to the Earth's surface during the time of emplacement. However, if we take into account that diatremes eroded to different degrees contain kimberlitic rocks allowing a gradual transition from breccia with fine porphyritic magmatic cement to massive coarse porphyritic varieties to be traced, then we cannot assign these bodies to stocks exclusively on the basis of these structural and textural features.

There is a striking difference between the above model and kimberlite pipes from the Bakwanga (Mbuji-Mayi) field, characterized by an abrupt widening in the near-surface section. Owing to this, a conical shape typical of the kimberlite pipes of this field is transformed into a mushroom shape (Fig. 2). An area of diatremes with surface funnels varies from 6.2 to 16.5 ha, with an average area for four pipes from the north-eastern group of 11.2 ha. Much smaller (one order of magnitude) are diatremes without funnels: their area is estimated at 0.4–1.6 ha, with an average for five pipes from the north-eastern group of 1.1 ha.

Two hypotheses were advanced to explain this mushroom shape, unusual for diatremes of the Bakwanga field. According to the first, the surface widening in pipes is considered to be karst cavities in the Bushimayi limestone, which were filled with weathered and displaced kimberlite mixed with fragments from the Mesozoic rocks. Pipe III provides a good example, its funnel being infilled with strongly altered kimberlite and terrigenous material while the vertical feeder is composed of dense kimberlite breccia. A distinct funnel occurs also in Pipe A, intruding sandstones overlying igneous rocks; carbonate rocks are absent from this part of the succession.

The second hypothesis ascribes the mushroom shape of these diatremes to a peculiar environment of their emplacement. The submarine eruption of kimberlites is supposed to have occurred concurrently with the deposition of Mesozoic sandstones of bed M-4, the Lualaba Series. The conditions were favorable for mixing and alternation of sedimentary and volcanic rocks and for the spreading of the latter among unconsolidated sediments over an area much larger than that of the vents themselves (Pipe V). It is noteworthy that kimberlite pipes have no mushroom-shaped widening in all sites where bed M-4 is eroded. They are represented by tubular bodies whose cross-section area does not change to a depth of several hundred meters (Pipes VII and VIII).

Kimberlite dykes are in fact known from all the areas of kimberlite volcanism. They are common in most regions and are exposed on the surface; however, in some areas dykes are more numerous than explosion pipes. Several dykes may form chains, and be arranged en echelon or fan-like. The following dyke types are recognized: (1) former feeders for diatremes and sills; (2) those emplaced along cracks in already solidified kimberlite rocks filling explosion pipes; (3) those developed in a shatter zone at diatreme exocontacts; (4) dykes not related genetically to explosion pipes or sills.

Dykes of type 1 are greater in thickness and more extensive than fissured kimberlite intrusions assigned to types 2 and 3. They are about 5 m, locally 10–15 m, thick and several kilometers long. These dykes often have "blows" subisometric or lenticular in shape, up to several tens of meters. When "blows" are filled with eruption breccia or kimberlite tuffisite breccia, they are proved to be associated with the root parts of explosion pipes. When massive kimberlites fill "blows" lenticular in shape, it becomes difficult to assign them to the root parts of diatremes because such a "blow" can be merely a local widening of a crack. Within distinct shatter and schist formation zones, dykes of this type often degenerate into swarms of steeply dipping branching small dykes and veinlets irregular in thickness and strike. Dykes which had intruded along fissures within already solidified kimberlites do not extend beyond the pipes. They often differ in composition from the kimberlite cement of breccia (toward higher contents of mica and olivine); they have apparently resulted from the intrusion of small batches of residual kimberlite melts. These dykes fill primary joints in kimberlite breccia, are often marked by a stepwise pattern in a section and have a small thickness.

Dykes occurring in shatter zones at contacts of kimberlite pipes are discussed in Chapter 4.

Dykes of type 4 are individual dykes situated at a great distance (several kilometers) from known explosion pipes, sills, and dyke swarms. Their thickness varies from less than 1 m to several meters, and they are several tens of meters to several hundreds of meters long. They have no "blows". They could have formed at the final phases of kimberlite magmatism when fissures were filled with melts whose low volatile content corresponds to periods of declining activity. However, they can also be attributed to poor exploration of the areas where dykes are present and there may still be undiscovered kimberlite diatremes and sills.

Sills constitute a small proportion of an overall number of the known kimberlite bodies, which can be explained by their rare occurrence and the difficulty in finding thin-bedded deposits among flat-lying strata in platform areas with poorly dissected topography. Kimberlite sills were originally known only from Africa (SAR, Zambia, Zimbabwe, Tanzania, Kenya), but recently they were also reported from the Russian platform (Stankovsky et al. 1979). Kimberlite sills display great variation in thickness and areal extent. For example, the Benfontein sill with a minimal average thickness (0.9–1.2 m) occupies an area of 5 km^2. The Wesselton Floors and Saltpetrepan sills from the same region can be classified as laccoliths on the basis of the ratio thickness (tens of meters) to horizontal dimensions (Hawthorne 1968).

In the late 1970's, kimberlite bodies of unknown morphological type were found in Yakutia in the vicinity of a large kimberlite pipe. They are exposed and have a very complex irregular pattern. Botkunov (1964), who had studied these bodies, wrote that in extent and depth they reach several tens to several hundreds of meters and their volume is estimated at hundreds to tens of thousands of cubic meters. These bodies are always associated with kimberlite dykes. According to visual observations, kimberlite of these bodies contains fragments derived from kimberlite dykes, which helps to trace their original occurrence.

Only detailed studies can give insight into the temporal and genetic relationships between these unique orebodies and associated dykes. As a working hypothesis we can propose a mechanism responsible for infilling of buried karst caves with kimberlite melts moving along radial cracks from a giant central diatreme or from deeper levels of the Earth's crust. Carbonate rocks enclosing these bodies show a tendency to karst formation, which implies their emplacement due to the infilling of karst caves. For example, a deep mantled karst was reported from the adjacent area, where it occurs in lithologically similar rocks (Filippov and Lelyukh 1980).

Relationship Between Shape, Size of Bodies, Structural Features of Rocks and Depth of Erosion

A geometric kimberlite diatreme model shows the increase in length and decrease with depth in a cross-sectional area. Diatremes different in shape and size and associated with a larger or smaller number of dykes occur in most kimberlite fields. However, it is difficult to find even two kimberlite fields identical in shape and size of diatremes; this holds true also for the relative occurrence, thickness, and length of dykes. Based on the above features, five groups of kimberlite fields can be recognized: (1) exclusive occurrence of tubular bodies; (2) predominant occurrence of pipes; (3) occurrence of dykes and pipes in similar proportion; (4) predominant distribution of dykes; (5) exclusive occurrence of dykes.

There are a few fields where kimberlites are exposed in the form of pipes, some of them can contain dykes not yet discovered (Verkhnyaya Muna field). Most kimberlite fields in Siberia and Africa belong to group (2). Subordinate kimberlite fields (Luchakan, Omonos-Kutuguna) contain an equal number of explosion pipes and dykes. Fields dominated by dykes are rare (Kao, Sefadu) and dyke fields are occasional (Segela).

There is a strong tendency to decrease in an average area and degree of isometry of diatremes from group (1) to group (5). Attempts have been made to attribute differences in morphology and average size of kimberlite bodies in various fields to the varied depth of erosion in certain parts of the territory (Kovalsky 1963). However, if a diatreme size at the level of the present exposure surface depended mainly on the depth of erosion, then all explosion pipes within the confines of a given district should have almost the same area, with the largest and smallest explosion pipes occurring respectively in watersheds and in river valleys. Nature does not provide such examples, and some closely spaced diatremes occurring at almost the same elevation differ in area by a factor of 100. Moreover, the average size of diatremes in different regions shows a correlation with the petrology and petrochemistry of kimberlites, as well as with the morphology, average size of crystals, and diamond content; the area of explosion pipes decreases regularly from the center toward the periphery in regions of kimberlite volcanism (Milashev 1965).

Those writers who relate the emplacement of kimberlite pipes with intermittent vents formed at the contact of the cover and platform basement must assume that the roots of all diatremes in a region are located at the same depth. Hence they do not take into account (1) important differences in pressure and supply of batches of kimberlite magma from the Earth's interior, (2) differences in geological and tectonic as well as hydrogeological environments that were not equally favorable for the formation of explosion pipes even in adjacent parts of the region during different periods of its geological history, and (3) voluminous original data, particularly on kimberlite geology, reliably indicating differences in the depth of emplacement of such adjacent diatremes as Kimberley and Saint Augustine, Mir and Sputnik.

Hawthorne (1975), in his account of some diamond-bearing diatremes in Africa, has constructed an ideal structural-petrographic model of a kimberlite pipe (Fig. 3). Despite the undoubted validity of the model, some important comments must be made. The diagrammatic presentation and description of the model suggest that the root of a hypothetic ideal pipe is assumed to be located at a depth of about 2 km, its original mouth being about 1 km in diameter. A cross-sectional area and a depth at the transition into dykes in all kimberlite diatremes smaller than those of the model are attributed to denudation processes which have exposed the middle and even lower levels of pipes. However, it cannot explain the important differences (700–800 m) in depth at which the Kimberley and Saint Augustine, Mir and Sputnik Pipes and other paired diatremes wedge out, because a possible denudation of some areas for many kilometers during the time elapsed between the formation of adjacent diatremes cannot be geologically confirmed.

These discrepancies do not arise if we ascribe changes in depth of diatreme emplacement to variation in the activity and, primarily, to the inevitable differences in pressure of kimberlite magma and the varying amount of volatiles involved in volcanic processes by diatreme emplacement both in different regions and within a single field. It is worth noting that the pressure of magma implied by the location of diamond subfacies kimberlites is several times higher than that of a melt which gave rise to pyrope-subfacies kimberlites (Milashev 1972, 1974a). The concentration of volatiles in magma and the proportion of vadose waters involved in volcanic processes by the emplacement of kimberlites of different fields and adjacent diatremes also show a strong variation.

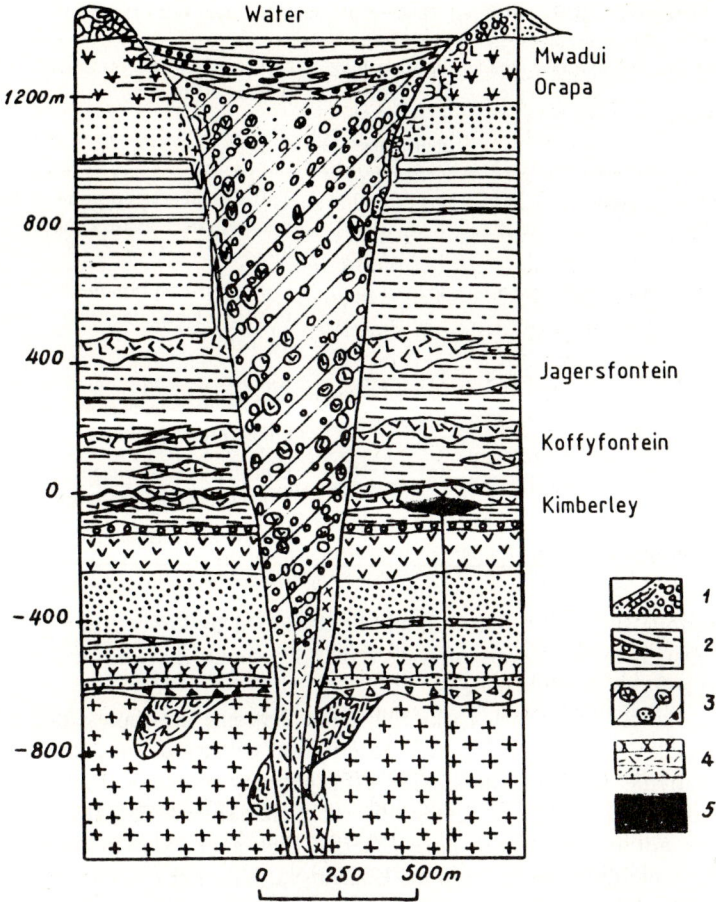

Fig. 3. Structural-petrographic model of a kimberlite pipe. (Hawthorne 1975). *1* Tuff cone; *2* fine- and coarse-grained sediments; *3* agglomerate and tuff; *4* intrusive breccia; *5* intrusive kimberlite (sills)

However, this does not diminish the importance of parameters for an ideal pipe presented by Hawthorne (1975) on the basis of several highly diamondiferous diatremes in Africa. Nevertheless, one should specify that the model does not fit any one kimberlite pipe in general, but can be applied to a very large kimberlite pipe which does not often occur in terranes of diamond-subfacies kimberlites. It is noteworthy that the depth of emplacement of the root of such a pipe (2 km) complies with the earlier values of 1.8–2.1 km obtained for diatremes in the fields of diamond-subfacies kimberlites (Milashev 1965). It was also emphasized that the release of pressure of the kimberlite magma inevitably leads to a decrease in the depth of emplacement and cross-sectional area of diatremes; average values were obtained for a possible depth of emplacement of explosion pipes in terranes where diamond and pyrope subfacies (0.7–0.8 km) and pure pyrope subfacies (0.5–0.6 km) kimberlites occur. Hence, most of the poorly diamondiferous kimberlites in the diamond, diamond-pyrope

Table 2. Morphology of kimberlite bodies at different stages of a kimberlite field vertical model

Stage (in descending order)	Relative number, %		Average area of diatremes (ha, m^2)	Short-lo-long axis ratio for typical diatremes	Structural and genetic groups of kimberlitic rocks
	Pipe	Dyke			
1	100	–	S	1:1 – 1:3	Tuffisite breccia, eruption breccia, rare massive kimberlites
2	61–99	1–39	(0.4–0.8) S	1:2 – 1:4	Kimberlite eruption breccia, tuffisite breccia, massive kimberlites
3	40–60	40–60	(0.15–0.4) S	1:3 – 1:6	Kimberlite eruption breccia, tuffisite breccia, massive kimberlites
4	1–39	61–99	< 0.15 S	1:4 – 1:10	Eruption breccia, massive kimberlites, tuffisite breccia
5	–	100	–	–	Massive kimberlites

and certainly in the pyrope surfacies terranes should be described by kimberlite pipe models which at a constant average slope of walls (82–84°) should have a smaller depth of emplacement and cross-sectional area.

According to the Hawthorne model, rocks filling the diatreme are represented exclusively by agglomerates and tuffs giving way to eruption (intrusive) breccias in the root parts of the diatreme. In Siberia, kimberlite tuff is not very common, accounting for about 4% of the total volume of kimberlitic rocks (Milashev 1965). It is of interest that tuff does not occur even in large Siberian diatremes extending, not passing into dykes, to a depth of above 1200 m (Udachnaya, Mir, and other pipes). We have no data to compare, so we do not know if the discrepancies are due only to different concepts of the genesis of identical rocks or to the features particular to kimberlite volcanism of Siberia and Africa. The former seems to be more likely.

In general, we think it is reasonable to associate tuffs, eruption breccia, and massive kimberlites with the higher, middle and lower horizons of explosion pipes, respectively. However, a complex trend caused, in particular, by the presence of structurally similar rock generations, as well as numerous unknown discrepancies, prevents using the structural and textural features of kimberlite rocks for the assessment of the erosion extent of diatremes.

The vertical variability in morphology and cross-sectional area, and the presence of dykes and sills at some depth from the earlier Earth's surface (present at the time of kimberlite volcanism) suggest the existence of primary vertical zoning (stages) of kimberlite fields. Therefore, it should be noted that each of the five groups of kimberlite fields, although differing in number of pipes and dykes, outline, and area of diatremes, contains fields similar in extent (not necessarily identical as to absolute depth) of denudation and, hence, can be used as an appropriate standard for an ideal model of a kimberlite field (Table 2). The original average area of a diatreme is determined by the kimberlite facies and, to a lesser extent, by the peculiar geological structure of a field. The depth of location of the root zone of an explosion pipe being directly proportional to the pressure of kimberlite magma, vertical dimensions of certain stages and a model as a whole would be, in general, maximal for kimberlite fields in the diamond subfacies and minimal for those in the pyrope subfacies.

In summary, we emphasize that vertical and horizontal dimensions of a general kimberlite diatreme model cannot be described in absolute metric units. Relative units should be used for linear parameters for a model to reflect major structural and petrographic features of most diatremes; such a model would illustrate the proportional pattern in changes of diatreme dimensions both vertically and horizontally independent of size.

CHAPTER 2

Internal Structure of Kimberlite Bodies

Endogenous Features

Endogenous features of single-step kimberlite pipes, dykes and sills, like other volcanic bodies, consist of components such as orientation of primary flow structures in magmatic rocks or banding in pyroclasts, original jointing of rocks, and distribution and orientation of long axes of porphyritic phenocrysts and xenoliths. The presence of two or more generations of kimberlitic rocks in a diatreme or dyke not only adds complexity to a body, but becomes its intrinsic structural feature. In multi-step bodies, apart from the above features of generation of each kimberlitic rock, one should study age relations, petrochemistry, physical and other features of different rock generations.

Conventional studies of petrochemistry cannot reveal structural features in rocks of single-step bodies due to negligible fluctuations in the content of the primary chemical components. On the contrary, petrochemical studies of multi-step diatremes become of importance because each kimberlitic rock generation is a major structural element having its own geochemistry. Moreover, the particular geochemistry of kimberlites, and, primarily, contrast differentiation and differentiation indexes are very important to assess the relative age and sequence of rock formation (for details see Chap. 7).

Linear- and flat-parallel flow structures reflected in a subparallel orientation of elongate flat inclusions and minerals are seen in hand specimens. Statistical assessment of numerous orientation of flattened-elongate inclusions in kimberlite breccia of one of the Yakutia pipes implies the presence of primary subvertical flow structures. Flattened inclusions in marginal parts dip toward the center at an angle of 70–80° with strikes parallel to the contact lines (Fig. 4). The data available suggest that the pipe becomes narrow with depth and is funnel-shaped with the vertex downward. The data additionally show that diatremes were filled with a fairly plastic melt, which makes comparison of a pipe formation mechanism with that of diapirs irrelevant.

The study of the joint system of kimberlite breccia from several explosion pipes shows three main types of cracks, viz. (1) horizontal and gentle with a slope less than 10–20°; (2) vertical and subvertical with a slope over 80°; (3) inclined with a slope of 30–80° (Figs. 5, 6).

Cracks of the first type are the most numerous. They extend perpendicular to flow lines and are therefore classified as transverse. Vertical and subvertical cracks are less common. They are perpendicular and parallel to the contact plane. The former radiate from the center of a pipe and coincide with flow lines of kimberlite, which

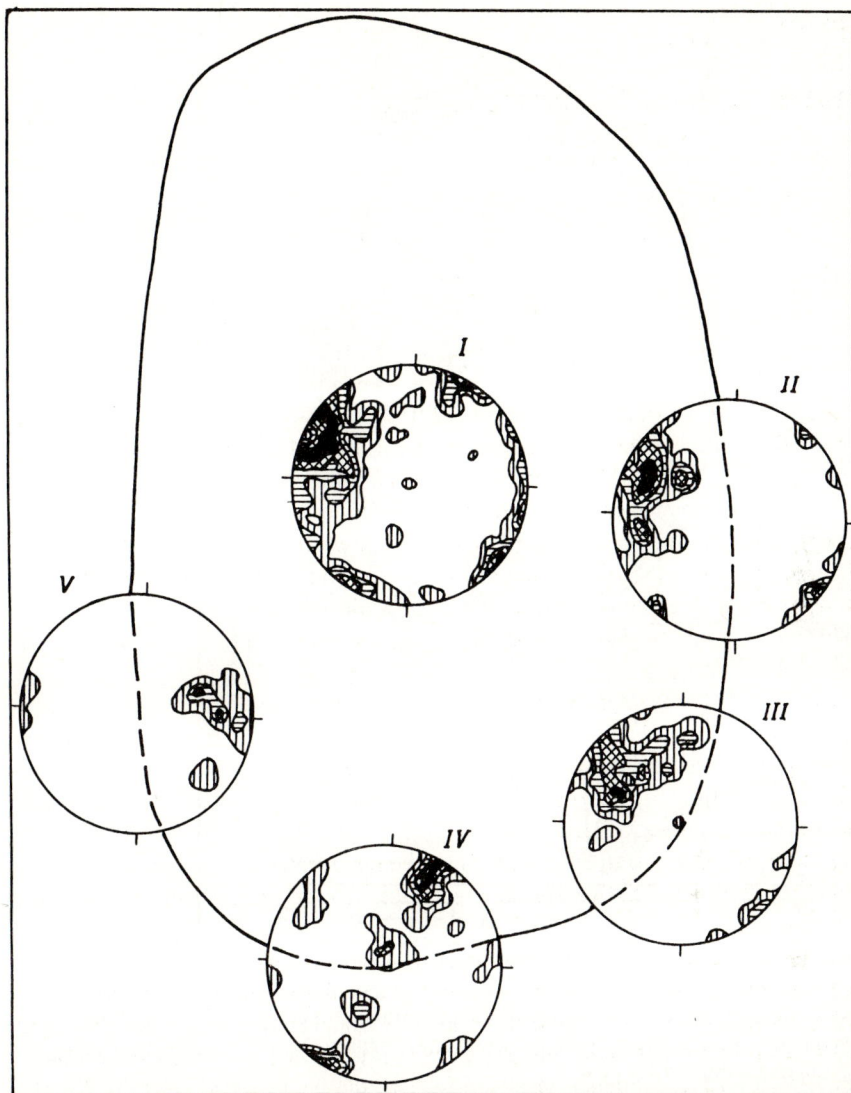

Fig. 4. Plan of the Obnazhennaya kimberlite pipe with oriented diagrams of normals to platy and flattened-elongate xenoliths and cognate inclusions. *I* Composite diagram (220 measurements); diagrams for different parts of the pipe; *II* eastern part (57); *III* south-eastern (56); *IV* southern (51); *V* western part (24)

makes them similar to longitudinal cracks of plutons. The latter are oriented concentrically to the vertical axis of a pipe and coincide with flat-parallel features of kimberlite and can be assigned to a joint of a bed. Inclined cracks (30–80°) are rare. Their strike mainly follows that of radiating cracks and can be considered as original oblique joints. Apart from these joints, superimposed cracks result from a regional deformation of the enclosing rocks after the formation of a diatreme (Milashev 1960).

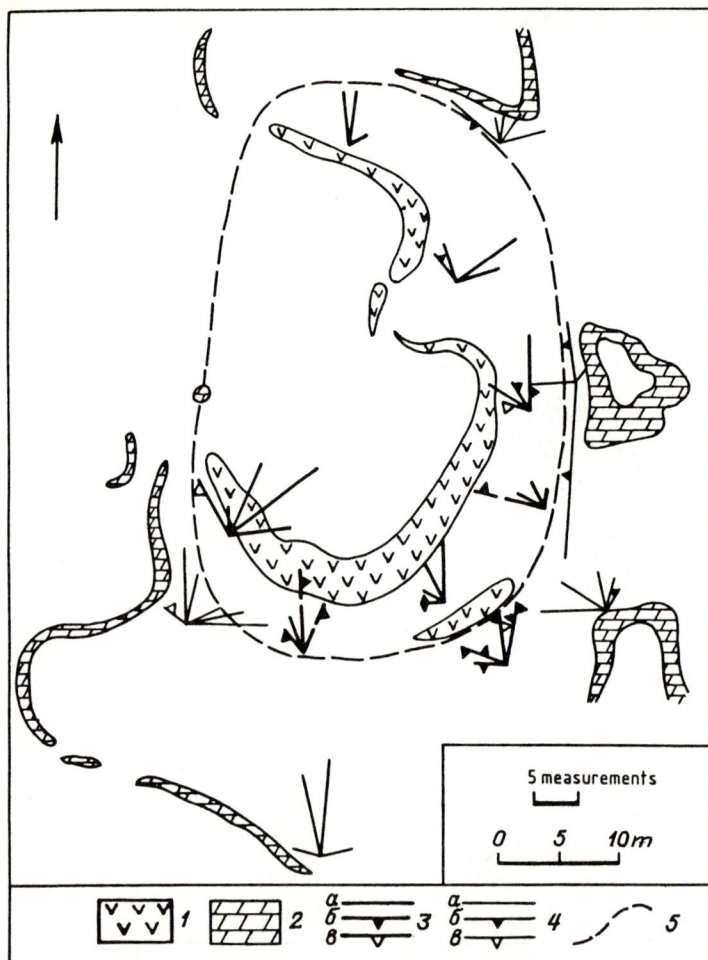

Fig. 5. Fissure system of the Obnazhennaya kimberlite pipe and enclosing dolomites of the Turkut Formation (Sinian). Bedrock exposures of: *1* kimberlite; *2* dolomite; fissures: *3* in kimberlite; *4* in dolomite [*a* vertical fissures (80–90°); *b* steeply dipping (45–79°); *c* gently plunging (10–44°)]; *5* inferred outline of the pipe

About 30% of diatremes and almost 100% of known bodies are composed of kimberlitic rocks of two and more generations, suggesting a much wider distribution of complex-structure diatremes. However, we cannot say that all pipes are multi-step because, as a rule, only the largest and diamond-rich pipes have been studied in detail. Some small pipes are undoubtedly composed of rocks of one generation. Nevertheless, it was proved that not only giant but also all large and most medium-sized diatremes have gone through a multi-step formation.

Successive generations of kimberlitic rocks in complex-structure diatremes differ in both structural and textural features, in content of porphyritic phenocrysts and xenoliths, chemical composition, and physical properties. The generations are separated

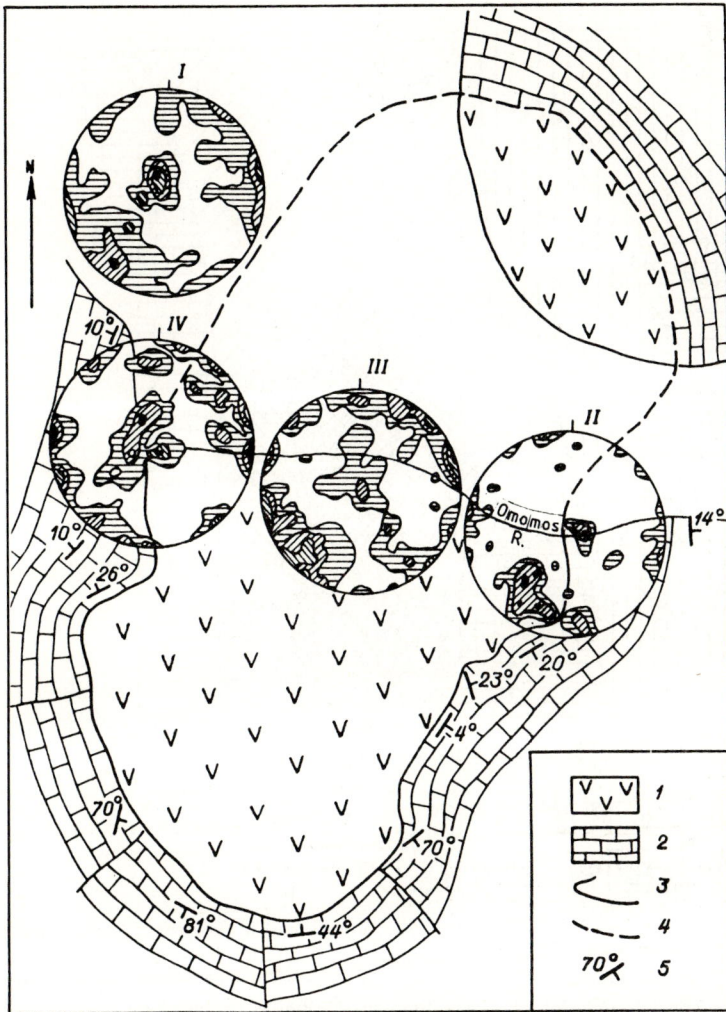

Fig. 6. Plan of the Leningrad Pipe and oriented diagrams of normals to joints in kimberlite breccia. (Milashev et al. 1963). *I* Composite diagram (600 measurements); diagrams for various parts of the pipe; *II* eastern part (100); *III* central part (300); *IV* western (200). *1* Kimberlite eruption breccia; *2* Cambrian limestone; outline of the pipe inferred from: *3* mining, *4* magnetic survey; *5* attitude for carbonate rocks

by sharp contacts. The position of the contact with respect to original flow structures in the rock generations studied and the presence of fragments of rocks of one generation in the form of xenoliths in the other allow the sequence of generations to be reliably established (Fig. 7). A distinct succession in the formation of structurally different kimberlitic rocks has been recognized, namely, massive kimberlites are always emplaced later than kimberlite breccia, tuffisite breccia being the earliest.

Diatremes whose difference in successive generations of rocks is reflected only in the abundance of xenoliths and unimportant variations in geochemistry of magmatic

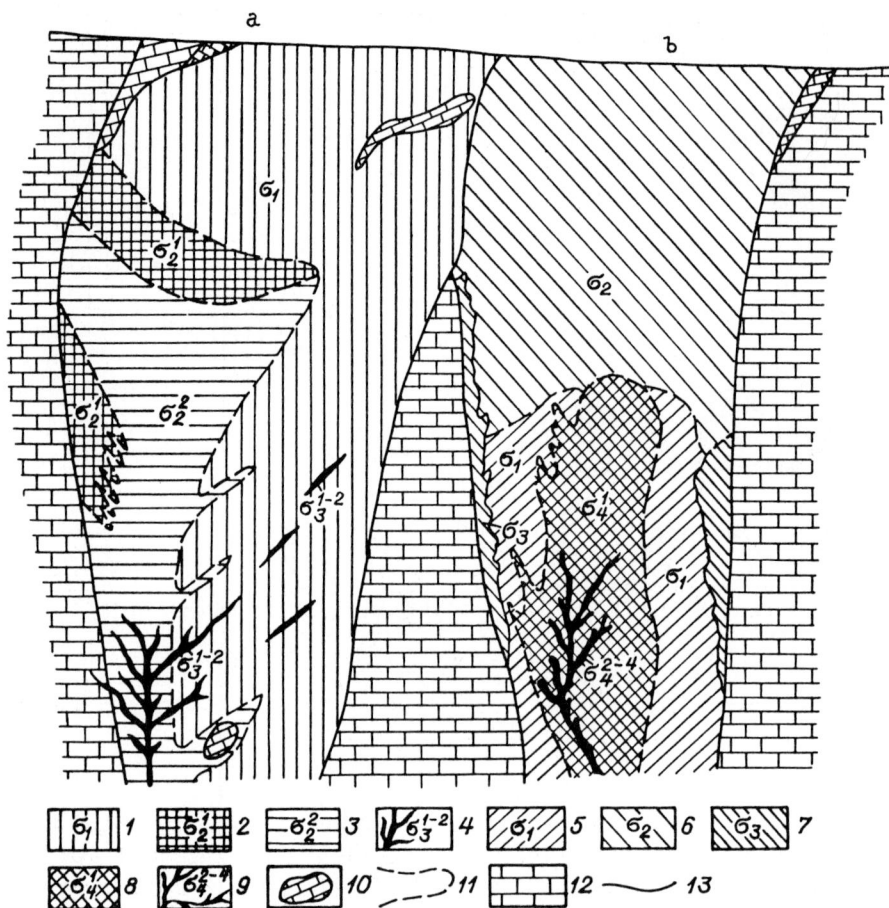

Fig. 7a,b. Diagrammatic section of **a** the Udachnaya-Zapadnaya and **b** Udachnaya-Vostochnaya pipes. (Marshintsev et al. 1975a,b). **a** Udachnaya-Zapadnaya. Kimberlite breccia of: *1* phase I; *2* phase II with primary flow structures; *3* phase III with massive cement; *4* veined kimberlite of the final phase. **b** Udachnaya-Vostochnaya. *5* Autolithic kimberlite breccia of phase I; *6* kimberlite breccia of phase II with massive cement; *7* kimberlite of phase III; *8* kimberlite breccia of phase IV; *9* veined kimberlite of phase IV; *10* large xenoliths of sedimentary rocks; *11* contacts of kimberlitic rocks of different generations; *12* enclosing carbonate rocks; *13* faults

products form a separate group. In fact, there are no contacts between such rock generations because instead of a plane there is a thick transitional zone (several tens of meters). For details the reader is referred to the paper by Zolnikov et al. (1979). The enclosing rocks of the diatreme that these authors describe contain Ordovician and Lower Silurian carbonate and terrigenous-carbonate deposits. Kimberlites and carbonate enclosing rocks are overlain by the Middle Carboniferous-Permian coal-bearing strata.

Borehole data clearly reveal the morphology of a diatreme, viz., a funnel-shaped widening to a depth of 300 m, columnar central and peripheral parts differing in rock

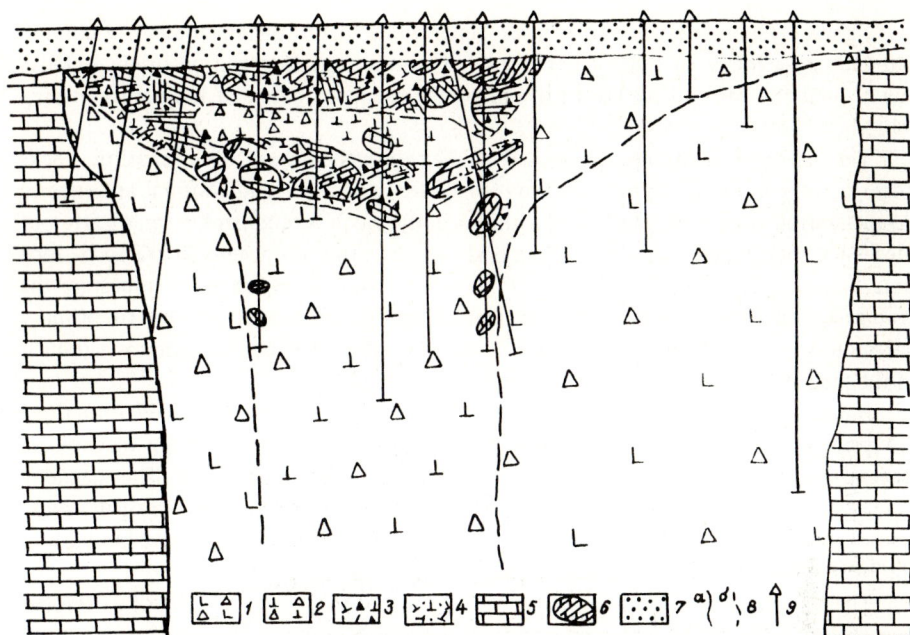

Fig. 8. Schematic section of a kimberlite pipe, Yakutia. (Zolnikov et al. 1979). Kimberlite breccia of: *1* variety I; *2* variety II; *3* variety II with high content of xenogenic material; *4* fine porphyritic to aphyric kimberlite rocks; *5* enclosing sedimentary carbonate rocks; *6* blocks of folded enclosing rocks; *7* overlying sedimentary rocks; *8* contacts (*a* sharp, *b* gradual); *9* boreholes

composition (Fig. 8). The pipe is filled with kimberlite breccias of type I and type II and hybrid kimberlitic rocks. The peripheral parts of the pipe are made up of type I breccia, while type II breccia (see Zolnikov et al. 1979) forms the central columnar part. A frequent alternation of type II kimberlite breccia, hybrid (fine porphyritic and aphyric) kimberlitic rocks and blocks of carbonate sediments is characteristic of a funnel-shaped neck. At a depth of 150 m hybrid rocks form vein-shaped bodies up to 1 m thick. The bodies dip westward at an angle of 20° and southeastward at 20–25° in the eastern and western parts of the funnel-shaped widening. In the central part they occur in the form of isometric, elongate, not distinct subvertical columns of different size within kimberlite breccia. Similar varieties of kimberlitic rocks occur also near "floating reefs" of sedimentary rocks.

The detailed study of borehole data and core material show that type I and type II breccia are separated by zones of mixed material several tens of meters thick. The upper part of the funnel-shaped widening contains isometric and angular blocks (up to 1 m by 1.5 m) of type I breccia within type II breccia, which enables their age relations to be reliably identified. There is a slight difference in chemical composition of type I breccia and that of type II, and observable small variations in contents of separate components can be attributed to impurity of the xenogenic material. Similar rock compositions and the presence of phenocrysts of one breccia in the other suggest their formation in the course of rapidly changing volcanic processes when not yet solidified breccia of type I was intruded by fluids and a melt and, as the result, breccia of type II was formed.

Fissure systems of the earlier rock generations in multi-step bodies are complicated because joints of later generations have been superimposed on original cracks due to pressure exerted by the intrusion of a subsequent batch of kimberlite magma (Milashev 1960).

The parts of diatremes composed of various generations of kimberlitic rocks differ greatly in the pattern and intensity of the magnetic field: normal and inverse magnetization occurs in such areas within a body. Such a remanent magnetization of successive rock generations of a certain body suggests that they were separated by long time intervals. It is worth noting that magnetic fields over kimberlite pipes are the youngest endogenic features of diatremes. Magnetic properties related to primary parameters of rock composition show up quite vividly by serpentinization when most of the magnetite is crystallized.

Based on magnetic field intensity, kimberlite bodies are subdivided into three groups, viz., weakly magnetized (several tens to 100–150 nT), average magnetized (several hundreds to 1000 nT), strongly magnetized (above 1000 nT). The magnetic field over kimberlite bodies mainly depends on magnetic susceptibility κ of the rocks varying within $n \cdot (10^{-4} \div 10^{-2})$ SI. As a rule, bodies composed of massive kimberlites exhibit the highest magnetic susceptibility and anomalies.

The occurrence of both normally and reversely magnetized kimberlites implies that the intensity of magnetic anomalies depends not only on magnetic susceptibility but on modulus and vector direction of the remanent magnetization as well. Kimberlitic rocks filling various bodies have different natural remanent magnetization J_n. According to the J_n/J_i ratio (where J_i is an induced magnetization), kimberlites are assigned to three groups with $J_n/J_i < 1$; $J_n/J_i = 1-1.5$; $J_n/J_i > 1.5$. Negative anomalies with vector J_n directed into the upper semispace and positive anomalies with vector J_n directed into the lower semispace are observed over kimberlite bodies of group 3. In this case, positive anomalies show intensity higher than could be expected on the basis of exclusively induced magnetization.

Positive magnetic fields are most common over kimberlite bodies, but negative anomalies and some fields of opposite sign are observed over individual parts of a diatreme. The differently magnetized kimberlitic rocks even within a single diatreme differ in color, petrology, density, and remanent and induced magnetization. However, any regular trend has not been established in the variations of the features in "negative" and "positive" kimberlites.

The difference in petrology of normally and reversely magnetized kimberlites filling the same body can be attributed to the nature and intensity of secondary processes and to the peculiar features of the original composition. For example, olivine is completely serpentinized and the matrix consists of hydromica with subordinate chlorite and serpentine in normally magnetized breccia of the Flazhok Pipe. The "negative" kimberlite from the same body contains some olivine in the chloritized phlogopite and serpentine groundmass. Positive anomalies over the Dvoinaya Pipe were observed in the parts composed of kimberlites whose groundmass consists mainly of phlogopite, talc, opaque minerals, perovskite, sphene, and leucoxene, while the total content of carbonate and serpentine does not exceed 25 vol. %. The reversely magnetized parts are composed of kimberlite breccia, kimberlite fragments similar to "positive" kimberlite accounting for 50%. The matrix is completely serpentinized.

The Pozdnyaya Pipe, that shows no difference in composition of normally and reversely magnetized breccias, in an exception.

In general, we can judge the internal structure of a kimberlite dyke from the distinctive zoning of its cross-section. This zoning is the result of quenching in selvage and primary flow structures away from the contacts. However, almost all dykes have contact chill zones (though not always visible), while primary flow structures occur only in some rocks. Primary flow structures are typical of intrapipe veins distinguished from enclosing kimberlite rocks by distinct orientation, zoning, size, and content of major minerals, such as olivine and phlogopite. The above features of dykes can be attributed to their formation from residual melts and fairly slow crystallization due to the intrusion in solidified, although not yet cold kimberlite breccia.

Because dykes are almost always composed of massive kimberlites, their flow structures are caused by the presence of zones rich in or depleted of porphyritic phenocrysts of particular minerals (e.g., olivine, opaques). When xenoliths from enclosing rocks occur in endocontacts of dykes, they are arranged parallel to the contacts (vein 73 in the Daldyn field).

The fact that kimberlite sills are related to zones of weakness at contacts of heterogeneous rocks suggests a mechanism similar to the intrusion of sills of different composition, e.g., dolerite sills. The study of the internal structure of sills also shows successive and in some cases repeated intrusions of kimberlite melts along the contact planes of rocks, differing considerably in mechanical properties. Kimberlite sills often exhibit horizontal stratification due to the alternation of differently colored layers varying, for example, in the composition and/or size of porphyritic phenocrysts or their amount to groundmass ratio. Individual laminae are several centimeters thick. The most frequent case is when rocks rich in olivine (or pseudomorphs after olivine) and opaque minerals are interbedded. Interbeds and lenses of the matrix devoid of porphyritic phenocrysts or containing a noticeable amount of apatite are also fairly common (Benfontein Sill).

The Wesselton Floors Sill, according to the peculiar features of kimberlitic rocks contained in it, is divided into the upper and lower parts separated by an obscured contact. Kimberlite of the upper part is fairly fresh (abundant relict olivine) and strong, being devoid of xenoliths. Small angular xenoliths from underlying shale occur in the middle of the sill. The rocks forming the lower part of the sill are strongly altered and rich in phlogopite, angular shale fragments are very abundant at the base of the intrusion. Such a structure of the sill may well be due to successive intrusions of two batches of kimberlite magma differing somewhat in composition (Hawthorne 1968).

Exogenous Features

Exogenous features of diatremes are reflected in the banded structure of their upper horizons, diapirism, and zoning resulting from weathering crust formation and hypergene carbonatization. Naturally, the development of weathering crusts and hypergene carbonatization does not depend on the intensity of denudation prior to these processes, although the banded pattern of the upper horizons is observed only in weakly eroded diatremes.

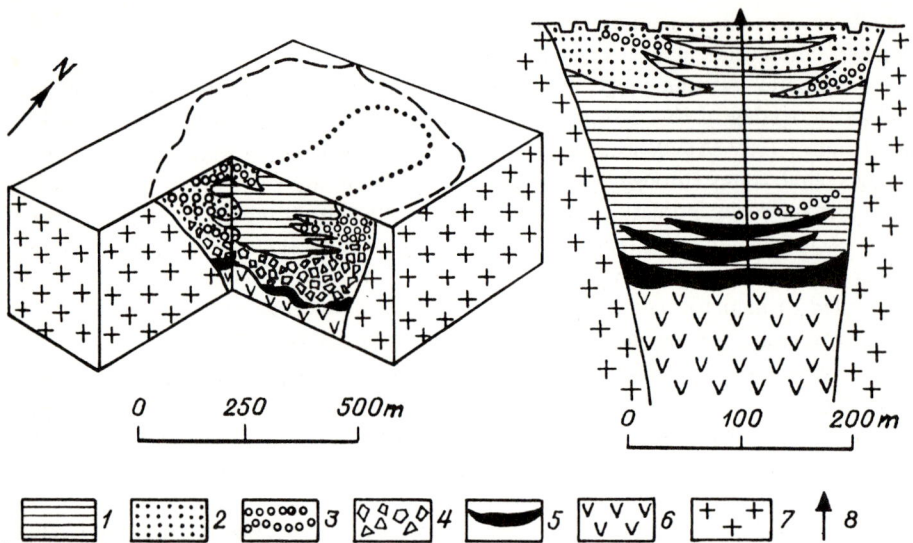

Fig. 9. Block-diagram of the Mwadui Pipe and geological section through the Koppieskraal Pipe illustrating crater morphology, composition, and occurrence of enclosing rocks and infilling rocks. (Edwards and Howkins 1966; Cornelissen and Verwoerd 1975). *1* Shale and mudstone; *2* sand; *3* pebblestone; *4* kimberlite-granite sedimentary breccia; *5* kimberlite "tuff"; *6* primary kimberlite pipe; *7* gneiss and granite; *8* boreholes

Two types of banding can be distinguished in the apical parts of kimberlite diatremes. The first type is marked by a fine intercalation of a mixture of secondary material and volcanic (kimberlitic) material extruded concurrently with sedimentation in mushroom-shaped enlargements of pipes in the Bakwanga field (see Fig. 2). Such a pattern, being primary in essence and time of formation, cannot be classified either as an endo- or exogenous feature and should be considered as intermediate in genesis. Banded structures of the second type are related to clastic formations filling vents or kimberlite diatremes to a depth of several hundred meters. For example, the crater of the world's largest kimberlite pipe, Mwadui (Tanzania), is filled with terrigenous Cretaceous or Paleogene lacustrine deposits to a depth of 366 m. Coarse clastics accumulated mainly along the crater walls and formed the shores of a lake, while fine-grained sediments were deposited in the center of the basin (Fig. 9).

The craters of most kimberlite pipes in Bushmanland are filled with sediments and breccia material. Vents are filled with shale, sandstone, and conglomerates, suggesting their deposition in crater lakes. However, beds of sedimentary rocks are often deformed by large blocks of enclosing rocks having fallen into the crater. A volcanic neck of angular gneiss blocks up to several meters in diameter erupted during repeated explosions is often seen within a crater (Cornelissen and Verwoerd 1975).

Kimberlites such as yellow ground, blue ground, and calcrete, known since the late 19th century, belong to the exogenous features formed in diatremes in the course of weathering and hypergene carbonatization. Yellow ground is oxidized kimberlite of yellowish color. Blue ground is unoxidized slate-blue or blue-green kimberlite, also known as hardebank. Calcrete is a term for kimberlite weathering products formed in the course of hypergene carbonatization.

Laboratory studies carried out by Shamshina (1979) have contributed greatly to the understanding of these processes. She showed that increased loss of magnesium up the section and other regular changes in chemical composition of kimberlitic rocks in the course of hypergenesis were accompanied by transformation of mineral parageneses, resulting in the formation of a structurally complex weathering crust. A complete section of weathering crust consists of four main zones: disintegration, leaching, hydrolysis, and final hydrolysis. Kimberlite bodies in Yakutia, like those from other provinces, are dominated by an incomplete weathering crust because the upper and often middle parts of the section have suffered denudation.

The lower, disintegration, zone of the weathering crust was found in all kimberlite bodies. It is represented by fractured primary rocks in which infiltrated aggregates of serpentine, carbonate, rare silica, and montmorillonite occur along cracks. Basically, the mineralogy and chemical composition of rocks in this zone are similar to those of kimberlitic rocks. The thickness of the disintegration zone is several tens of meters.

Above comes the leaching zone marked by loss of magnesium from serpentine minerals leading to the alteration of serpentine into saponite. This zone is characterized by minerals of the talc and hydromica type, and a mixed mineral of hydromica-saponite type resulting from alteration of serpentine. Phlogopite is unaltered. About half the magnesium is lost in the leaching zone, and the rock density decreases from 2.5–2.7 to 1.9–2.0 g cm^{-3}. The thickness of the zone reaches 10 or even 15 m. A horizon of hypergene carbonatization about 2.3 m thick is often observed in the upper part.

In the hydrolysis zone, trioctahedric montmorillonite gives way to the dioctahedric variety owing to continuing loss of magnesium. When conditions become unfavorable for the loss of silica, a horizon of hypergene silicification is formed and newly formed minerals are now dominated by nontronite and beidellite. The hydrolysis zone is marked by an intense chloritization of phlogopite and in the presence of calcium ion vermiculite or a mixed layer mineral of vermiculite-phlogopite type is formed after phlogopite. As the result of phlogopite chloritization, a mineral of chlorite-phlogopite type is formed.

Under conditions favorable for silica mobility and its partial loss, kaolinite-gallusite minerals are formed at the expense of montmorillonite minerals. In this case the iron in montmorillonite is oxidized and occurs in the form of hydroxides. In general, the hydrolysis zone is characterized by an extreme loss of chemical components, totalling 70–80% of the original rock mass. The thickness of the zone does not exceed 10 m; it is either completely removed by denudation or occurs as relics in kimberlite bodies in most areas of Yakutia and other provinces.

The complete zone of final hydrolysis is not known in Yakutia. Its typical parageneses are various iron oxides and hydroxides along with subordinate aluminum hydroxides. It is several meters thick on kimberlite bodies in Yakutia.

An intense loss of matter by weathering crust formation is accompanied by a considerable decrease of the final product, viz., residual crust as compared to original rock. Naturally, not only kimberlites but also enclosing rocks are subject to weathering. A decrease in volume of the final product by the formation of weathering crusts

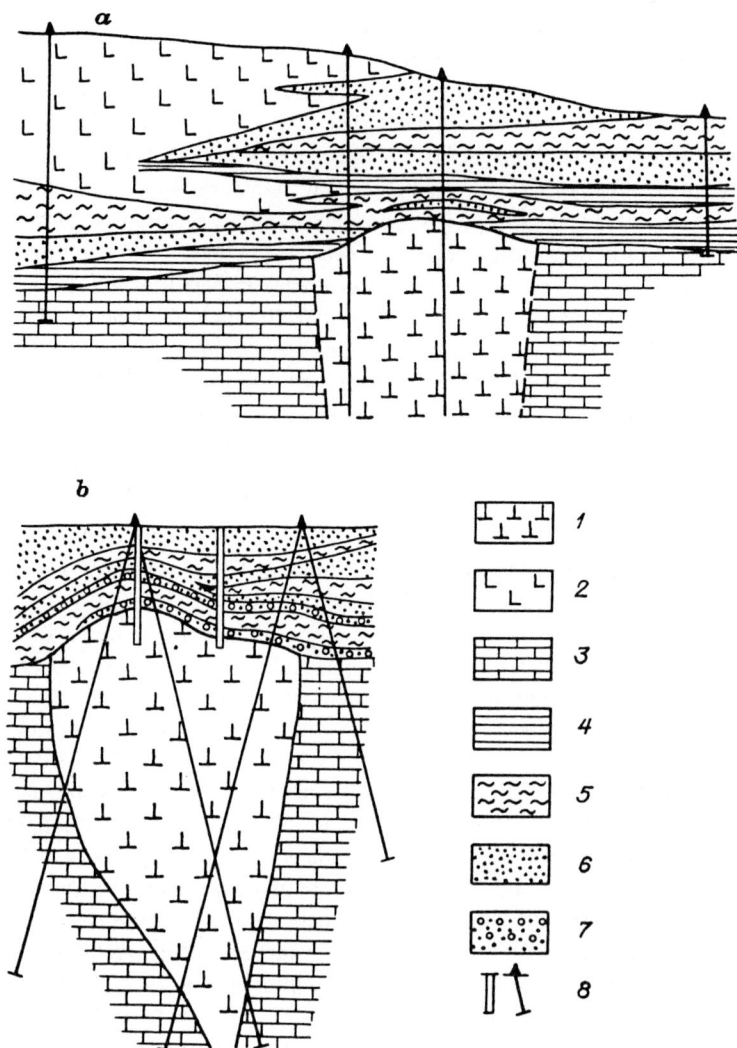

Fig. 10a,b. Schematic section of the Vostok (**a**) and Imeni XXIII S'ezda KPSS (**b**) pipes. (Kharkiv 1978). *1* Kimberlite; *2* dolerite; *3* carbonate rocks; *4* shale; *5* siltstone; *6* sandstone; *7* conglomerate; *8* pits and boreholes

on kimberlites and enclosing rocks depends on their composition and regional (including climatic) regime under which exogenic processes are operating. For example, in the kimberlite fields of Africa, the weathering crust formation resulted in the presence of negative landforms over some diatremes. In Siberia, there are rare depressions over kimberlite bodies and some kimberlite "diapirs" pierce overlying younger rocks.

Early and mature stages of "diapirism" can be distinguished. Visually, the "diapirism" of the early stage is reflected in a middle part of a kimberlite pipe higher than that of the enclosing carbonate rocks when overlying younger terrigenous rocks and traps are folded ("enveloping folding") (Fig. 10). The elevation of kimberlite above country rocks was found to be 12 m. The surface slope of bulging parts of kimberlite bodies varies from 5° to 25°. Slopes and apexes of kimberlite bulges are composed of strongly schistose ochreous clay mass.

Apart from xenoliths of dense limestone, the clay mass contains grains of picroil-menite pyrope and flared scales of chloritized phlogopite. Pyrope, as a rule, has no kelyphitic rims. The arrangement of rock fragments and minerals in yellow ground suggests the direction along which they were "dragged out". The trends coincide with distinct schistosity, which in turn strictly follows the contact between the kim-berlite surface and that of overlying rocks. Schistosity becomes less intensive with depth and disappears completely at 5–7 m below the contact.

The mature stage "diapirism" of kimberlites is observed when the latter pierce overlying younger formations. The Moskvichka Pipe provides a good example: its kimberlitic rocks have pierced the Upper Paleozoic strata and a dolerite sill had thus formed a concentric structure with ring zones of volcanic and semimentary rocks at the level of the present exposure surface (Fig. 11).

Dolerites and pierced rocks are separated by shatter and fold zones. The former is filled with fragments of Permian claystone, shale, and tuffstone, the latter consist-ing of upthrown beds of rocks dipping steeply to vertically. Within a 0.5–2.0 m boundary, kimberlites are crushed and ferruginated. A zone of contact shattering in the dolerites is also about 2 m thick. The shatter zones in the country rocks do not occur at less steep contacts; the Upper Paleozoic rocks rest there directly on schistose kimberlitic rocks whose schistosity parallels the contact plane.

All investigators who studied the problem of kimberlite "intrusion" in the overly-ing younger rocks agree that the above textural features have formed after the deposi-tion of overlying deposits. It was even emphasized that the thickness of the sediments above pipes does not decrease (Kharkiv 1978). However, this observation is not supported by sections of diatremes shown in Fig. 10, because some beds wedge out approaching the top of a kimberlite bulge (Vostok Pipe) or its vertical projections into the overlying rocks (Imeni XXIII S'ezda KPSS Pipe). Without original geological data, it is difficult to decide if the above features of the sections have resulted from erosion (second siltstone horizon of the Imeni XXIII S'ezda KPSS Pipe), or may be ascribed to the "growth" of kimberlite diatremes which took place synchronously with sedimentation.

Most researchers attribute kimberlite diapirism to an increase in volume of kim-berlitic rocks during serpentinization or their tectonic extrusion (Kharkiv 1978). Serpentinization being a typically postmagmatic (Milashev 1962, 1963b) or partly late magmatic process (Bobrievich et al. 1964) terminates, in general, prior to a dia-treme burial beneath younger sediments. In this context diapirism cannot be ascribed to the volume effect of olivine and volcanic glass serpentinization. Moreover, detailed study of kimberlite serpentinization shows that this process is not associated with important changes in rock volume (Milashev 1962; Milashev et al. 1963), so that kimberlite diapirism cannot be explained by the volume effect of serpentinization even independently of its time of manifestation.

Statements which ascribe kimberlite diapirism to tectonic effects are geologically contradictory and critically poor and groundless. Let us mention the main arguments

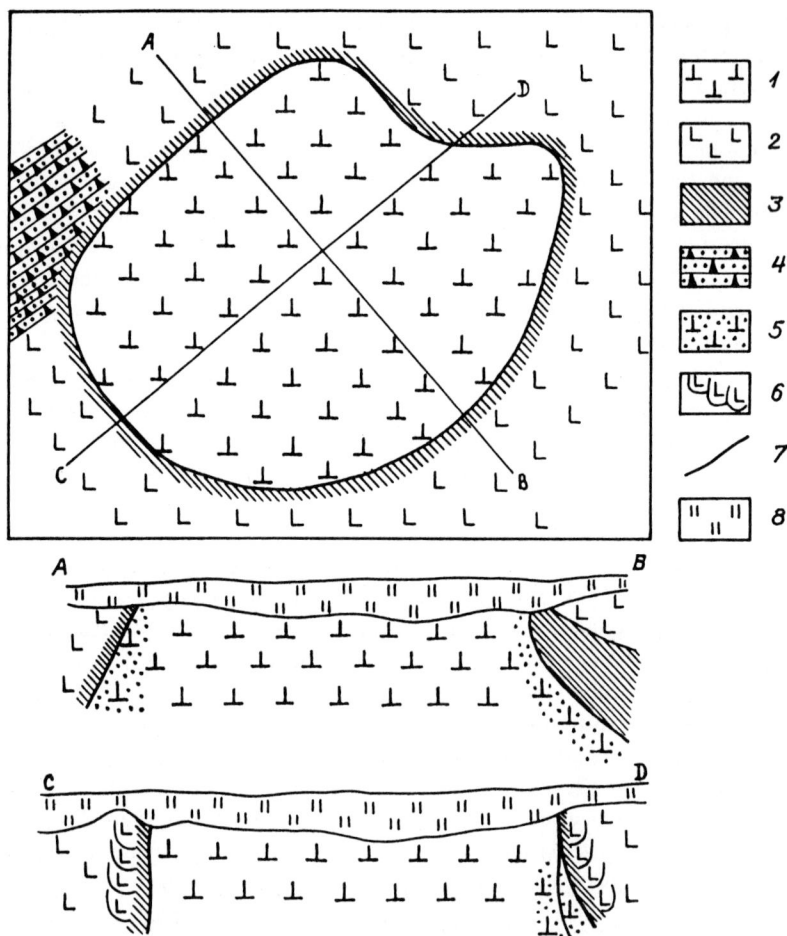

Fig. 11. Schematic plan and section of the Moskvichka kimberlite pipe. (Shamshina 1979). *1* Kimberlite; *2* dolerite; *3* shale, mudstone; *4* tuffstone; *5* shatter and sheeted zones of kimberlite; *6* shatter zones of dolerite; *7* deformation with slickensides; *8* eluvial-deluvial deposits

against this concept. The above authors do not specify the character and scale of tectonic stress. If we assume that there are some regional tectonic stresses then it would be difficult to explain the rare occurrence of kimberlite diapirism and the abundance of buried diatremes with subhorizontal upper surfaces. We can hardly ascribe kimberlite diapirism to the effect of local tectonic stresses, there are also no arguments for kimberlite schistosity, as this occurs only at contacts of the uppermost part of a diatreme, while there are no deformations in rock structure at a depth below several tens of meters.

The above facts suggest that a source of stresses responsible for kimberlite diapirism is located in higher levels of diatremes. If we take into account that the near-surface parts of kimberlite bodies (at any depth of denudation) differ from deeper levels in the presence of rocks always weathered but to a different degree, then we can infer that diapirism may be the direct result of exogenic processes. The same conclusion was drawn by Shamshina (1979), who, studying the mineralogy, found several varieties of montmorillonite in weathering crusts of most diatremes. We know that the volume of montmorillonite can increase greatly owing to its ability to absorb water.

By absorbing water and increasing volume, montmorillonite defeats high external pressure. When the ambient pressure is higher than that of montmorillonite, then the mineral will not expand even at excessive water content. In this case, montmorillonite will act like a highly compressed spring: when the external pressure drops below the critical one, the mineral starts to absorb water, increasing its volume, and hence the volume of the rock.

In term of the concept stated, we can formulate major geological features pertinent to kimberlite diapirism, namely, that (1) the ability of weathered varieties of kimberlitic rocks to increase their volume depends directly on the montmorillonite content; (2) the potential ability of montmorillonite-bearing weathered kimberlitic varieties can be realized only in cases when a swelling mineral defeats resistance (pressure and mechanical strength) of overlying strate; (3) conditions favorable for the formation of montmorillonite in kimberlites and resultant increase in rock volume (often not synchronous with weathering processes) in the geological history of the Siberian, African, and other platforms took place repeatedly, although they were not ubiquitous. The consistency of the statements and geological observations allow us to use this hypothesis to explain the causes and mechanism of kimberlite diapirism.

The occurrence of montmorillonite exclusively in exogenically altered kimberlite varieties and the increase of its content up the weathering crust section, with maximal concentration in the hydrolysis zone, comply with the observations that the source responsible for kimberlite diapirism is located in the higher levels of a diatreme. The higher levels of all the buried kimberlite bodies are composed to a certain extent of altered rocks related to various, mainly middle, zones of residual weathering crusts. Despite the noticeable occurrence of montmorillonite in the weathering crusts of seven well-known bodies, diapirism was reported from only five of them (Vostok, Moskvichka, Pobeda, Internatsionalnaya, and Imeni XXIII S'ezda KPSS Pipes).

The thickness of the overlying formations in four of these five diatremes is less than 30 m at a diapir amplitude of 12 m (excluding the erosionally exposed Moskvichka Pipe). The Pobeda Pipe is buried under 60-m-thick strata of terrigenous rocks and traps and it seems reasonable that the kimberlite extends for only 5 m from the surface of the Lower Paleozoic carbonate rocks. For example, the upper surface of weathered kimberlites is in fact horizontal in the Sytykanskaya and Podtrappovaya Pipes covered by thick (above 100 m) traps.

The eighth of the well-known buried diatremes, the Lira Pipe, is covered by Upper Paleozoic terrigenous rocks and traps to a total thickness of about 20 m. In the geology and rock composition of the diatreme, two peculiar features are to be stressed: (1) despite the small thickness of the kimberlite-overlying deposits, diapirism was not observed; (2) the weathering crust consists here of intensely leached and highly (45–70%) carbonatized kimberlite devoid of montmorillonite (Shamshina 1979). This is strong evidence for the validity of the hypothesis (using the rule of contraries) of the relationship between kimberlite diapirism and montmorillonite swelling.

Detailed studies of buried diatremes of the Siberian platform carried out by the Yakutian geologists suggest not only several epochs of crust formation, but much younger associated kimberlite diapirism. For example, within the Markha-Alakit field, the formation of buried weathering crusts on kimberlite diatremes was dated as Middle Paleozoic (at the Early/Middle Carboniferous boundary), while diapirism was manifested no earlier than Late Triassic. Within the Malaya Botuobiya field, weathering crusts on buried kimberlite bodies are Early Mesozoic (pre-Early Jurassic), and diapirs are probably post-Early Jurassic in age.

CHAPTER 3

Distribution of Xenoliths and Minerals in Kimberlite Pipes

Distribution of Xenoliths

Kimberlite breccia set in a magmatic (eruption breccia) and pyroclastic (tuffisite breccia) matrix dominates the kimberlitic rocks in most regions. Kimberlite breccia accounts for 80% of the rocks infilling pipes and dykes of the Central Siberian kimberlite province. Xenoliths of country rocks and those of rocks pierced in deeper levels comprise up to 80 vol. % or average 20 vol. % of the kimberlite breccia; they provide a very important record of the mechanism and regime of diatreme formation.

Understanding xenolith distribution in the kimberlite breccia is somewhat limited because all the best-studied diatremes fall into the category of bodies showing a complex structure, as they consist of several rock generations with a wide variety of features, including xenogenic rock content. The boundaries between sites composed of various generations of kimberlite breccia are different even within the outcrops. Consequently, the drawing of boundaries between successive rock generations in the deeper levels of the diatreme is a subject of controversy, because in most cases drill cores are the only source of information.

Any generation of kimberlite breccia displays a fairly even distribution of xenoliths; however, local clusters are not unusual in the interior parts of the diatremes, and high concentrations are common in the endomorphic zones. A high proportion of xenoliths occurs in the narrow contact zones of the diatremes composed even of massive kimberlite varieties. As a rule, different generations of kimberlitic rocks within a diatreme vary considerably in their content of xenogenic material, the maximal concentrations being typical of the portions of the diatremes with the least steeply dipping walls (Fig. 12). The tendency observed in xenoliths to cluster mainly in the relatively flattened portions of the diatremes and in narrow endomorphic zones as a whole persists also in those rare cases where the late generations of kimberlite breccia seem impounded in the products of earlier, more violent eruptions (see Fig. 7).

As to the vertical distribution of xenogenic material in the kimberlite breccia, the vast majority of scientists consider that the concentration of xenoliths decreases with increasing depth, and this opinion is consistent with the notion of the "draining off" action of kimberlite liquids at the bottom of the diatremes and the transport of xenoliths to higher levels. However, even if the theory of a regular decrease in the concentration of xenoliths with depth were basically true, currently available data are not sufficiently strong to permit any firm conclusion. Moreover, some evidence appears to be not in full agreement with one single rule in the vertical distribution of

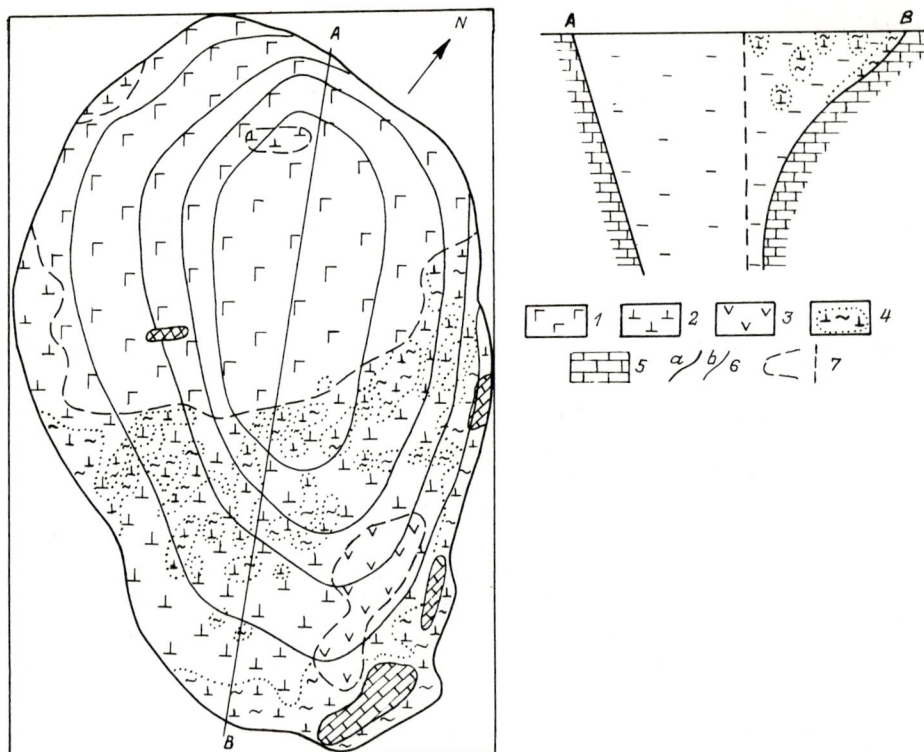

Fig. 12. Plan and cross-section of the Mir Pipe. (Marshintsev et al. 1975a). *1–3* Kimberlite breccia of varieties I–III; *4* sites rich in xenogenic material; *5* large xenoliths of host rocks; *6* contours of the pipe at the surface (*a*) and at different levels (*b*); *7* boundaries separating kimberlite breccia varieties. Section *A–B*

xenoliths, common for any diatreme. It should, however, be emphasized that more thorough study of that body of data which, at first sight, do not agree with the theory of decrease in concentration of xenoliths with depth, results in a more logical explanation on the basis of different geological and petrographic evidence.

For example, detailed investigations by Marshintsev et al. (1975a) with subsequent machine computations of the concentration of xenoliths in drill cores show that in the Udachnaya-Vostochnaya Pipe at a depth of 500–510 m below the present exposure surface the concentration of xenoliths in the kimberlite breccia spasmodically rises from 10% to 30%. The workers explain such a drastic increase in the number of xenoliths in the lower levels "by the close vicinity of the pipe contact to the country rocks" (p. 138). But this explanation appears to be unsatisfactory as one of the two boreholes studied (Borehole 222) penetrates the zone which is almost equidistant from the contact (at depths of 0–500 m and 500–700 m the mean distances measure, respectively, 110 m and 80 m with the minimal distance being 60 m), where a spasmodic change in the concentration of xenoliths seems improbable in a single generation of kimberlite breccia. It can be said to this point that we have at our disposal firm data suggesting a multiphase structure of the Udachnaya-Vostochnaya

Pipe and the replacement of kimberlite breccia filling its higher level by different kimberlite rock varieties (different generations) at a depth of a few hundred meters. The amount of xenoliths occurring at deeper levels in one of the varieties (sometimes named "The Second") exceeds that in the first variety by almost three times (Marshintsev et al. 1975b), which fully complies with data from Borehole 222.

Of the many aspects of study of xenoliths set in the kimberlite breccia, that of xenoliths transported vertically for a long distance is of particular interest. The xenoliths are represented by rock fragments both derived from greater depth and those which subsided within the diatremes below the level of their primary occurrence. Xenoliths captured by the kimberlite magma from the upper mantle and from different levels of the Earth's crust are much less abundant than fragments of the diatreme country rocks. The former comprises no more than 1 vol. % and averages about 0.5 vol. % of the kimberlite breccia. Unlike xenoliths belonging to this group, inclusions of enclosing kimberlitic rocks are constantly present not only in the kimberlite breccia, but also in massive kimberlite varieties.

The study of xenoliths of the mantle and crustal rocks leads to a more valid judgement on the depth of the genesis of the kimberlite magma, the chemistry of the lower mantle, and the thermodynamic conditions dominating in it; the results obtained are also helpful in the interpretation of geophysical fields, etc. Some authors recognize "plutonic" and "metamorphic" rocks in xenoliths of these groups (Marshintsev et al. 1975a), but this does not appear appropriate either terminologically or essentially, since "plutonic" rocks include metamorphogenic diamond-bearing eclogites as well, while "metamorphic" rocks contain granites and some other igneous rocks occurring in the Earth's crust. It seems more expeditious to distinguish between xenoliths of the mantle and crustal rocks and to recognize subgroups on petrographic grounds.

A point of much more essential controversy is the genesis of apparent plutonic inclusions, whose mineralogy displays a continuous series of holocrystalline ultrabasic rock varieties such as garnet, dunite, peridotite, and pyroxenite. These inclusions and their individual mineral components (pyrope, spinel, chrome diopside, and diamond) are permanently present in kimberlites, and have thus been known as "cognate inclusions" since the end of the last century. Some authors believe that crystallization of cognate rocks takes place at abyssal depths in the subcrust and consider them as fragments of the upper mantle rocks (Sobolev 1974 and others); others assume that these inclusions are partly of clastic origin and relate them to the segregation of intratelluric phenocrysts from the kimberlite magma during the subcrustal stage of its evolution (Milashev 1972a and others).

No decisive evidence for either viewpoint has been found as yet, so that both receive similar recognition. It should only be noted that variations observed in the chemistry and diamond content of kimberlites from different diatremes, related to the thermodynamic regime of formation, differentiation, and other alterations in the kimberlite magma, may be better explained by the "segregation" hypothesis of cognate inclusions as early crystallization products; it theoretically analyzes different aspects of kimberlite magma evolution and helps to construct models of the upper mantle of platforms (Milashev 1972, 1974a).

Mantle- and crust-derived xenoliths, as well as cognate inclusions, are unevenly distributed in kimberlite breccia. As a rule, individual generations of kimberlitic rocks within even a single pipe differ markedly in the ratio of abundance of xenoliths to cognate inclusions. It should be noted that in diatremes of simple structure, plutonic xenoliths and inclusions are concentrated chiefly in the central (axial) parts, while xenoliths of country rocks are more abundant along the periphery.

Xenoliths set in the kimberlite breccia of any diatreme and occurring substantially below the level of their primary occurrence should be examined to recognize fragments of the rocks which are present in, or absent from, the modern section of the territory. Xenoliths of the rocks observed in the country rocks surrounding the diatreme permit determination of the depth of sinking of rock fragments and, hence, a minimal depth to an active cavity of the pipe prior to its filling with kimberlite magma. Of the xenoliths reliably located in the section, the greatest depth of sinking is recorded for fragments of the Dwyka black shales. They are cut by the Kimberley Pipe at the 15.5 m level, and in the kimberlite breccia they are traced down to 769 m from the present surface (Williams 1932). A point worthy of mention is that the transition of the Kimberley Pipe into the feeding dyke is established at a depth of 1073 m.

Fragments of rocks absent from the present section can be found in kimberlite breccia; they contain a very important record not only of kimberlite volcanism, but of the geologic history of the whole region. The study of the xenoliths ensures a more reliable determination of the composition and age of deposits developed during diatreme formation, but removed by later erosion. Moreover, the feedback principle enables us to draw a more accurate age boundary of kimberlite volcanism and to determine the denudation depth of pipes and dykes.

One of the best-known xenoliths from the group in question is a belemnite rostrum *Pachyteuthis (Acroteuthis)* sp. (?) cf. *Subrectangulata* Blüthg, found in the kimberlite breccia of the Obnazhennaya Pipe, according to V.N. Saks, typical of the Upper Volgian-Valanginian (Milashev and Shulgina 1959; Milashev et al. 1963). At the present exposure surface the kimberlitic rocks of the pipe occur in dolomites of the Turkut Formation (Sinian). No deposit younger that Early Cambrian has been found in the region, and the nearest outcrops of Late Jurassic and Cretaceous rocks occur 150 km to the north and 160 km to the east. Estimates suggest that the rostrum sank down the pipe cavity to a depth of a few hundreds of meters; at the same time this is the minimal depth of diatreme erosion, since the rostrum was found in the present near-surface level.

In Yakutia one of the pipes studied in detail contains deverse xenoliths yielding fossils which help to definitely establish the age of the country rocks. The xenoliths contain rocks both building up the diatreme walls and removed by erosion from the area adjacent to the pipe. The youngest preserved deposits pierced by kimberlites are Early Silurian (Middle Llandoverian); their apparent thickness is about 60 m. The Middle Ordovician carbonate deposits occur below. Table 3 presents a list and elevations of the fossiliferous xenoliths from the kimberlite breccia of the pipe. Analysis of the above material and the thickness of deposits composed of xenoliths in the adjacent areas led Brakhfogel et al. (1979) to the conclusion that no less than 200–300 m of the pipe has been eroded since the time of formation. The depth of sinking of xenoliths containing fossils of Late Llandoverian (sp. 66), Middle and Late Devonian (sp. 62 and 93/239) age is estimated at 400–600 m.

Table 3. Elevation of fossiliferous xenoliths from kimberlite breccia in one of the pipes in Yakutia. (Brakhfogel et al. 1979)

Specimen	Fossils[a]	Age	Depth of occurrence m
50	*Stenojonotriletes* aff. *formosus* Naum., *Leiotriletes microrugosus* (Jbr.) Naum.	D_3 fr(?)	91
53	*Stenojonotriletes* aff. *definitus* Naum., *Calamospora microrugosa* (Jbr.) Naum., *Leiotriletes* sp.	D_3 fr(?)	103
93/239	*Leiotriletes nigratus* Naum., *Trachytriletes* sp., *Retusotriletes* sp., *Spinosporites* sp., *Dictyosporites* sp., *Granisporites* sp., *Turrisporites* sp., *Cyclobaculisporites trichacantus* (Lub.) Lub., *Archaeoperisaccus*? Naum.	D_3 fr(?)	239
62	*Striatoproductus tungusensis* Nal., *Productella* sp., *Gypidula* sp., *Devonogypa* aff. *spinulosa* Navl., *Dechenella* sp.	D_2 (gv?)	310
67	*Novakia* sp.	$D_1 z_1$-D_2 ef	67.5
75	*Paranovakia* sp., *Knoxiella* sp., *Baschkirina* sp.	$D_1 z_1$	155
69	*Leiociamus* aff. *circularis* Abush., *Hatangeus* sp., *Costaegera* sp., *Nichamnella*? sp., *Cytherellina* aff. *oviformis* Abush., *Pseudorayella* sp.	$S_1 e_3$-v	103
60	*Lenatoechia elegans* Nikif.	$S_1 l_3$	56.4
66	*Sibiritia eurina* Abush.	$S_1 l_3$	196
11	Brachiopods	$S_1 l$	64–78
20	Tabulate corals	$S_1 l$	222
21	Brachiopods	$S_1 l$	234
63	Ostracods, bryozoans	$S_1 l$	68.5
68	Trilobites	$S_1 l$	93

[a] The table presents data only on xenoliths from the eroded deposits and from the youngest, still preserved, deposits pierced by kimberlites.

Similar depths of downward movement of xenoliths (about 550 m) were recorded for pipes in the Kuranakh kimberlite field, where the Universitetskaya Diatreme, located in Middle Cambrian deposits, was found to contain xenoliths of: humic fusovitrain coal and tuffstones yielding *Equisetites* sp. and a pollen assemblage suggesting a Middle Carboniferous-Early Triassic age for the xenoliths (Brakhfogel et al. 1975).

Sinking of fragments of sedimentary rocks to a depth of 1 km was established in the kimberlite pipes in North America, where diatremes in the basement of the platform contain xenoliths derived from sedimentary strata varying in age from Ordovician to Cretaceous and Eocene inclusive (Hearn 1968; McCallum et al. 1975).

Distribution of Diamond

Even in rich primary deposits diamond amounts to only a few hundred thousandths of a percent. Some deposits with a diamond content of $(2-3) \times 10^{-6}\%$ of the rock are mined. Only about 2.5% of the total number of kimberlite bodies are mined throughout the world. The remainder of the pipes and dykes even within the diamond-subfacies kimberlite terranes are composed of medio- and low-diamondiferous rocks where the diamond content is a hundred to a thousand times lower than in mined deposits.

The content of diamond within any individual diatreme varies to a much lesser degree, although there could also be sections with either high or low content of this valuable mineral. Variations in diamond content of the rocks filling a body may be due to: (i) the formation of weathering crusts rich in diamond on the kimberlite bodies; (ii) the presence of two or more generations of kimberlitic rocks in complex-structure bodies differing in original diamond content; (iii) the uneven distribution of xenoliths of country and other foreign rocks.

Weathering crusts may often be much richer in diamond than primary rocks. For example, in the weathering crust of the Premier Pipe, the amount of diamonds several times exceeded that of the underlying kimberlite (Williams 1932). Even under tropical climate conditions a zone of exogenic enrichment in diamonds usually does not exceed several tens of meters; as a rule, the lower boundary of the zone is complex in outline, and the depth of penetration may vary greatly through different sections of a diatreme.

Various generations of kimberlitic rocks in complex-structure bodies greatly differ in their proportion of diamond. In the Mir Pipe the content of diamond in the richest variety of kimberlite breccia is 2 to 2.5 times higher than in the poorest variety (Botkunov 1964). The western portion of the Kimberley Pipe has never been mined, as the kimberlite breccia is poor in diamond there (Williams 1932).

The depletion caused by the presence of foreign rock fragments affects the concentration of diamonds in the kimberlites; this is not only intuitively perceived but also supported by the results of statistical processing of a wealth of data for the Mir, Udachnaya-Zapadnaya, and Udachnaya-Vostochnaya Pipes (Marshintsev et al. 1975a; Zolnikov and Kovalsky 1976). However, the proportion of fragments of foreign rocks and minerals generally totals about 50% of the rock so that the mixture of xenogenic material infrequently leads to a decrease of more than half the diamond content.

The relationships between diamond content and heavy-mineral fraction concentration as a whole, and the content of individual minerals obtained through data processing (Zolnikov and Kovalsky 1976) confirm the idea that the content of diamond and other heavy-mineral fractions varies in different generations of kimberlitic rocks.

These causes affect the variability of diamond content across horizontal and vertical sections of diatremes. Nevertheless, some authors describe a regular decrease in diamond content in the kimberlite pipes with depth and relate this to "crystallization of diamond in an intermittent magma reservoir". As the understanding of a law of diamond distribution in kimberlites is of theoretical and practical significance, we must first consider evidence for "a steady and regular" decrease in diamond content with depth. A very low content of gems in the kimberlite necessitates volumi-

carat/t

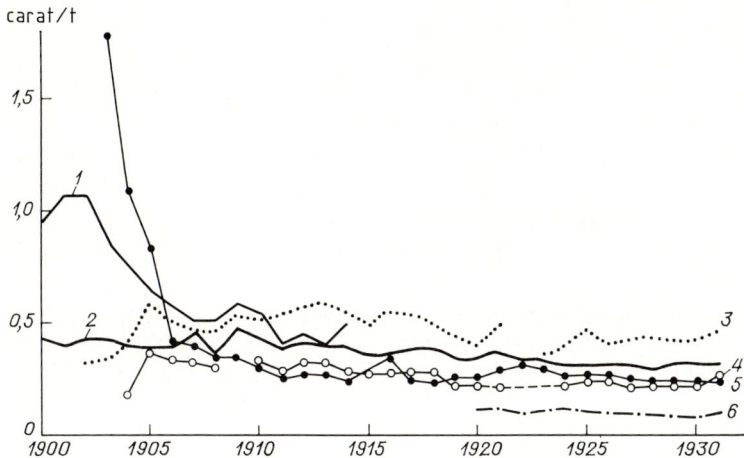

Fig. 13. Diamond content determined from long-term mining operations in the most productive mines of Southern Africa. (Williams 1932; Stutzer 1935). Pipes: *1* De Beers and Kimberley; *2* Wesselton; *3* Bultfontein; *4* Dutoitspan; *5* Premier; *6* Jagersfontein

nous sampling to ascertain the presence of diamond, and depends on the average size of crystals in the deposit studied. Williams (1932) states that it is not unusual to obtain diamonds far surpassing medium-sized crystals for which many hundreds of tons of ground without diamonds must be sorted. He provides an example of a mine yielding an average of about 0.3 carat of diamond per ton of kimberlite and notes that if, for example, a 400-carat diamond were discovered in a load of ground (as could happen), over 1300 tons of ground (without diamonds) has to be processed.

Even larger volumes of samples are required to reliably determine the diamond content in kimberlites. Meager samples taken from boreholed cores probably cannot give very accurate estimates. Therefore, Botkunov's conclusion (1964) of a 2.5 decrease in diamond content at the deeper levels of the Mir Pipe should be checked, because the variations in diamond content across the section, recorded by Botkunov, comply with those in different generations of the kimberlite breccia filling the pipe.

To determine the character of diamond distribution across the vertical section of kimberlite bodies, we may rely on the mineral content data accumulated over many years, during which hundreds of thousands and millions of tons of kimberlitic rocks from successively deeper levels of diatremes have been mined and processed. Figure 13 shows the diamond distribution pattern as a function of depth in six of the most productive mines of South Africa, based on data presented by Williams (1932) and Stutzer (1935).

In analyzing the above data, one must remember the thick weathering crust, highly rich in diamond, developed on the Premier Pipe so that the diamond contents determined from the outputs recorded in the early years of operation after its discovery (1902) are not indicative of diamond concentration in the kimberlite. In the period 1903 to 1906, increasingly deep levels of the residual weathering crust were developed. Since 1906 diamonds have been extracted from a rather poorly weathered kimberlite. However, owing to the irregular penetration of weathering processes to

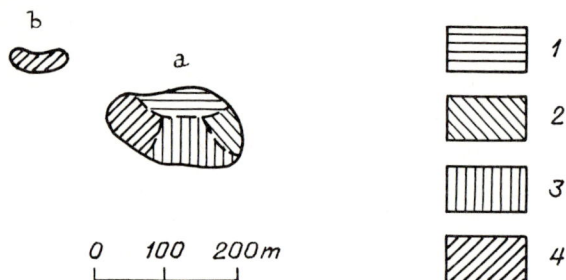

Fig. 14a,b. Distribution of diamonds in **a** the Kimberley and **b** St. Augustine Pipes. Kimberlite breccia containing: *1* abundant bort; *2* edged colorless octahedrons; *3* numerous crystal clasts and pale yellow octahedrons with characteristic obtuse crystal edges and rounded faces; *4* unpayable content of diamond with dominating brown and smoke-colored octahedrons and numerous twin crystals

various depths at different sections of the diatreme, a gradually diminishing amount of exogenic enriched material contributed to it for further 2 years. From 1908 to 1970 the diamond content in the kimberlitic rocks of the Premier Pipe did not vary considerably, and amounted to 0.3 carat per ton.

The data on the De Beers and Kimberley consolidated mine cover the last 14 years of the 25 years of its operation, but the De Beers Pipe was mined until 1908. The data reflect a variability in initial content of diamonds in the rock, since the residual weathering crust and sections containing apophyses at depth had been exploited in the early years of operation. Despite this fact, half the amount of diamond in the kimberlites was extracted during the above period. In interpreting the data, one should allow for a possible causal relationship between the observed phenomenon and a progressive increase in the proportion of subdiamondiferous generations of kimberlite breccia in lower regions of the diatreme, apart from the assumption of nonsteady vertical distribution of diamonds in a single rock variety.

Detailed studies of higher regions of the Mir Pipe show that kimberlite breccia varieties differing substantially in diamond content but similar in crystal habit may display monotonous petrographic structure, unidentified macroscopically in some cases. Williams (1932) clearly indicates the presence of an uneconomic complex of kimberlitic rocks in the north-western De Beers Pipe and of the kimberlite breccia poor in diamond in the western Kimberley Pipe. Figure 14 shows a complex diamond distribution pattern in the higher levels of the Kimberley Diatreme.

It is quite possible that generations of kimberlite breccia similar to rich varieties in habit, but substantially inferior in diamond content, may have contributed to the formation of deeper levels of the Kimberley and De Beers diatremes. This assumption is indirectly evidenced by the fact that these pipes so far represent the only recorded case of decrease in diamond content as a function of depth in rocks similar in the extent of weathering.

Of the other four diatremes described by Williams (1932) and Stutzer (1935) (see Fig. 13) the Wesselton and Jagersfontein mines show no substantial variations in diamond content, but at Bultfontein and Dutoitspan the higher levels were even somewhat lower in diamond as compared to underlying levels. Stutzer (1935) relates the data on the Dutoitspan mine to an increase of original diamond content in the kimberlites with depth.

Thus, highly representative data obtained through the study of the multiyear production of some primary deposits suggest that, despite a slight difference in the original content of diamonds in five of the six best-known mines, it does not vary substantially with depth. Hence, irrespective of the cause of the observed drop in diamond content in the kimberlites from one of the six mines, this may be considered rather as an exception to a general rule. The above data imply a fairly even original distribution of diamond, both horizontally and vertically, across diatremes filled with kimberlitic rocks of one generation.

Persistent diamond distribution in any kimberlite variety and considerable change in mineral concentration in rock generations differing in composition and thermodynamic regime indicate a certain relationship between the amount of diamonds, the morphology of their crystals, and the content of representative elements in the kimberlites. A method of evaluating the diamond-bearing potential of kimberlite was proposed earlier by the author (1965), based on the thesis that the diamond-bearing properties (DBP) of kimberlite are the result of two inter-related factors: (1) the chemical potential for diamond formation (CPD), and (2) the degree of preservation of the crystals (DPC). The CPD is expressed by the formula:

$$CPD = \frac{Fe : Ti}{\log (Fe + Ti) + 1/2 \log (Al + K + Na)} . \tag{1}$$

The substitution of weight contents of proper elements into the formula gives some abstract positive numbers which vary from ones to tens. The correlation between the CPD and DBP in deposits formed under similar PT conditions indicates that the DBP increases with increasing CPD.

The argument behind (1) is that a magma with the highest Fe/Ti ratio would theoretically be the most favorable for crystallizing diamond, whereas Al, Na, and K will be "diluting" factors. The degree of preservation of diamonds is held to be the result of two factors: (a) the PT conditions of crystallization of a particular body of kimberlite magma, which may be either favorable or unfavorable to diamond crystallization, and (b) the degree of resorption of the diamond in the magma when the initial PT conditions change. Both these factors are responsible for the resulting morphology of diamonds, and the present author (1965) proposed that they may be estimated by the proportion of various morphological types of diamond within a kimberlite body. The DPC of a body is expressed by a formula:

$$DPC \cong O + 0.45i + \frac{0.45d}{1 + \lg d} \tag{2}$$

where O = the percentage of flat-faced octahedral crystals, d = percentage of rhombic-dodecahedra; and i = percentage of transitional forms (octahedral-dodecahedra).

In the search for a mathematical expression for the relationship between the DBP and the CPD and DPC, the author assumed that both factors are equally important, since even in a kimberlite with a high CPD, only the correct PT conditions will produce diamonds, whereas kimberlite crystallizing under the correct PT conditions will produce no diamond if the degree of resorption is too high. A mathematical entry of these factors into the equation of diamond-bearing properties of kimberlite is dictated by the necessity to obtain a "response" in metric units (at best, in mg m^{-3}). Such an equation, however, may be applied to evaluate the diamond-bearing potential of kimberlite only when the mean mass of crystals equals 7 mg. For the deposits where the mean mass of diamond crystals is above or below 7 mg, as well as for diatremes composed of kimberlite breccia, corrections for a difference in mean mass of crystals and for the presence of xenoliths should be introduced. With due regard for the corrections, the following equation of diamond-bearing properties (DBP) of kimberlite was derived:

$$DBP = (CPD + \frac{CPD^3}{120} - 1.75\sqrt{CPD})\ \frac{DPC^2\ \sqrt[3]{G}\ \cdot(100 - XC)}{13{,}750}, \tag{3}$$

where DBP is the diamond-bearing properties of kimberlite in mg m^{-3}; G is the mean mass of crystals in a deposit studied in mg; XC is the content of xenoliths in kimberlite in vol. %.

The difference between the evaluated and assayed diamond-bearing potential in all relatively well-known pipes is small, thus proving the actual cause of different diamond-bearing properties in various bodies. Certainly, new data can further specify the relationship, although it has already been used for the prospective evaluation of the diamond-bearing properties of some primary deposits. On the strength of a few chemical analyses of the rock, without resorting to costly assaying, this technique permits rejection of kimberlite pipes whose diamond-bearing properties appear uneconomic (Milashev 1974c).

Distribution of Other Minerals

The major primary mineral in kimberlite is olivine, which ranges from 15 to 60% and averages about 30% of the rock. All other primary minerals are much lower, only phlogopite in kimberlite, subject to micatization (mineral generation formed by metasomatism), locally constitutes up to 20% of the rock.

Olivine in the form of phenocrysts of different generations or as relics in available kimberlite is far from being pervasive. In the rocks filling the higher levels of most pipes and dykes, olivine is completely replaced by secondary products, and only serpentine and carbonate pseudomorphs are present. However, due to the fact that serpentine-carbonate pseudomorphs after olivine stand out sharply against the matrix, it is easy to determine their sizes and amount. Olivine relics in some of pseudomorphs and their peculiar outlines allow the composition of the primary mineral to be reliably established.

Olivine, the main rock-forming mineral of kimberlite, is rather evenly distributed in one rock generation, but locally its concentration drastically decreases in narrow endomorphic zones of pipes, dykes, and sills. Detailed petrographic studies and data processing for the Udachnaya-Zapadnaya and Udachnaya-Vostochnaya Pipes (Barashkov et al. 1976) have confirmed the lateral and vertical persistence of olivine distribution in kimberlite, along with its fairly low content in the contact aureoles.

With olivine evenly distributed in a particular kimberlite generation, its content varies greatly in different kimberlite generations building up a diatreme and in rocks of various bodies and terranes. It is noteworthy that the variations are evident not only in the variability of total olivine content, but in the abundance ratios of individual generations of the mineral.

Three main olivine generations crystallized at intratelluric, plutonic, and hypabyssal phases of the intrusive stage of kimberlite formation are recognized. The intratelluric phase embraces the time interval between the beginning of the rise of the kimberlite magma by the mechanism of zone melting in the upper mantle and its emplacement following the general lines of weakness in the lithosphere. From this time on a new — plutonic — phase of evolution of the kimberlite magma starts and continues until the magma rises to the higher levels of the Earth's crust. The peculiar features pertinent to the magma evolution display a separate — hypabyssal — phase in kimberlite formation.

Olivine I of the intratelluric generation is a part of intratelluric segregations (cognate inclusions) and forms occasional megacrysts in kimberlite. Olivine I accounts for less than 1% of the mineral in the rocks discussed, but it contributes greatly to the energy and chemistry balance of subcrustal differentiation of the kimberlite magma (Milashev 1972a). The distribution and abundance ratios of olivine euhedra of the plutonic and hypabyssal generations yield new insight into the evolution of the kimberlite magma during transit to its present position. Magmatic processes operative there predetermine to no inconsiderable extent the character of the explosion stage resulting in the formation of explosion pipes; we shall now deal with these crucial processes.

The mechanism of the movement and the energy balance of kimberlite magma in the uprising following the shattering of the Earth's crust differ essentially from the radiate transport of the molten magma in the upper mantle. For example, when the magma is transported due to zone melting, its volume and heat increase through most of the intratelluric phase, but the volume of the kimberlite magma (porphyritic crystals, or phenocrysts) remains practically constant, and the heat decreases in the course of the upward movement along fractures.

During the rise along the zones of weakness in the lithosphere, the kimberlite magma loses energy, chiefly to the surrounding rocks. However, despite the heat loss, the temperature of the molten magma remains optimal owing to partial crystallization accompanied by a considerable heat release. Most phenocrysts crystallizing during the plutonic phase consist of olivine of the second generation, hence, the anions of orthosilicic acid, cations of magnesium and, in part, iron represent in fact "fuel" which in "burning" maintains the required magma temperature. Consequently, the amount of olivine II phenocrysts depends on the magma heat balance and hence may be used as a measure of the relative variability of the thermodynamic regime at the plutonic phase of kimberlite formation. The amount of olivine II in the Yakutian kimberlites varies widely, suggesting considerable changes in the physicochemical conditions of their formation (Table 4).

The kimberlite and picrite magmas which followed in their strikes the general local zones of weakness reach the higher levels of the Earth's crust. Although the geology alone cannot definitely distinguish between the plutonic and the hypabyssal

Table 4. Mean content of olivine of plutonic (II) and hypabyssal (III) generations in kimberlites of major fields of the Central Siberian province

Field	Number of diatremes studied	Mean content[a], vol. %	
		Olivine II	Olivine III
Malaya Botuobiya	6	21.6	14.0
Daldyn	22	9.2	22.8
Markha-Alakit	9	8.5	19.9
Verkhnyaya Muna	8	17.3	17.2
Omonos-Ukukit	6	3.1	21.7
Luchakan	8	15.4	15.2
Kuranakh	6	15.8	13.1
Motorchuna	2	19.7	18.0
Chomurdakh	5	18.2	17.8
Nizhnyaya Kuonamka	9	6.2	12.7
Srednyaya Kuonamka	8	6.4	13.6
Dzhyuken	8	14.8	11.6
Nizhnii Ukukit	1	15.8	17.9
Merchimden	9	15.4	20.8
Verkhnyaya Molodo	5	13.0	22.7
Kuoika-Beenchime	5	24.9	11.6

[a] The content of olivine in each pipe was calculated as an arithmetic mean for three to five, or, rarely, for seven to ten thin sections; the value obtained always denotes the volume of kimberlites, since xenoliths were omitted when calculating.

rocks, the mineralogy and petrology convincingly suggest the break in crystallization and, hence, in the regime of magma formation. Phenocrysts of olivine II, ilmenite and phlogopite I, crystallized in the plutonic phase, are fused and corroded. They are larger than olivine III, ilmenite, and phlogopite II, which are always idiomorphic and grow locally on the corroded phenocrysts of the earlier generations, making them idiomorphic. Their optical properties suggest that both olivine generations are similar in composition, and that ilmenite and phlogopite are probably somewhat more ferruginous (Milashev et al. 1963; Blagulkina et al. 1975). Abundance ratios of different generations of the above minerals in individual rock varieties differ essentially, the morphology of the crystals remaining, however, unchanged in all the kimberlitic and picritic rocks. A decrease in the temperature of mineral crystallization (melting) due to drop in pressure, rather than heating of the ascending magma, leads to the fusion of phenocrysts.

Total heat losses and the amount of crystallized phenocrysts of olivine II and III, controlled by heat losses, are related not only to the temperature difference between the country rocks and the magma, but also to the length of time they were in contact. By virtue of the fact that the areas of kimberlite fields usually do not exceed a few thousand square kilometers, the temperature difference within a single field terrane may be taken to be similar for certain sections of the Earth's crust. Hence, a relative variability of heat regime in the course of intracrustal magma evolution which resulted in the formation of kimberlitic and picritic rocks of each individual terrane depends practically only upon the time necessary for the upward movement of the molten magma, i.e., upon the rate of the movement.

As a result, we can turn from the content and abundance ratio of olivine of different generations and the rocks of a particular diatreme to the estimation of the relative rate of rise of the melt in the plutonic and hypabyssal phases and during its intracrustal evolution. The relative rate of the upward movement, calculated from the content of olivine II and III, and an approximate excessive pressure of the magma, estimated from the size of an explosion pipe, enable us to roughly determine the apparent permeability of the weakened crustal zones during the plutonic and hypabyssal phases and the average permeability within the zones from the lower to higher levels in the Earth's crust. Strictly speaking, differences in viscosity of the magmas filling different diatremes should also be allowed for the calculations. However, slight changes in the chemistry even of the extreme kimberlite varieties from different provinces (Milashev 1965, 1974a and others) suggest no essential differences in viscosity of the magmas filling pipes and dykes within a single area.

The present author has already dealt with the above problems at length (Milashev 1973) and derived a formula to approximately estimate the relative crustal permeability v from the content of second and third olivine generations (Ol_{II} and Ol_{III}) in kimberlites with regard to diatreme size S:

$$v \approx (200/Ol_{II} + 100/Ol_{III})\,(1/\log S). \tag{4}$$

Formula (4) was further transformed to calculate the crustal permeability in the deeper and near-surface levels where the plutonic and hypabyssal phases of kimberlite formation proceed (Milashev 1979):

$$v_p \approx 200/(Ol_{II} \log S); \tag{5}$$

$$v_h \approx 100/(Ol_{III} \log S). \tag{6}$$

Apart from pervasive primary phlogopite, metasomatic mica occurs locally in kimberlite. Naturally, the difference in genesis is reflected in the distribution pattern and proportion of these mineral varieties. Primary phlogopite occurs as euhedra of two generations. Phlogopite crystals of the first generation are a few centimeters in diameter and average 1–5 mm in size. On a plane (001) they are oval and rounded. Phlogopite of the second generation occurs in the form of idiomorphic crystals, 0.3–0.5 mm across, evenly distributed throughout the matrix. On the contrary, metasomatic phlogopite laths are different in size, and form intricately shaped sections in kimberlitic rocks.

Ilmenite is evenly distributed in one kimberlite generation, and no important changes have been recorded in its content. Only in rock varieties which were subject to maximal alterations at the post-magmatic stage is much of the ilmenite replaced by rutile and anatase which obscures the primary content and distribution pattern of ilmenite. As a rule, some kimberlitic varieties within one single or separate diatremes of the same area, along with rocks of different areas, differ greatly in the amount of ilmenite from a few hundredths of a percent to a few percent of the rock.

In kimberlitic rocks, pyrope garnet and spinel show a similar pattern of distribution. However, the weight content of ilmenite is a few times and a few tens of times, respectively, higher than those of garnet and spinel; moreover, unlike ilmenite, they could hardly have been post-magmatically altered.

CHAPTER 4

Contact Effects of Kimberlite

Mechanical Effects

Like other igneous rocks, kimberlite produces a contact effect on the enclosing rocks and xenogenic material therein. In character and consequences the effects fall into two major groups: (1) mechanical; (2) thermal and chemical.

The term explosion pipe which has been applied to kimberlite diatremes since the end of the last century means that kimberlite volcanism was accompanied by violent explosions which affect the pipe-enclosing country rocks. At first sight this is also suggested by the occasional deformation of pipe-overlying deposits, although in practice, as shown above, they should be classified as exogenic structures. Without dwelling on the folding (exogenic in nature) of diatreme-overlying deposits and on the resultant sheeted and shatter zones, we shall consider deformations which affect the enclosing country rocks during diatreme formation.

Mechanical actions operative in the process of formation of kimberlite bodies give rise to contorted bedding and heavy fracturing of the enclosing rocks. Contorted bedding is very common in the blocky structure of exocontact zones, where some blocks dip toward the contacts of the diatreme, and others to the opposite side (see Fig. 6). Vertical displacements of the blocks are low in amplitude, about 1 m, infrequently reaching 3–4 m (Poturoev 1976). Within a distance of a few meters to tens of meters from the diatreme contacts there are no deformation effects and the sedimentary kimberlite-enclosing deposits lie subhorizontally, i.e., in the mode typical of platforms.

Kimberlite pipes and dykes may occur at hinges of brachyanticlines. A causal genetic relationship between these geologic features remains uncertain, and may be accepted only in terms of structural control. An inverse relationship of events if the intrusion of kimberlite were a reason and the folds were a consequence seems quite improbable because of the incommensurability in sizes of folds and coincident kimberlite bodies.

Strata containing a kimberlite body are cut by numerous fractures, some of them developing throughout the geological history of a certain crustal block[1], while others

---

[1] The concepts of planetary and regional fractures in rocks apply to planetary and regional investigations, but detailed studies of fissures developed at some localities of the territory suggest the existence of relatively small crustal blocks with areas of several hundred to several thousand square kilometers cut by typical and persistent fissure systems differing from those developed in some adjacent blocks in density and orientation of fissures.

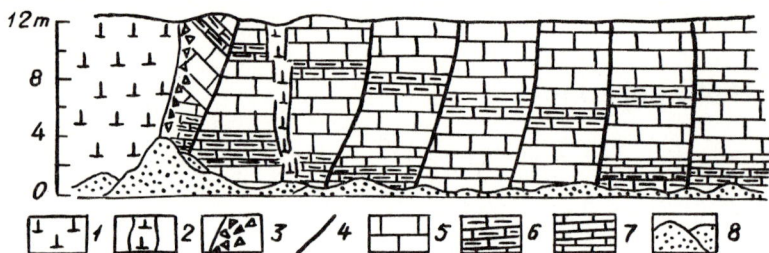

Fig. 15. Structure of a near-pipe space (detail). (Nikitin 1980). *1* Kimberlite breccia; *2* kimberlite vein (out of scale); *3* shatter zone of country rocks; *4* faults (shear joints); *5* dolomite; *6* clay limestone; *7* limestone; *8* talus

are related to the formation of the kimberlite bodies. The fact that some fractures occur near the kimberlite bodies and are absent at some distance from the diatremes, along with fracture orientation relative to the contacts of pipes, dykes, and sills, permit their assignment to the second group.

Kimberlite volcanism causes the formation of fissure systems, subperpendicular and subparallel to the diatreme contacts. Subparallel fissures are located mainly in relatively narrow zones at a short distance from the contacts. For example, in the dolomite surrounding the Obnazhennaya Pipe, apart from regional fractures, there are sets of fissures subperpendicular and subparallel to the body boundary (see Fig. 5). Structural analysis indicates the funnel shape of the Obnazhennaya Pipe with a slope of the contact surfaces of about $70°$ to the horizontal (Milashev 1960).

Systems of fissures subparallel to the contacts in the pipe-enclosing dolomite also dip toward the diatreme center at angles of $60-75°$. The fissures are rectilinear, occasionally slightly flexuous, with rugged surfaces. They are distributed unevenly: at a distance of $5-10$ m from the contact one can observe two or three badly cracked zones, $0.5-1.0$ m thick, with single fissures of this type in between. In each zone the fissures are spaced $1-3$ m apart. In structural position and genesis the fissures are similar to marginal cracks; their presence indicates the vertical direction of force during the emplacement of the kimberlite pipe.

Some large and slightly eroded diatremes show much greater deformation of the country rocks, which commonly collapse in the near-pipe space along concentric shear joints with crescent-shaped cross-section. The total amplitude of displacement along all the fractures may reach many tens of meters over an area extending from the pipe contacts for a distance many times larger than its radius (Figs. 15 and 16). In places some diatremes are characterized by dome structures with subsidence features in place of negative structures and rare thrusts (Nikitin 1980).

The above systems of fissures are thus oriented fairly regularly toward the diatreme contacts; but sedimentary deposits in some narrow zones of the country rocks, mainly at contacts with the deeper parts of the diatremes, may be pierced by such a dense network of erratically oriented fissures and locally even be impregnated with kimberlite veinlets, that they are practically indistinguishable from endomorphic breccia richer in xenoliths than in kimberlite material. There are common slickensides, traces of carbonate, sulfide (mainly pyrite) and sulfate (gypsum, celestite)

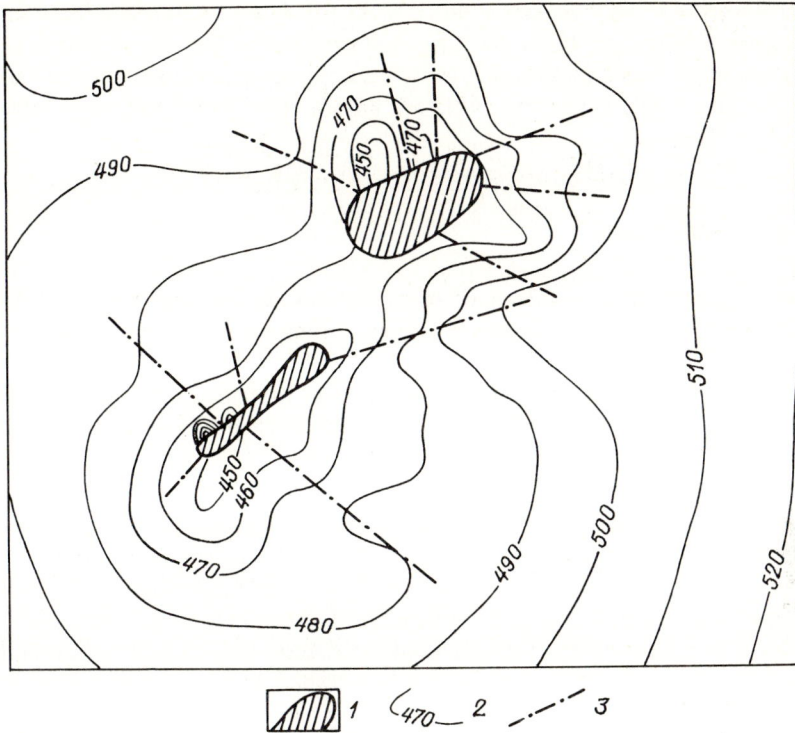

Fig. 16. Negative structure of the sedimentary cover rocks surrounding kimberlite pipes. (Nikitin 1980). *1* Pipe area, plan view; *2* contour lines for the base of Llandoverian succession (Lower Silurian), m; *3* inferred faults

mineralizations with rare pockets and veinlets of bitumen. Most probably the occurrence of mineralizations and hydrocarbon, not typical of kimberlite, in the contact aureoles of the diatremes may be ascribed to a strong hydraulic permeability of disintegrated rocks.

These contacts which are, in essence, transitions from endo- to exomorphic zones of diatremes can be found at Mir (Frantsesson 1962) and some other primary diamond deposits thoroughly studied and exposed by minings. Xenoliths of carbonate rocks derived from the host rocks increase in number and size from the center of the diatreme toward its contacts. At the same time the breccia filling the pipe varies in composition: kimberlite cementing fragments of carbonate rocks in breccia is replaced by broken carbonate material. Apophyses and veinlets of kimberlite breccia impregnated with limestone and dolomite fragments cut large xenoliths, erratic carbonate rocks, and the enclosing strata. Kimberlite filling the apophyses and veinlets is almost fully carbonatized. It contains "kimberlite" minerals such as pyrope, mica, ilmenite, and diamond. The thickness of this transitional zones varies from a few centimeters to 20 m.

The thickness of shatter zones shows no distinct relationship to the dip of contact surfaces, although relatively flattened parts of the contacts of a particular diatreme are generally composed of badly broken country rocks. No direct relationship has been established between the scale of the local fracture pattern and the size of diatremes, but thick zones of fractured rocks and series of erratically oriented (commonly radial) lengthy faults filled with kimberlite dykes are known to surround only very large pipes.

Thermal and Chemical Effects

As a rule, thermal and chemical effects of kimberlite on the host rocks and xenoliths are minor. Carbonate and particularly argillaceous carbonate rocks are most subject to such effects. Although the country rocks and xenoliths of some adjacent diatremes are similar in composition, the scale and nature of their thermochemical action vary greatly, thus suggesting a significant difference in the physical and chemical conditions of their emplacement and in the composition of the kimberlite magma filling the diatremes even within a single field.

At contacts with kimberlite pipes, dykes, and sills, the enclosing carbonate rocks show evidence of some relatively weak contact metamorphism. This is supported by the presence of thin (less than 1 m) zones of dense recrystallized limestones and dolomites at places weakly chloritized, serpentinized, micatized, or pyritized. Only locally do the host rocks undergo high-grade metamorphism parallel with the generation of garnet and apatite. A good example is provided by the Egientei Pipe in northern Yakutia: at its western contact the enclosing limestone is altered to banded inequigranular garnet-phlogopite-apatite carbonate rock showing heterogranonematoblastic texture. The rock is composed of xenomorphic grains of carbonate (50%), laths of phlogopite (20%), apatite (25%), small idiomorphic grains of garnet (grossular) and minor chlorite. Apatite forms bed-by-bed aggregates of erratically oriented laths. Garnet is also distributed unevenly, in some bands its content increases. Large grains of carbonate contain apatite, garnet, and phlogopite (Milashev et al. 1963).

Apart from unaltered fragments of carbonate rocks which have sharp contacts with the embedding kimberlite, almost all pipes generally contain fragments of rocks which had been metamorphosed to a certain degree. For some pipes this may be ascribed to specific processes of alteration, probably related to the composition of kimberlite magma and the composition and temperature of postmagmatic solutions. The metamorphism of carbonate rocks leads to their partial or complete recrystallization, accompanied by garnetization, apatitization, micatization, magnetitization, epidotization, serpentinization, chloritization, and silicification; moreover, in some diatremes neogenic monticellite and cuspidine occur in argillaceous carbonate xenoliths. Occasionally the effects of many processes are recorded in a single specimen. A point worthy of mention is that limestone xenoliths embedded in kimberlite breccia of the earlier generations are more badly altered than those enclosed in the later rock generations. Fragments of cryptocrystalline and fine-grained carbonate rocks are altered, mainly along the rims, to medium- and coarse-grained rocks showing a serrate or mosaic granoblastic texture.

According to Kharkiv (1967c) the high-temperature metamorphism of xenoliths in clayey carbonate rocks is typical of the Verkhnyaya Muna field (Novinka, Komsomolskaya, Zimnyaya Pipes). The fragments have a zonal structure: the center is composed of fine calcite crystals set in an almost unaltered clayey carbonate, while cuspidine and garnet of grossular-and-radite series form a peripheral zone up to several centimeters thick. Fine euhedra of monticellite, together with occasional grains of garnet and cuspidine, occur in the transitional zone. Some xenoliths exhibit pronounced banding, probably sedimentary in origin. Bands differ not only in color, but in mineralogy: some bands are composed of garnet and cuspidine, but the others are essentially of garnet or garnet-cuspidine-monticellite composition. The space between the bands is filled with unaltered clayey carbonate. Banding disappears exactly at contacts with kimberlite, and a fine-grained aggregate of cuspidine and garnet appears (Kharkiv 1967c).

Only in a few diatremes are carbonate xenoliths garnetized. Here we can observe fine (up to 0.1 mm) rhombic dodecahedrons or irregular microcrystals of colorless and weak yellow garnet-grossular, along the rims or throughout the groundmass. Garnet occurs mainly in highly chloritized and serpentinized limestones.

Apatitization of carbonate xenoliths was reported only for Egientei. Columnar needles of apatite (up to 30%) and laths of phlogopite (up to 15%), and minor clinopyroxene, opaque minerals, chlorite, and serpentine occur along the outer zone (several millimeters wide) of recrystallized limestones having serrate granoblastic texture.

Micatization of carbonate fragments in kimberlite breccia results in the development of erratically oriented brown mica laths mainly around the periphery of xenoliths. In monticellite picritic porphyry of the Originalnaya Pipe, fine carbonate fragments are wholly altered to fibrolepidoblastic serpentine-phlogopitic rocks with relict stratification.

Magnetitization of carbonate xenoliths is found in kimberlite breccia of a few bodies (Nadezhnaya, Aerologicheskaya). Partly recrystallized or chloritized limestone has contact rims of microflaky phlogopite; limestone is rich in magnetite powder.

Epidotized limestone fragments were reported from the Lvinaya Lapa Pipe. Isotropic serpentine, along with epidote microcrystals, fills the space between carbonate grains. At Egientei, radially fibrous aggregates of clinozoisite filling rounded interstices are found beyond the micatized rim along the margins of some limestone fragments.

Serpentinization of carbonate fragments is a fairly common feature of kimberlite breccia. Usually serpentine coexists with chlorite or with some other minerals. Microflaky light yellow to greenish isotropic serpentine fills in joints, pores, and interstices of the rock and occurs among carbonate grains, locally with separation of opaque dust. Serpentine spots are common in the rocks; besides, serpentine regularly replaces the carbonate material of the xenoliths. Locally, serpentine forms concentric zones, 3−5 mm wide, or fills the center of the inclusion. Serpentine mainly replaces cement in brecciated carbonate varieties. Primary serpentinization characterizes both the center and periphery of xenoliths.

Chloritization of carbonate leads to the development of chlorite microflakes primarily around the periphery of fragments, where chlorite rims, up to several millimeters wide, are present. Minor chlorite flakes are commonly scattered throughout the xenolith groundmass. Occasional fine fragments are completely altered to carbonate-chlorite rocks. Marlstone is the most strongly chloritized rock; bands rich in clay are preferably replaced.

Locally xenoliths of carbonate rocks are subject to intense sulfide mineralization. In places where numerous pyrite-galenite-sphalerite-carbonate veinlets occur, sulfide constitutes about 50–60% of some fine xenoliths of limestone (Frantsesson 1962).

Concentric structure of xenoliths of bituminous limestone and dolomite may be attributed to the redistribution of bitumen. The alternation of dark (rich in bitumen) and light (almost free of bitumen), mainly external zones is a peculiarity of the xenoliths.

Kimberlite breccia contains few silicate xenoliths, hence the difficulty in understanding their contact alteration. Of particular importance for the solution of the above problems is the knowledge of very high temperature metamorphism and, in particular, of the alteration of syenite xenoliths to sanidinite hornfels observed in kimberlite of Guinea (Vladimirov et al. 1971), as well as fusion rinds on the surface of schist xenoliths at contacts with kimberlite at Zapolyarnaya and Novinka (Kharkiv 1978).

CHAPTER 5

Size of Diatremes and Distribution of Bodies of Variable Sizes in Kimberlite Fields

At the present exposure surface or beneath the cover of young deposits, kimberlite diatremes vary in size from 0.01 to 141.6 ha (but to 54.5 ha on the Siberian platform). In all fields having numereous (more than 10) kimberlite exposures, the area of the largest pipe is a hundred times larger than that of the smallest diatreme. Data available for hundreds of pipes recognized in 16 major fields over the Central Siberian province conclusively indicate a wide variation in size of explosion pipes within each field (Fig. 17). However, we cannot as yet recognize natural clustering of diatremes by using their size. The distribution of diatremes over the Central Siberian province and, probably, of those known from other kimberlite provinces of the world with respect to size (area in plan) is approximated by the logarithmic-normal law (Fig. 18).

The results thus obtained are inconsistent with the idea of the essentially discrete distribution of diatremes relative to size and, hence, with the consequent conclusion regarding the occurrence in each kimberlite field of one or, less commonly, two large (primary) pipes with minor diatremes radiating from them. A regular distribution pattern of kimberlite bodies throughout the fields and, in particular, the tendency toward the distribution of diatremes of different size in the fields are pertinent to a

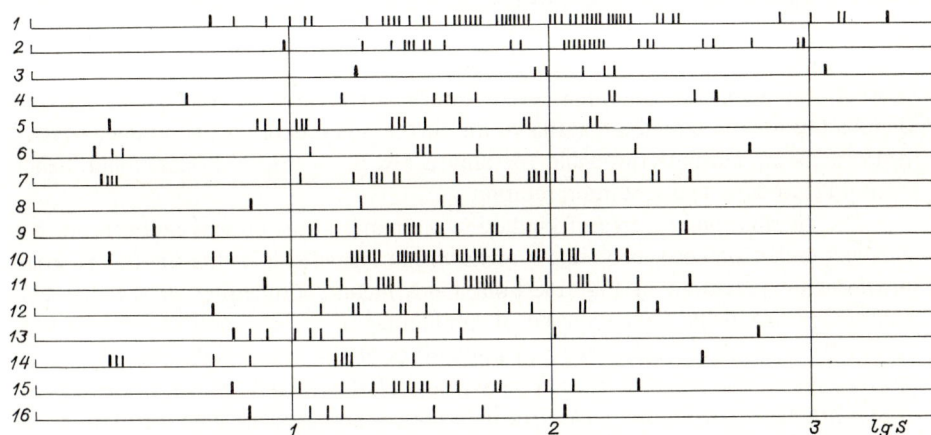

Fig. 17. Area of kimberlite diatremes S (ha, m²) of the major kimberlite fields, Central Siberian province. Fields: *1* Daldyn; *2* Markha-Alakit; *3* Malaya Botuobiya; *4* Verkhnyaya Muna; *5* Chomurdakh; *6* Kuranakh; *7* Omonos-Ukukit; *8* Omonos-Sukhan; *9* Luchakan; *10* Nizhnyaya Kuonomka; *11* Srednyaya Kuonamka; *12* Merchimden; *13* Verkhnyaya Molodo; *14* Kuoika-Beenchim; *15* Dzhyuken; *16* Nizhnyaya Ukukit

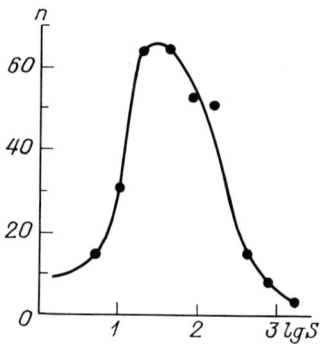

Fig. 18. Frequency of occurrence of kimberlite diatremes of varying sizes, Central Siberian province (for reference see Fig. 17)

wide range of genetic questions of kimberlite volcanism; in practice they help to locate prospective large pipes in the already known kimberlite fields.

Frantsesson (1976) and Frantsesson et al. (1976) discussed in detail the spatial distribution of explosion pipes of different size in kimberlite fields. They believe that any kimberlite field is a central-type structure having one center of "volcanic pressure". A field includes one primary, or central, pipe (rarely two pipes) and a series of smaller "satellite" pipes around the primary pipe, i.e., central vent. In all cases, the central bodies are believed to be the largest, isometric, relatively long-living multistep pipes filled with kimberlites of several generations. The central bodies have lengthy vents of peculiar morphology: they become progressively narrower and remain isometric in shape with depth. The central pipes are high in diamond content and many of them prove to be economic. Within a field the "satellite" pipes are characterized by kimberlite of more variable composition, although each pipe consists of only one or two rock generations. These pipes disappear within a short distance and turn into dykes. As a rule, they are low in diamond or merely barren.

The papers cited have not provided examples from real kimberlite fields. Therefore, it is interesting to correlate the above pattern with the distribution of diatremes, which differ in the above features, in some kimberlite fields in Yakutia. It should, however, be borne in mind that even though comprehensive and reliable data on diamond-bearing potential are available for all the pipes, data on the number of rock generations are currently available for only a few pipes studied in detail and, hence cannot be used for a correct interpretation.

The distribution and area of kimberlite diatremes at the present exposure surface and just under the cover of younger rocks within eight fields of the Central Siberian province are presented in Fig. 19. Four fields include kimberlites of the diamond subfacies; three fields are composed of kimberlites of the diamond and pyrope subfacies; the Dzhyuken field consists of kimberlites of the pyrope subfacies. Figure 19 shows pairs of the adjacent fields (Daldyn and Markha-Alakit; Nizhnyaya and Srednyaya Kuonamka) with due regard for their mutual arrangement as to orientation and distance (to scale) from each other. Morphological boundaries of each field are also drawn by marginal kimberlite outcrops. These boundaries do not provide independent geological information and may be specified in the case of the discovery of new kimberlite bodies outside the field, as the boundaries are merely outlines of the area

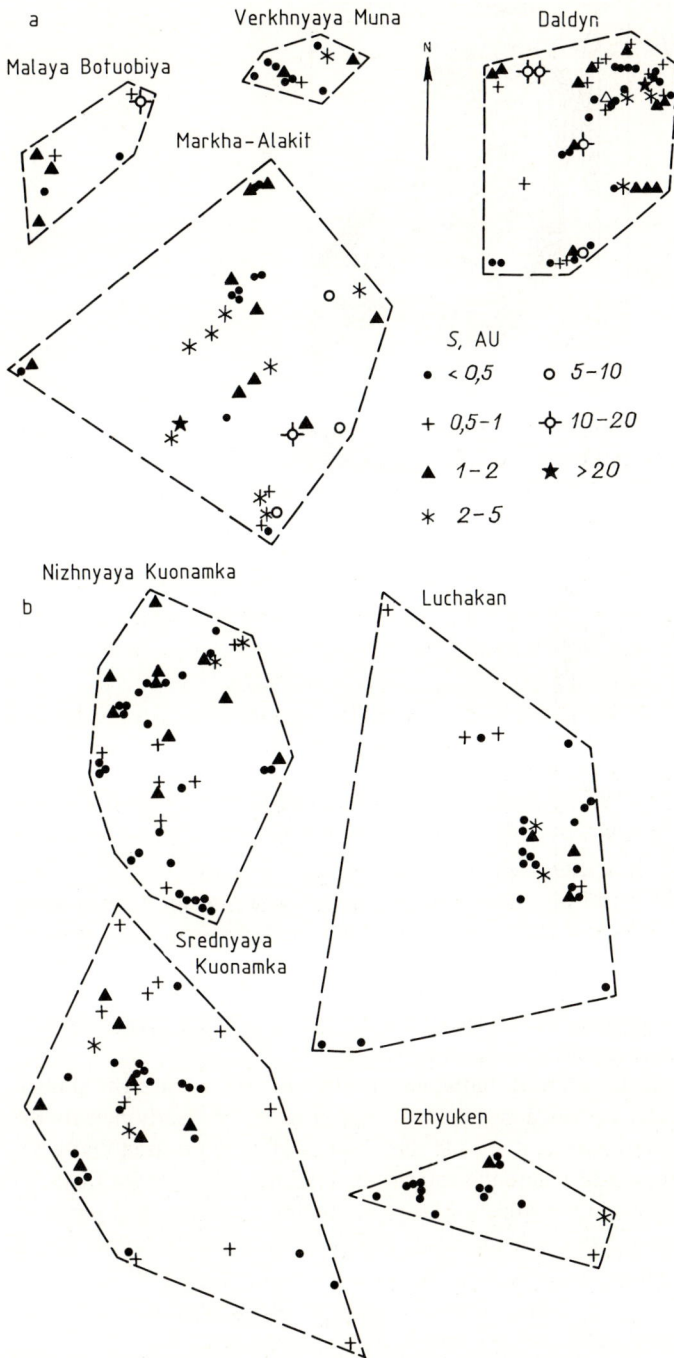

Fig. 19a,b. Distribution of diatremes of varying sizes in the fields of: central (**a**) and north-eastern (**b**) Central Siberian kimberlite province. *Dashed lines* show the morphological (marker) boundaries of the kimberlite fields

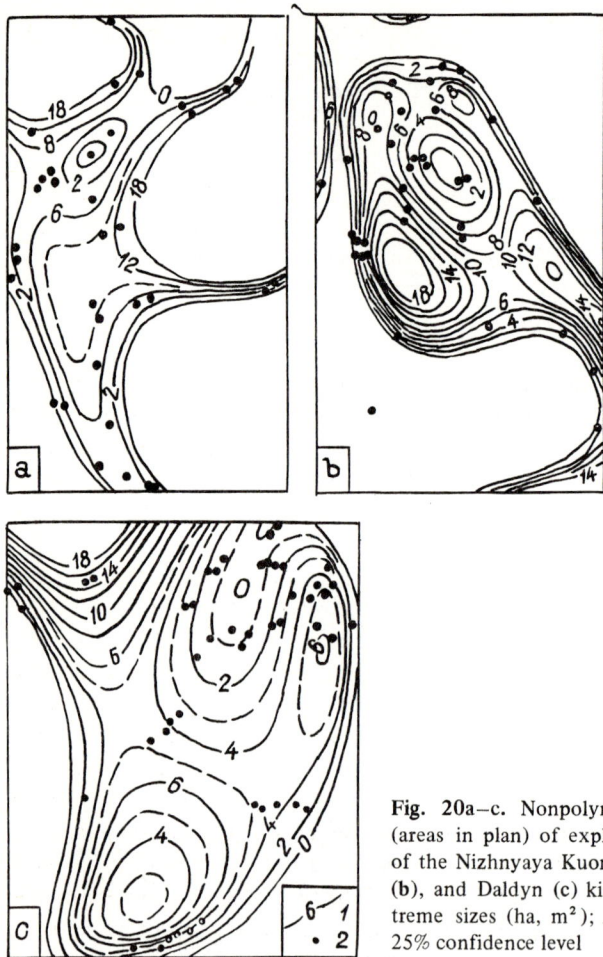

Fig. 20a–c. Nonpolynomial trend surfaces for sizes (areas in plan) of explosion pipes within the confines of the Nizhnyaya Kuonamka (a), Srednyaya Kuonamka (b), and Daldyn (c) kimberlite fields. *1* Isolines of diatreme sizes (ha, m²); *2* kimberlite pipes; trend at the 25% confidence level

of known kimberlite exposures within each field and give a graphic presentation of the field incorporating clusters and individual diatremes.

Analysis of spatial distribution of diatremes of different size in the eight kimberlite type fields of Yakutia reveals a certain tendency. Figure 19 clearly depicts the presence of very large diatremes and a wide distribution of medium-sized diatremes only in the fields composed of diamond-subfacies kimberlites, while in the fields incorporating diamond-pyrope and pyrope-subfacies kimberlite, medium-sized pipes act as large ones, with most diatremes being small and fine. Reduction in mean size of diatremes in the fields where kimberlites of the diamond and pyrope subfacies occur together, as compared to the "pure" diamond-subfacies fields, previously inferred from mean values (Milashev 1965, 1972a), now becomes evident from Fig. 19.

Figure 19 also shows the distribution pattern of the largest pipes in a field. In four fields (Malaya Botuobiya, Verkhnyaya Muna, Nizhnyaya Kuonamka, Dzhyuken) the largest diatremes are marginal ones, and in two fields (Daldyn and Markha-Alakit) they are adjacent to the morphological boundaries of the fields. In the Srednyaya

Kuonamka field, one of the two largest diatremes is located close to the boundary of the field, while the other is situated in the vicinity of its central part. The two largest diatremes of the Luchakan field lie actually in the center of the arcuate kimberlite zone. This implies that the distribution of large and largest pipes within kimberlite areas shows no regular trend.

Analysis of Fig. 19 gives a negative answer to the question as to whether the kimberlite fields are too extensive subdivisions of kimberlite volcanic provinces and whether large and very large diatremes occur in the center of lesser sites, such as clusters of pipes, presumably correlating them with volcanic centers. It is worthwhile to note that despite the fact that two to more than ten clusters and linear groups of diatremes can be recognized in the eight field types presented in Fig. 19, the largest (of the bodies of each group) pipes occur in the central part of only three such groups, the Novinka and Zapolyarnaya Pipes in the southern and northern linear groups of the Verkhnyaya Muna field and the Zarnitsa Pipe in the cluster of kimberlite bodies in the extreme north-east of the Daldyn Field. The absence of a consistent distribution pattern for diatremes of variable size within kimberlite fields is confirmed by trend analysis based on three of the eight field types (Fig. 20).

Thus the model of the internal structure of a kimberlite field proposed by Frantsesson and coworkers, with a large diamondiferous pipe in its center and poor or barren diatremes radiating from the pipe, seems invalid. Neither analysis of graphic presentation (Fig. 19), nor data processing (trend analysis, Fig. 20) reveal a concentric arrangement of diatremes of varying extent within the best-known fields of Yakutia. The internal structure of not only these fields of the Central Siberian province, but also of the remainder of the fields in this, as well as in all other kimberlite provinces of the world, cannot be approximated by such an ideal model. It might well be pointed out that the discrepancy holds true for all the main statements: (1) gradual transitions from the largest to the smallest diatremes are observed in any adequately studied field (see Fig. 17); (2) one or two of the largest diatremes generally occur on the margins, but not in the center of the field; (3) in most fields the highest, including economic, diamond concentrations are present primarily not in the largest, but in medium-sized pipes.

It should also be said that no volcanic edifices of different size show a concentric arrangement in volcanic provinces which would merit observation and comprehensive study. It should be emphasized that parasitic craters in the lower and middle parts of some very high stratovolcanoes are formed as the result of magma outburst through an effusive-pyroclastic cone above its "socle". These craters cannot be considered as prototypes of "satellite" kimberlite pipes surrounding the central vent, since all kimberlite pipes occur in the host rocks (of the cover and basement of the platform) and, with rare exceptions, are not connected with each other at an observable depth, i.e., at a distance of $1-2$ km from the paleosurface.

Moreover, the search for a genetic volcanic cell, minimal in size, but providing important information about the main trends of the process, indicates that no one single volcano is such a unit, since it does not incorporate all the major characteristics of volcanism. A genetic volcanic cell, acccording to Masurenkov (1979), is a volcanic center consisting of a set of volcanoes united by an endogenic matter and energy flux, local in space and persistent in time, which generates a magma; structurally, the flux is embodied in a ring dome-caldera complex. There is no zoning in the distribution of volcanic edifices of variable size within volcanic centers.

In kimberlite provinces, an analogous genetic unit is refered to as a kimberlite field, which, by the only definition known in literature (Milashev 1972b) is an individual site of kimberlite occurrence, formed under similar thermodynamic and structural conditions within a narrow age range (up to several tens of millions of years).

CHAPTER 6

Size of Diatremes and Genetic Features of the Rocks

It is important to establish a possible relationship between the size of diatremes and genetic rock features to solve the problems of the mechanism and conditions of diatreme formation. Chapter 5 deals with a relatively apparent relationship between the mean area of a pipe and the facies of kimberlite in each field: the largest diatremes are found in fields composed of diamond-subfacies kimberlite; medium-sized diatremes occur in fields of diamond- and pyrope-subfacies kimberlites; and very small diatremes occur mainly in fields of pyrope-subfacies kimberlite and picrite porphyries.

The mean size of diatremes composed of rocks of differing facies shows a regular trend for each field marked by varied mineralogical facies of ultrabasic volcanism, for the fields composed of diamond- and pyrope-subfacies kimberlites in association with porphyritic ultrabasic rocks of the picrite facies, as well as for the fields of pyrope-subfacies kimberlite, generally associated with comagmatic picrite rocks. Appropriate information for the north-eastern Central Siberian province is reviewed by Tabunov (1971b). In the fields containing diamond- and pyrope-subfacies kimberlites, the mean size of diatremes (0.44 ha) exceeds those of picrite and picrite-porphyritic pipes (0.29 ha) by 1.5 times, while the maximal mean areal extent (0.67 ha) is recorded for diatremes composed of diamond-bearing kimberlite. It is of interest that in the fields of the group discussed the mean size of kimberlite diatremes, including diamondiferous pipes, shows a clear tendency to decrease as the proportion of kimberlite and, in particular, diamond-bearing pipes drops. However, the mean area of picrite and picrite-porphyritic pipes increases.

By contrast, in the fields of pyrope-subfacies kimberlite, the mean size of kimberlite pipes (0.29 ha) is 1.5 times less than in those of diatremes filled with picrite and picrite-porphyritic rocks (0.47 ha). The mean area of kimberlite diatremes also decreases as their proportion drops and that of picrite-bearing pipes increases. In the north-eastern Siberian platform, diatremes filled with monticellitic and melilitic picrites have minimal mean sizes (Tabunov 1971b).

Together with the thermodynamic regime of formation, the crucial feature of kimberlite, intimately related to the conditions of diatreme emplacement, is the chemistry of the rocks. However, most kimberlitic and comagmatic porphyritic ultrabasic rocks in the picrite facies are subject to heavy alteration due to postmagmatic processes leading to the loss of many important components. Therefore, chemical analyses do not show, as a rule, original contents of silicon, magnesium, calcium, and other element oxides. Traditional methods used in petrochemical analysis, taking all major elements into consideration, seem inefficient to determine the original chemical content of these highly altered, i.e., essentially metasomatic, rocks (Milashev 1962, 1964).

Consequently, preference was given to the content of so-called representative elements, such as iron, titanium, chromium, potassium, and some others, which had suffered no essential redistribution and loss during the postmagmatic stage. To reveal the original chemistry of kimberlite one may use either proportions of major representative elements followed by a statistical estimation of their discrepancies, or the contrast differentiation index (CDI) and differentiation index (DI) of a magma, both estimated from the content of the above elements (Milashev 1965, 1972a). Theoretical premises and the deduction of approximate formulae to calculate CDI and DI of a kimberlite magma have already been discussed elsewhere (Milashev 1972a). The equations may be written as

$$\text{CDI}_{\text{Ti}}^{(n)} = \frac{(I_{k1}-I_{s1})Ti_s}{I_{s1}(Ti_k-Ti_s)} + \ldots + \frac{(I_{k\,n-1}-I_{s\,n-1})Ti_s}{I_{s\,n-1}(Ti_k-Ti_s)} + 1, \tag{7}$$

where $\text{CDI}_{\text{Ti}}^{(n)}$ is the contrast differentiation index of a kimberlite magma calculated from n elements (including titanium) provided that assimilativity of titanium = 1 (hereafter referred to as the titanium contrast differentiation index from n elements); $I_{k1}, \ldots, I_{k\,n-1}$ are the contents of individual representative elements in kimberlite; $I_{s\,1}, \ldots, I_{s\,n-1}$ are the contents of the same representative elements in substratum; Ti_k and Ti_s are the contents of titanium in kimberlite and substratum, respectively;

$$\text{DI}_{\text{Ti}}^{(n)} = \left[\frac{(I_{k\,1}-I_{s\,1})Ti}{I_{s\,1}(Ti_k-Ti_s)} + \ldots + \frac{(I_{k\,n-1}-I_{s\,n-1})Ti_s}{I_{s\,n-1}(Ti_k-Ti_s)} + 1\right] \cdot \frac{Ti_k}{Ti_s}, \tag{8}$$

where $\text{DI}_{\text{Ti}}^{(n)}$ is the differentiation index of a kimberlite magma calculated from n elements (including titanium) provided that the assimilativity of titanium equals unit (hereafter referred to as the titanium differentiation index from n elements); for other designations see Eq. (7).

If we calculate the titanium differentiation index from five major representative elements using the garnet-peridotite model of the upper mantle matter (Milashev 1965, 1972a) the equation takes the form

$$\text{DI}_{\text{Ti}}^{(5)} = \left[\frac{(Fe_k - 5.50) \cdot 0.036}{5.50(Ti_k - 0.036)} + \frac{(Al_k - 1.25) \cdot 0.036}{1.25(Ti_k - 0.036)} + \right.$$

$$\left. + \frac{(K_k - 0.12) \cdot 0.036}{0.12(Ti_k - 0.036)} + \frac{(Cr - 0.13) \cdot 0.036}{0.13(Ti_k - 0.036)} + 1\right] \cdot \frac{Ti_k}{0.036} . \tag{9}$$

The absolute value of $\text{DI}_{\text{Ti}}^{(n)}$ obtained from Eq. (8) depends on the I_{si} content controlled by the above model. Because DI is a relative, but not an absolute characteristic, comparison of relative $\text{DI}_{\text{Ti}}^{(n)}$ values for kimberlites of individual diatremes, fields, and provinces is valid only in the case of the values being calculated using a single substratum model.

Of the specific contrast differentiation indexes of a kimberlite magma which are used in the above equations, the specific index for iron records the most complete genetic information. This index shows the heat balance of kinematic differentiation

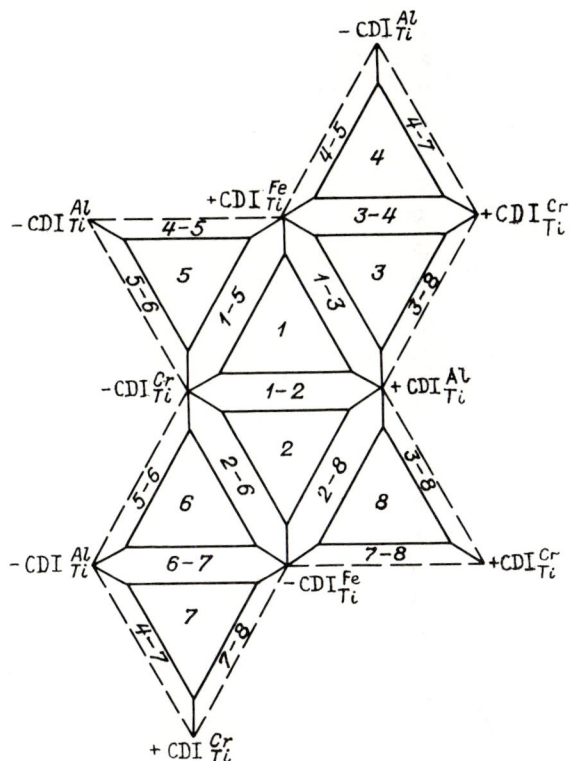

Fig. 21. Basic and intermediate types of kimberlite magma differentiation and appropriate petrochemical rock types recognized with regard for relative values of contrast differentiation index with respect to iron, chromium, and aluminum

of ultrabasic magmas and hence characterizes, in particular, the energetic trend in their evolution (Milashev 1972a).

When calculating the contrast differentiation and differentiation indexes of a kimberlite magma with respect to the five most important representative elements, calculations are performed in practice with respect to only four elements, since in all cases the assimilativity of titanium is taken to equal 1. The degree of final potassium abundance in the rocks is generally determined from the content of phlogopite which allows the recognition of micaceous (> 5% of phlogopite) and nonmicaceous kimberlite varieties. With respect to the remaining three elements, contrast differentiation of a kimberlite magma can be expressed as positive and negative values.

Different methods can take into account the variability of three specific contrast differentiation indexes with sign reversal, and thus recognize appropriate types of kimberlite magma differentiation. The graphic presentation using eight isosceles triangles seems most illustrative. To save space and for convenience, the triangles may be developed as an octahedron corresponding to the basic and intermediate types of magma differentiation and adequate petrochemical rock types (Fig. 21).

Table 5 shows mean areas of diatremes filled with rocks distinguished by CDI_{Ti}^{Fe} and $DI_{Ti}^{(5)}$ values and belonging to different petrochemical types. The data presented in Table 5 show no evidence of any relationship between the mean size of the diatremes and CDI_{Ti}^{Fe} value either for a single field or for all the fields of the Central Siberian province. The absence of correlation does not seem surprising if we assume

Table 5. Mean areas of diatremes (ha) infilled with kimberlitic and picritic rocks of various petrochemical groups and types in major fields of the Central Siberian province

Petrochemical groups and types of kimberlitic rocks		Fields of diamond-subfacies kimberlites			
		Malaya Botuobiya	Markha-Alakit (without Pipe N 1)	Daldyn	Verkhnyaya Muna
Rock groups recognized with respect to relative CDI_{Ti}^{Fe} values	$+ (50 \div 100)$	—	—	0.8	0.1
	$+ (10 \div 50)$	6.3	3.3	0.4	2.4
	$(-10) \div (+10)$	1.1	0.6	4.2	0.1
	$-(10 \div 50)$	0.6	1.6	1.5	0.4
	$-(50 \div 100)$	0.7	3.3	0.4	—
Rock types: diatreme areas are given in accordance with $DI_{Ti}^{(5)}$ values: $\dfrac{<25\ ;\ 25-62}{63-100;\ >100}$	1	$\dfrac{5.0\ ;\ -}{-\ ;\ -}$	$\dfrac{-\ ;\ 3.1}{1.4\ ;\ -}$	$\dfrac{0.3\ ;\ 0.4}{0.4\ ;\ -}$	$\dfrac{0.1\ ;\ -}{-\ ;\ -}$
	1–2	$\dfrac{1.0\ ;\ 0.9}{-\ ;\ -}$	$\dfrac{0.3\ ;\ 0.4}{-\ ;\ -}$	$\dfrac{1.0\ ;\ 1.2}{-\ ;\ -}$	$\dfrac{-\ ;\ 0.1}{-\ ;\ -}$
	1–3	—	$\dfrac{1.0\ ;\ -}{-\ ;\ -}$	$\dfrac{0.3\ ;\ 1.3}{-\ ;\ -}$	$\dfrac{-\ ;\ 1.0}{-\ ;\ -}$
	1–5	$\dfrac{-\ ;\ 12.0}{-\ ;\ -}$	$\dfrac{-\ ;\ 0.1}{-\ ;\ -}$	—	—
	2	$\dfrac{0.8\ ;\ -}{-\ ;\ -}$	$\dfrac{2.4\ ;\ 0.7}{-\ ;\ -}$	$\dfrac{1.0\ ;\ 1.2}{-\ ;\ -}$	$\dfrac{0.4\ ;\ -}{-\ ;\ -}$
	2–6	$\dfrac{0.7\ ;\ 0.1}{-\ ;\ -}$	$\dfrac{-\ ;\ 1.4}{1.6\ ;\ -}$	$\dfrac{6.7\ ;\ 0.2}{-\ ;\ -}$	—
	2–8	$\dfrac{1.4\ ;\ -}{-\ ;\ -}$	—	$\dfrac{1.4\ ;\ 0.3}{-\ ;\ -}$	$\dfrac{0.4\ ;\ -}{-\ ;\ -}$

3	1.4 ; 5.8 / −;−	−;0.8 / −;−	−;1.8 / −;−	—
3–4	−;2.0 / −;−	−;0.9 / −;−	−;3.7 / −;−	—
3–8	−;0.1 / −;−	−;0.3 / −;−	—	—
4	−;6.6 / 0.1;−	—	4.4 ; − / −;−	—
4–7	—	—	—	—
5	−;5.6 / −;−	—	−;0.2 / −;−	−;1.0 / −;−
5–6	0.3 ; 0.7 / −;−	−;5.3 / −;−	—	—
6–7	1.4 ; 3.7 / −;−	—	—	—
7	−;2.6 / −;−	—	—	—
7–8	4.4 ; − / −;−	—	—	—
8	1.3 ; 0.3 / −;−	−;1.0 / −;−	—	—
Averaged data for all rock types	1.6 ; 2.4 / 1.0 ; −	2.8 ; 1.1 / 0.4 ; −	1.3 ; 1.4 / −;−	1.8 ; 4.4 / −;−

Table 5 (continued)

Petrochemical groups and types of kimberlitic rocks	Fields of diamond- and pyrope-subfacies kimberlites							
	Chomur-dakh	Omonos-Ukukit	Omonos-Sukhan	Luchakan	Kuranakh	Srednyaya Kuonamka	Nizhnyaya Kuonamka	Motor-chuna
Rock groups recognized with respect to relative CDF_{Ti}^{Fe} values:								
$+(50 \div 100)$	0.1	1.0	—	0.5	0.1	0.8	0.4	—
$+(10 \div 50)$	0.6	1.0	0.2	0.7	1.0	0.6	0.6	6.1
$(-10) \div (+10)$	0.1	0.6	0.3	0.4	—	0.9	0.6	4.2
$-(10 \div 50)$	0.9	0.6	—	—	—	—	0.8	—
$-(50 \div 100)$	—	1.2	—	—	—	—	—	—
Rock types: diatreme areas are given in accordance with $DI_{Ti}^{(5)}$ values: $\frac{<25 \; ; \; 25-62}{63-100 \; ; \; >100}$								
1	$\frac{0.1 ; 0.9}{0.7 ; 0.1}$	$\frac{- ; 0.3}{0.8 ; 3.1}$	$\frac{- ; -}{0.4 ; -}$	$\frac{- ; 0.3}{0.6 ; 0.2}$	$\frac{- ; 1.7}{<0.1 ; -}$	$\frac{0.2 ; 0.5}{1.3 ; 0.2}$	$\frac{1.1 ; 0.7}{0.3 ; -}$	$\frac{- ; 4.9}{2.6 ; -}$
1–2	$\frac{- ; 0.1}{- ; -}$	$\frac{- ; 0.6}{0.9 ; -}$	$\frac{- ; 0.3}{- ; -}$	$\frac{- ; 0.3}{- ; -}$	—	$\frac{- ; 1.1}{- ; -}$	$\frac{1.2 ; 0.2}{0.3 ; -}$	$\frac{- ; 4.2}{- ; -}$
1–3	$\frac{- ; 0.1}{- ; -}$	$\frac{- ; 1.1}{0.1 ; -}$	$\frac{- ; -}{0.1 ; -}$	$\frac{- ; 0.8}{3.2 ; 0.2}$	$\frac{- ; 0.6}{<0.1 ; -}$	$\frac{0.6 ; -}{1.4 ; 0.2}$	$\frac{- ; 0.5}{0.6 ; -}$	—
1–5	—	$\frac{- ; 0.6}{- ; -}$	—	—	—	$\frac{- ; 0.6}{- ; -}$	$\frac{0.6 ; -}{- ; -}$	—
2	$\frac{1.4 ; -}{0.3 ; -}$	$\frac{1.2 ; 0.2}{- ; -}$	—	—	—	—	$\frac{- ; 1.0}{- ; -}$	—

	1	2	3	4	5	6	7	8
2–6	—	—	—	—	—	—	—	—
2–8	—	0.9 ; 1.0 / —;—	—	—	—	—;1.2 / —;—	—	—
3	—	— / 0.2;—	—	—;0.4 / 1.1;—	—	—	—;0.9 / —;—	—
3–4	—	—	—	—	—	—	—	—
3–8	—	—	—	—	—	—	—	—
4	—	—	—	—	—	—	—	—
4–7	—	—	—	—	—	—	—	—
5	0.5;— / —;—	—	—	—	—	—	—	—
5–6	—	—	—	—	—	—	1.0;— / —;—	—
6–7	—	—	—	—	—	—	—	—
7	—	—	—	—	—	—	—	—
7–8	—	—	—	—	—	—	—	—
8	—	—	—	—	—	—	—	—
Averaged data for all rock types	0.7 ; 0.4 / 0.5 ; 0.1	1.1 ; 0.6 / 0.5 ; 3.1	—; 0.3 / 0.2 ; —	—; 0.5 / 1.9 ; 0.2	—; 1.2 / 0.1 ; —	0.4 ; 0.8 / 1.4 ; 0.2	1.1 ; 0.7 / 0.4 ; —	—; 4.6 / 2.6 ; —

Table 5 (continued)

Petrochemical groups and types of kimberlitic rocks	Fields of pyrope-subfacies kimberlites and picrites					For all the fields (bracketed values refer to a rock type)
	Dzhyuken	Nizhnii Ukukit	Merchimden	Verkhnyaya Molodo	Kuoika-Beenchime	
Rock groups recognized with respect to relative CDI_{Ti}^{Fe} values:						
$+ (50 \div 100)$	0.5	—	1.5	—	0.1	0.5
$+ (10 \div 50)$	0.6	0.4	0.6	0.3	0.2	1.5
$(-10) \div (+10)$	0.4	—	0.2	1.7	0.3	1.0
$-(10 \div 50)$	0.4	—	—	0.1	1.4	0.8
$-(50 \div 100)$	—	—	—	—	—	1.4
Rock types: diatreme areas are given in accordance with $DI_{Ti}^{(5)}$ values: $\dfrac{<25 \; ; \; 25-62}{63-100; >100}$						
1	$\dfrac{-\,;\,0.5}{0.2\,;\,-}$	$\dfrac{-\,;\,0.4}{0.4\,;\,-}$	$\dfrac{1.1\,;\,0.5}{0.7\,;\,0.2}$	$\dfrac{-\,;\,0.4}{0.3\,;\,-}$	$\dfrac{0.1\,;\,0.2}{0.1\,;\,-}$	$\dfrac{1.0\,;\,1.1}{0.7\,;\,0.8}$ (0.9)
1–2	$\dfrac{-\,;\,0.4}{-\,;\,-}$		$\dfrac{-\,;\,0.4}{0.1\,;\,-}$	$\dfrac{-\,;\,3.4}{-\,;\,-}$	$\dfrac{0.1\,;\,0.1}{-\,;\,-}$	$\dfrac{0.7\,;\,0.9}{0.4\,;\,-}$ (0.7)
1.3	$\dfrac{-\,;\,0.5}{-\,;\,-}$	$\dfrac{-\,;\,0.4}{0.2\,;\,-}$	—		$\dfrac{-\,;\,0.1}{-\,;\,-}$	$\dfrac{0.6\,;\,0.6}{0.8\,;\,0.2}$ (0.6)
1–5	$\dfrac{-\,;\,0.7}{-\,;\,-}$		$\dfrac{1.3\,;\,-}{-\,;\,-}$	$\dfrac{0.1\,;\,0.1}{-\,;\,-}$	—	$\dfrac{0.7\,;\,2.4}{-\,;\,-}$ (1.5)

2	$\dfrac{0.3\;;\;0.4}{-\;;\;-}$	—	$\dfrac{-\;;\;0.1}{-\;;\;-}$	$\dfrac{3.9\;;\;-}{-\;;\;-}$	$\dfrac{1.4\;;\;0.7}{0.3\;;\;-}$ (0.8)
2–6	—	—	—	$\dfrac{-\;;\;0.3}{-\;;\;-}$	$\dfrac{3.7\;;\;0.5}{1.6\;;\;-}$ (1.9)
2–8	—	—	—	—	$\dfrac{1.3\;;\;0.8}{-\;;\;-}$ (1.0)
3	—	—	—	$\dfrac{0.2\;;\;0.4}{-\;;\;-}$	$\dfrac{0.8\;;\;1.6}{0.6\;;\;-}$ (1.0)
3–4	—	—	—	—	$\dfrac{-\;;\;2.2}{-\;;\;-}$ (2.2)
3.8	—	—	—	$\dfrac{1.6\;;\;-}{-\;;\;-}$	$\dfrac{1.6\;;\;0.2}{-\;;\;-}$ (0.9)
4	—	—	—	—	$\dfrac{4.4\;;\;6.6}{0.1\;;\;-}$ (3.7)
4–7	—	—	—	$\dfrac{0.3\;;\;-}{-\;;\;-}$	$\dfrac{0.3\;;\;-}{-\;;\;-}$ (0.3)
5	—	—	—	$\dfrac{0.3\;;\;0.1}{0.7\;;\;-}$	$\dfrac{0.4\;;\;1.7}{0.7\;;\;-}$ (0.9)
5–6	—	$\dfrac{-\;;\;0.1}{-\;;\;-}$	$\dfrac{0.2\;;\;-}{-\;;\;-}$	$\dfrac{0.2\;;\;-}{-\;;\;-}$	$\dfrac{0.4\;;\;2.0}{-\;;\;-}$ (1.2)
6–7	—	—	—	—	$\dfrac{1.4\;;\;3.7}{-\;;\;-}$ (2.5)
7	—	—	—	—	$\dfrac{-\;;\;3.6}{-\;;\;-}$ (2.6)

Table 5 (continued)

Petrochemical groups and types of kimberlitic rocks	Fields of pyrope-subfacies kimberlites and picrites					For all the fields (bracketed values refer to a rock type)
	Dzhyuken	Nizhnii Ukukit	Merchimden	Verkhnyaya Molodo	Kuoika-Beenchime	
Rock types: diatreme areas are given in accordance with $DI_{Ti}^{(5)}$ values: <25 ; 25–62 / 63–100; >100						
7–8	—	—	—	—	—	$\dfrac{4.4\;;\;-}{-\;;\;-}$ (4.4)
8	—	—	—	—	—	$\dfrac{1.3\;;\;0.6}{-\;;\;-}$ (1.0)
Averaged data for all rock types	$\dfrac{0.3\;;\;0.5}{0.2\;;\;-}$	$\dfrac{-\;;\;0.4}{0.3\;;\;-}$	$\dfrac{1.2\;;\;0.3}{0.4\;;\;0.2}$	$\dfrac{0.2\;;\;1.0}{0.3\;;\;-}$	$\dfrac{0.8\;;\;0.2}{0.4\;;\;-}$	$\dfrac{1.0\;;\;1.2}{0.7\;;\;0.8}$ (1.0)

that CDI_{Ti}^{Fe} reflects mainly the energetic trend of subcrustal evolution of kimberlite magmas and so cannot greatly contribute even to the differentiation index of a melt. Alternatively, when the properties reflecting the variability of kimberlite magma composition are considered as an entity, we obtain a distinct relationship between the $DI_{Ti}^{(5)}$ of kimberlite, on the one hand, and the mean area of diatremes on the other. Table 5 shows that of 17 fields examined, the maximal mean area was recorded for: diatremes filled with fairly differentiated rocks (DI = 25—62) in nine fields; diatremes of slightly differentiated kimberlites (DI <25) in five fields; diatremes of highly dffferentiated rocks (DI = 63—100) in two fields; and diatremes composed of drastically differentiated rocks (DI >100) in only one field.

Study of the mean sizes of the diatremes filled with petrochemically different kimberlites reveals a similar trend: the area of the pipes shows no evidence of a particular dependence on the rock types, but among the pipes composed of kimberlite of one type, diatremes, filled with fairly to slightly differentiated kimberlite, are maximal in area. Because false correlations can result from the values given in Table 5, a different representativeness of the initial data should be taken into account. Thus the maximal (2.2—4.4 ha) and the minimal (0.3 ha) areas of the diatremes are obtained for rare petrochemical types known to occur in one or more pipes (Fig. 22). The scatter of their sizes relative to areas (0.6—1.5 ha) calculated as arithmetic mean values from much more representative samplings is bound to occur and this naturally precludes any assumption of geologic events being responsible for the deviation.

The variability in the size of diatremes filled with rocks of uniform petrochemical type shows that 17 major kimberlite fields of the Central Siberian province contain rocks belonging to 18 basic and intermediate types and four kimberlite types are represented by rock varieties differentiated to a similar extent. In the kimberlites of each of the remaining 14 petrochemical types the maximal mean area of diatremes is (i) most commonly (eight types) reported from the fairly differentiated rock varieties; (ii) less commonly (five types) determined in the poorly differentiated rocks; and (iii) observed in only one type (1—3) of the highly differentiated varieties. The resulting relationship between the mean area of the diatremes and the extent of kimberlite differentiation both within an individual field and in the rocks of one petrochemical type from different fields is to be anticipated, since analysis of a great quantity of bodies practically nullifies the probability of random error.

Although the extent of coverage by prospecting of the 17 fields is different, and large and medium-sized diatremes are usually found at early stages, while smaller bodies are discovered at later stages of prospecting, this cannot explain such a correlation between the mean area of diatremes and the extent of kimberlite differentiation. As a result, three fields, prospected in detail, do not fall into one single group containing no other fields, but into the two larger groups, one including nine, and the other five fields. The most numerous group of fields, in which the pipes composed of fairly differentiated kimberlite have the maximal mean area, includes only two thoroughly prospected fields (Malaya Botuobiya and Markha-Alakit). The group of fields next in abundance, characterized by slightly differentiated kimberlites in the diatremes of maximal mean area, includes only one field (Daldyn) investigated in some detail.

We see no substantial increase in mean area of pipes filled with kimberlites of one, two, or three, of 18 basic and intermediate petrochemical types over the 17 fields of

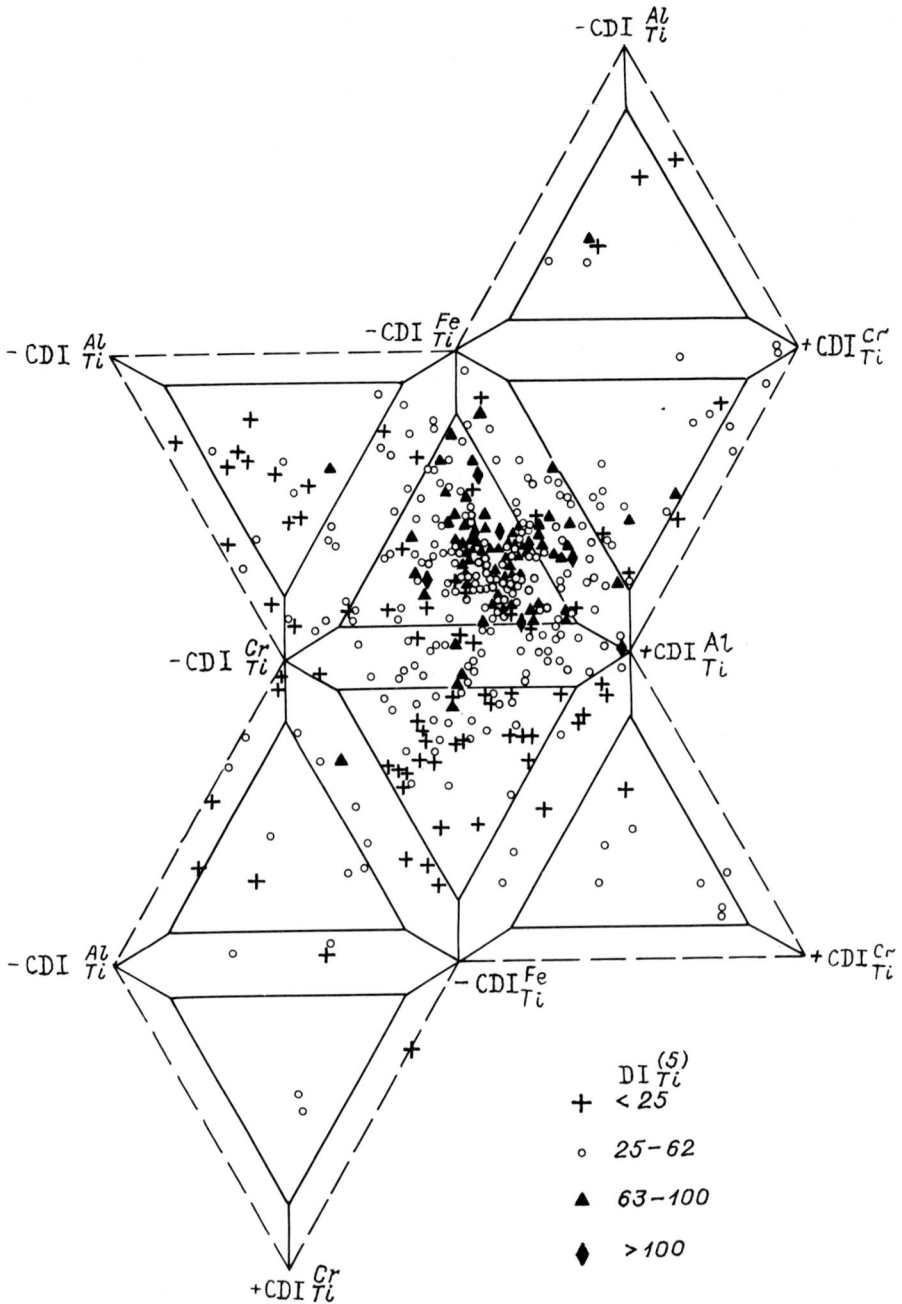

Fig. 22. Abundance of different petrochemical types of rocks infilling explosion pipes in 17 major fields, Central Siberian kimberlite province

the Central Siberian province, hence the conclusion that variability in rock composition with respect to the four major representative elements has not greatly affected the emplacement of the diatremes. This conclusion is evidently not ambivalent because the differentiation of kimberlite magmas with respect to titanium, iron, chromium, and aluminum, as well as the relative enrichment or depletion of magmas in these elements, are unlikely to have affected the intensity of volcanic processes to any extent.

A closer relationship should exist between the extent of volcanic activity and the differentiation index of kimberlite, which takes into consideration not only the contrast differentiation, but also the amount of substratum reworked by a magma. In the course of this process, which proceeds by the mechanism of zone melting, a kimberlite magma loses infusible and accumulates fusible components, primarily volatiles (Milashev 1972a), whose role in volcanic processes is well known. This question, as applied to the emplacement of kimberlite and other diatremes, is considered at length in Part III.

CHAPTER 7

Chronology of Kimberlite Formation

Age of Kimberlite and Methods of its Determination

An analysis of data concerning all areas where effects of kimberlite volcanism have been recorded shows that the formation of the best-known kimberlite provinces spans several epochs, each lasting for about 250 m.y. The duration of the epoch during which any province is formed may be determined from the difference in age of the kimberlite occurring in the center and around the periphery of the province (Milashev 1974a). These estimates are generally based on the geological age of the rocks containing and overlying the kimberlite. K-Ar ages of kimberlite usually yield higher values as compared to the geological data on rock formation. The reason for the difference in age has been discussed elsewhere (Milashev 1968c).

Attempts at dating kimberlite from fission tracks on zircon (Komarov and Ilupin 1978) yield dubious results in terms of geology. Lakatos and Miller (1973) studied uncertainties in fission-track datings of rocks and minerals on muscovite. The authors ascribed the uncertainties, which are probably also valid for zircon, to the following: (a) uneven distribution of uranium in the mineral; (b) variation in uranium abundance during the existence of the mineral; (c) the regime of track annealing; (d) incomplete revelation of partly annealed tracks. The authors reported the results from the experimental study of the effect of the last two factors. They studied how all the above factors affect the results obtained, and arrived at the conclusion that fission-track dates cannot be relied upon.

Zircon grains in kimberlite have been also used to determine the age of the rock emplacement by the U-Pb method (Davis et al. 1980). The authors emphasize the high reliability of the dates obtained, and advise using the U-Pb method widely in prospecting for diamond deposits. However, considering the significance of foolproof methods of kimberlite datings for a more complete understanding of the geology of diamond deposits and for the problems treated here, we must now subject some of the conclusions that the above authors arrived at to a somewhat closer scrutiny.

U-Pb zircon datings of kimberlite are based on the assumption that in the course of the magmatic stage ($>1000°C$) all lead produced by uranium decay had diffused from crystals and the amount of lead determined from the analysis had accumulated in zircon after the kimberlite cooled down. An opposite assumption of the survival of all or much of the radiogenic lead in zircon would eliminate the idea of using the amount of lead to date kimberlite, as grains of this mineral may have been picked up by the kimberlite magma from the plutonic (mantle?) rocks.

The assumption of a complete diffusion of lead from zircon at about $1000°C$ rules out the above constraints, but raises, in turn, a number of serious questions. Diffusion of lead from zircon suggests a substantial difference in the abundance of this component in zircon and in enclosing kimberlite, as diffusion, i.e., penetration of one component into another due to thermal particle motion, occurs in the direction of the decreasing concentration gradient and establishes a uniform concentration of a component in all parts of the system (i.e., diffusion equilizes the chemical potential of the system). Ultimately, it is not diffusion motion that is of importance, but the equalization of the component concentration thus established in the primary heterogeneous medium. In the course of the process, particles leave the places of high component concentration rather than those of low concentration. A unilateral matter flux − the diffusion flux − passes via unit area in heterogeneous medium per unit time in the direction of decreasing concentration gradient.

From Fick's law, the diffusion flux is directly proportional to the concentration gradient and the diffusion coefficient, which increases as temperature rises and decreases as molecular weight of matter increases. With concentration gradient equaling zero, the intensity of diffusion flux also equals zero at any, even the highest temperature. The content of lead in kimberlite ranges from 6 to 43 g per ton (Borodin et al. 1976). In two cases (Amakinskaya Pipe and one sample from the Internatsionalnaya Pipe), anomalously high lead amounts (430 and 1000 g per ton, respectively) were recorded; the values are correlative with a high zinc content and are prpbably due to hydrothermal mineralization (Ilupin et al. 1978).

Unfortunately, Davis et al. (1980) report no data on the bulk content of lead and its individual isotopes in zircon with low uranium content. The authors indicate only that ". . . due to a very low ^{207}Pb content in low uranium zircons the best data reproducibility with the least analytical uncertainty was obtained for the $^{206}Pb/^{238}U$ ratio" (p. 176). This implies that the total amount of lead in these zircons is relatively low and differs only slightly from its abundance in kimberlite. It should be borne in mind that during Paleozoic and Mesozoic time the lead content in zircon was lower than at present, since by now its concentration has been substantially increased by uranium decay.

The above leads to two important conclusions. First, at the magmatic stage of kimberlite formation (at $T \geqslant 1000°C$) in some cases the lead concentration gradient between zircon and enclosing kimberlite was close to zero and, hence, from Fick's law, the diffusion lead flux between them was also zero. Second, and more important, lead concentration in zircon could not drop below that in the enclosing rocks, as diffusion establishes the uniform concentration of a component throughout the system.

Thus, the assumption of a complete diffusion of lead from zircon in kimberlites at $T \geqslant 1000°C$ is inconsistent with the laws of physics and thus appears erroneous. This implies that both total lead amount and ^{206}Pb content, recorded in the zircon, are higher than the amount of radiogenic lead which had formed furing the time interval from kimberlite emplacement to the present. Therefore, a time interval obtained from the $^{206}Pb/^{238}U$ ratio cannot show the age of kimberlite. The value thus obtained could be considered as the age of zircon, if we could prove that the mineral belongs to the intratelluric or plutonic phenocrysts of the kimberlite.

If, however, according to Sobolev (1974), all zircons contained in kimberlite, including low uranium mineral varieties, were caught by the uprising kimberlite magma from the basement and upper mantle, then the results of $^{206}Pb/^{238}U$ recalculations have a very wide bearing even on the age of zircon itself. In fact, the zircon of the oldest (obviously pre-kimberlitic) rocks of the upper mantle must have contained a comparatively high amount of radiogenic lead, whose excess, had diffused in the kimberlite magma with lower lead concentration. Of particular interest for our study is the equalization of lead concentration in all zircon clasts in the magma which captured them.

As a result, the $^{206}Pb/^{238}U$ "age" of low uranium zircon in any pipe would show only a slight scatter. Close estimates would also be obtained for low uranium zircon from kimberlites which differ slightly in the total amount of lead and in the time of emplacement. In the general case it is anticipated that the $^{206}Pb/^{238}U$ "age" of low uranium zircon, all things being equal, would be older in lead-bearing kimberlite varieties.

One of the factors which is difficult to take into account, but which may have a pronounced effect on the $^{206}Pb/^{238}U$ estimates, is the rate of the ascent of the kimberlite magmas from subcrustal depth to the Earth's surface. In the case of sufficiently rapid upward movement and cooling of the melt, chemical potentials of lead in the zircon clasts and in the magma had not always managed to equalize, and the excess ^{206}Pb which had survived in zircon further increases the inaccuracy of the results obtained.

Hence, the genesis of low uranium zircon in kimberlite is of crucial importance for valid geological interpretation of the recorded $^{206}Pb/^{238}U$ ratio. We cannot therefore agree that the inaccuracy of kimberlite datings performed by Davis et al. (1980) is ± 2 m.y. It would be better to note the high sensitivity of the analytical method of uranium detection (to 0.2 g per ton) used by the authors. The $^{206}Pb/^{238}U$ value obtained through this method cannot be strictly geologically derived.

Some writers believe that according to Dirac's hypothesis of the change of gravity constant in proportion to the age of the Universe, the rate of radioactive decay should vary with time. Such being the case, all radiometric ages are very far from showing the actual duration and absolute dates of geological processes and events.

Common geological data permit the dating of a few kimberlite fields within provinces. The age range thus obtained is estimated at tens of millions of years. For instance, the formation of kimberlite of the Markha-Alakit field continued from the Middle Devonian to the Early Carboniferous inclusive; the intrusion of kimberlite composing the Motorchuna field took place between the Early Triassic and the beginning of the Jurassic. Any information on such lengthy (tens of millions of years) continuous volcanic processes within any field is lacking, but at the same time we should also exclude a single-pulse eruption of the ultrabasic magmas even over limited (on average, 1000 km^2) areas, recognized as separate kimberlite fields on the basis of a variety of data.

Available, but unfortunately meager data lead to the conclusion that the period between the initial and final stages of volcanic processes resulting in the formation of any kimberlite field lasted for many millions of years. Investigation of different rock generations in complex-structure diatremes suggests that major pulses of fresh magma may have intruded at recurring intervals. Younger kimberlite is found to have pierced the early rock generations which had time to solidify and suffer high-grade autometamorphism. The supposition of a considerable time gap in the formation of individual kimberlite generations in the complex-structure bodies is also supported by paleomagnetic studies: in some diatremes various kimberlitic rock generations are reversely magnetized, thus suggesting a substantial (0.25–10 m.y.) difference in time of formation. Hence, the process of formation of any kimberlite field includes many outbursts of ultrabasic magmatism, considerably dispersed in time, with the result that separate pulses of fresh magma formed separate pipes and dykes or pierced earlier diatremes and turned them into complex-structure bodies.

To our knowledge, no method has been reported in literature which would allow establishment of the true order of succession of the emplacement of kimberlitic rocks comprising various diatremes within a single field. Radiometric datings cannot be used for this purpose, because uncertainties (tens and hundred of millions of years) are obvious. A reliable succession in the formation of different kimberlite generations in each complex-structure pipe cannot be used to determine the relative ages of the rocks which composed even adjacent bodies. The use of data on kimberlite magnetization also gives no desirable results.

Petrochemistry and Succession in Eruption of Kimberlite

The assumption of the origin of explosion pipes and dykes of each kimberlite field in the course of prolonged ultrabasic magmatism at a certain site of the territory implies that successive batches of magma may differ in composition. Substantial differences in texture and structure, in color and other physical properties, as well as in chemistry and mineralogy, have been recorded in various kimberlite generations of all complex-structure diatremes. Geological and structural relationships between the rocks of various structural-genetic groups in well-studied explosion pipes indicate conclusively the following succession of formation; tuffisite breccia — eruption breccia — massive varieties of kimberlite or picrite.

The contrast differentiation and differentiation indexes should be used in petrochemical analysis of kimberlites. The most informative for our study seems to be CDI_{Ti}^{Fe} which is more closely related to the energetics of initiation and evolution of certain pulses of fresh magma and, hence, to the regime and timing of origin as a result of subcrustal processes during the epochs of kimberlite volcanism. Table 6 presents major petrological data on successive kimberlite generations from the best-studied diatremes in Yakutia. Even at first sight, Table 6 reveals obvious differences in diachronous rocks of one and the same diatreme. Attention is drawn to the fact that in most diatremes the earlier rock generations are represented, as a rule, by eruption breccia or tuffisite breccia, but massive kimberlite varieties dominate the later generations. There are pipes in which both rock generations are represented by brecciated (Festivalnaya, Podsnezhnaya, Yanvarskaya) or, less commonly, massive (Pipe An. 135/63) varieties. No reverse situation (i.e., the first generation being represented by massive and the second by brecciated rocks) has been reported.

Table 6. Petrology of various kimberlitic rock generations in some complex-structure diatremes of Yakutia

Ord. No (see Fig. 23)	Diatreme	Rock generation (see Table 7, Figs. 7, 23)	Rock	Type of differentiation (see Fig. 21)	CDI^{Fe}_{Ti}	$DI^{(5)}_{Ti}$
1	Udachnaya-Zapadnaya	1	Eruption breccia of micaceous microlitic kimberlite	2−6	−0.0152	17.45
2	Udachnaya-Vostochnaya	2	Nonmicaceous nonmicrolitic kimberlite	5−6	−0.0002	26.05
3	Aerogeologicheskaya	1	Eruption breccia of micaceous nonmicrolitic kimberlite	1−2	0.0021	42.03
		2	Eruption breccia of nonmicaceous nonmicrolitic kimberlite	1	0.0052	64.16
4	Iskorka	1	Eruption breccia of nonmicaceous microlitic kimberlite	2−6	−0.0036	73.35
		2	Nonmicaceous microlitic kimberlite	1	0.0125	64.52
5	Flogopitovaya	1	Eruption breccia of micaceous nonmicrolitic kimberlite	5−6	−0.0001	50.05
		2	Micaceous nonmicrolitic kimberlite	1	0.0073	73.62
6	Dalnyaya	1	Eruption breccia of micaceous nonmicrolitic kimberlite	1−2	−0.0009	43.60
		2	Nonmicaceous nonmicrolitic kimberlite	1−2	0.0008	52.97
7	Sibirskaya	1	Eruption breccia of nonmicaceous nonmicrolitic kimberlite	1−2	−0.0009	27.76
		2	Nonmicaceous nonmicrolitic kimberlite	3	0.0036	39.85
8	"An. 48"	1	Tuffisite breccia of micaceous nonmicrolitic kimberlite	1−2	−0.0023	67.95
		2	Tuffisite breccia of micaceous microlitic kimberlite	1−2	0.0003	61.64
9	Festivalnaya	1	Eruption breccia of nonmicaceous microlitic kimberlite	6	−0.0074	29.85
		2	Eruption breccia of nonmicaceous nonmicrolitic kimberlite	5−6	0.0006	33.26

Table 6 (continued)

Ord.No (see Fig. 23)	Diatreme	Rock generation (see Table 7, Figs. 7, 23)	Rock	Type of differentiation (see Fig. 21)	CDI_{Ti}^{Fe}	$DI_{Ti}^{(5)}$
10	Chomur	1	Eruption breccia of micaceous nonmicrolitic kimberlite	1	0.0039	37.09
		2	Micaceous microlitic kimberlite	1	0.0064	74.47
11	Kuranakh-skaya	1	Micaceous nonmicrolitic kimberlite	1−3	0.0083	42.10
		2	Micaceous microlitic kimberlite	1−3	0.0113	49.71
12	Universitet-skaya	1	Eruption breccia of micaceous nonmicrolitic kimberlite	1−3	0.0102	29.33
		2	Micaceous microlitic kimberlite	1	0.0103	38.03
13	"An.135/63"	1	Micaceous nonmicrolitic picritic porphyry	1	0.0427	33.62
		2	Monticellite picrite	1−5	0.0790	25.63
14	Senkyu Severnaya	1	Tuffisite breccia of micaceous nonmicrolitic kimberlite	1−3	0.0049	43.32
		2	Micaceous nonmicrolitic kimberlite	1−3	0.0105	60.68
15	Podsnezhnaya	1	Eruption breccia of micaceous microlitic kimberlite	1	0.0221	21.41
		2	Eruption breccia of nonmicaceous microlitic kimberlite	1	0.0376	16.16
16	Geofiziche-skaya	1	Eruption breccia of nonmicaceous nonmicrolitic kimberlite	2	−0.0085	21.10
16	Geofiziche-skaya	2	Nonmicaceous nonmicrolitic kimberlite	1	0.0082	26.51
17	Molodezh-naya	1	Eruption breccia of nonmicaceous nonmicrolitic kimberlite	2	−0.0128	26.77
		2	Nonmicaceous microlitic kimberlite	4	0.0035	23.52
18	Sputnik	1	Eruption breccia of micaceous nonmicrolitic kimberlite	2	−0.0331	21.06
19	Dyke between Sputnik and Mir Pipes	2	Micaceous nonmicrolitic kimberlite	6	−0.0075	34.44

Table 6 (continued)

Ord. No (see Fig. 23)	Diatreme	Rock generation (see Table 7, Figs. 7, 23)	Rock	Type of differentiation (see Fig. 21)	CDI_{Ti}^{Fe}	$Di_{Ti}^{(5)}$
20	Mir	3	Eruption breccia of non-micaceous nonmicrolitic kimberlite	1–5	0.0042	27.08
		4	The same as generation 3	1–5	0.0077	23.57
		5	Nonmicaceous nonmicrolitic kimberlite	5	0.0111	26.70
21	Snezhinka	1	Eruption breccia of micaceous nonmicrolitic kimberlite	7–8	−0.0144	24.64
		2	Micaceous nonmicrolitic kimberlite	3–4	0.0014	30.21
22	Yanvarskaya	1	Tuffisite breccia of micaceous nonmicrolitic kimberlite	2	−0.1156	9.22
		2	Eruption breccia of non-micaceous nonmicrolitic kimberlite	6–7	−0.0647	7.80

Successive rock generations which belong to different groups evidenced by contrast differentiation may have a different $DI_{Ti}^{(5)}$ parameter, although this shows a tendency to increase in relatively later generations. If both rock generations in a diatreme do belong to a single group on the basis of CDI_{Ti}^{Fe}, then the earlier rock generations are characterized by lower $DI_{Ti}^{(5)}$ values (Dalnyaya, Chomur, Senkyu-Severnaya and other pipes) or show no substantial (over ± 20%) discrepancies with respect to younger rocks.

Much more conspicuous relationships which may be classified as regularities are recorded between relative ages of kimberlitic rocks and CDI_{Ti}^{Fe}. Values of the index are without exception lower in the earlier rock generations than in later rock generations, whose time of emplacement has been reliably determined by geological observation. The CDI_{Ti}^{Fe} values may vary to a great extent in the rocks forming one pipe, thus suggesting considerable differences in the type of parent magma differentiation (Udachnaya, Geofizicheskaya, Molodezhnaya, and some other pipes). However, even where both rock generations belong to one single type of differentiation, the CDI_{Ti}^{Fe} values are markedly higher in the later than in earlier rock generations within the same pipe (Dalnyaya, Chomur, Podsnezhnaya, and other pipes).

A graphic idea of the above tendencies is conveyed in Fig. 23. Structural relations between various rock generations in individual diatremes are schematically represented

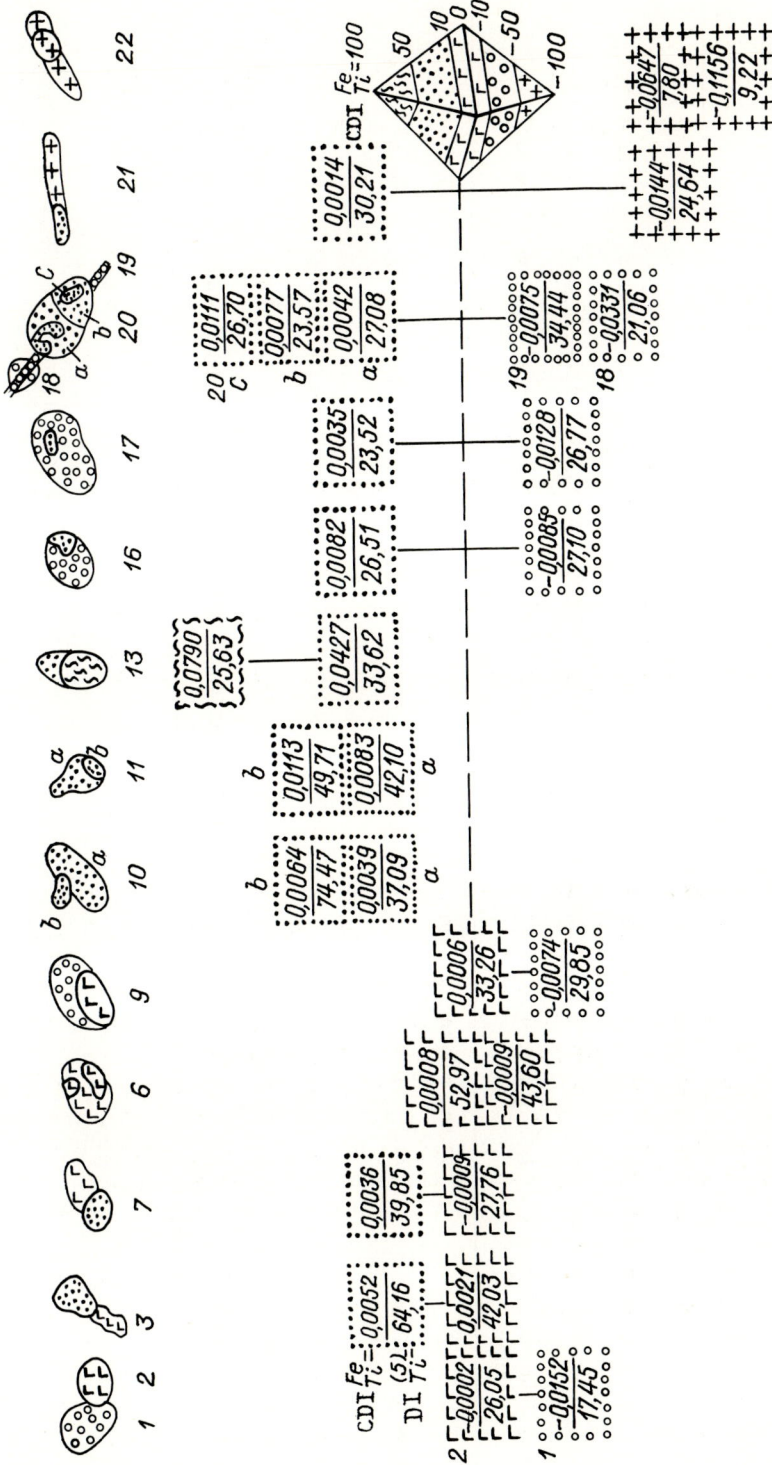

Fig. 23. Age relationships (inferred from geological data) and petrochemistry of different generations of kimberlitic rocks infilling some complex diatremes, Central Siberian province. Distribution of rectangular cells with respect to zero (*dashed line*) shows correlation between $CDI\frac{Fe}{Ti}$ and relative age of the rocks (see Table 6,1–22)

Table 7. Petrology of different kimberlitic rock generations across the section of the adjacent Udachnaya-Zapadnaya and Udachnaya-Vostochnaya Pipes (see Fig. 7)

Phase of emplacement	Rock	Number of analyses	Content, wt %					Specific contrast differentiation indexes				$CDI_{Ti}^{(5)}$	$CDI_{Ti}^{(5)}$	Type of differentiation
			Fe	Ti	Al	K	Cr	CDI_{Ti}^{Fe}	CDI_{Ti}^{Al}	CDI_{Ti}^{K}	CDI_{Ti}^{Cr}			
Udachnaya-Zapadnaya Pipe														
1	Kimberlite breccia	89	4.35	0.53	1.26	0.50	0.075	−0.0152	0.0006	0.2308	−0.0308	1.1854	17.45	2–6
2	Kimberlite breccia	14	4.26	0.57	1.37	0.61	0.075	−0.0152	0.0065	0.2753	−0.0249	1.2417	19.66	2
3	Massive kimberlite (veins)	9	4.02	0.76	1.47	0.59	0.082	−0.0134	0.0088	0.1948	−0.0184	1.1718	24.74	2
Udachnaya-Vostochnaya Pipe														
1 (4)	Autolithic kimberlite breccia	14	4.89	0.74	1.18	0.89	0.062	−0.057	−0.0029	0.3281	−0.0269	1.2926	26.58	2–6
2 (5)	Kimberlite breccia	32	5.48	0.87	1.07	0.41	0.068	−0.0002	−0.0062	0.1044	−0.0204	1.0776	26.06	5–6
3 (6)	Kimberlite	5	5.33	0.99	0.99	0.48	0.062	−0.0012	−0.0078	0.1132	−0.0199	1.0843	29.82	5–6
4 (6)	Kimberlite	13	6.12	0.91	1.05	0.81	0.068	0.0046	−0.0066	0.2712	−0.0195	1.2497	31.59	1

at the top of the figure; to the right is given a "key" developed into an octahedron (see Fig. 21). Upper and lower apexes of the octahedron correspond to 100% of positive and negative CDI_{Ti}^{Fe} values, respectively, while on the horizontal plane passing through the other four apexes ($\pm\ CDI_{Ti}^{Cr}$ and $\pm\ CDI\ _{Ti}^{Al}$) the values equal zero. To ensure more vivid presentation and to show rock differentiation with respect to CDI_{Ti}^{Fe} in greatest detail, the "key" presents, along with the zone $\pm\ 10\%$, zones of positive and negative index values more than 50%.

Naturally, the relationship between the CDI_{Ti}^{Fe}, $DI_{Ti}^{(5)}$ and the timing of formation of various kimberlitic rock generations is revealed not only in plan, i.e., at the present exposure surface, but to a depth of many hundreds of meters. An excellent example is provided by two conjugate pipes, Udachnaya-Zapadnaya and Udachnaya-Vostochnaya (see Figs. 21, 23; Tables 6, 7). The conspicuous structural relations of kimberlites in the pipes make it possible to reliably establish the order of succession in the formation of each pipe, while the conjugate volcanic pipes and the petrology of the rocks enable the consideration of both diatremes as a single complex-structure body, and assess the age of all seven kimberlite generations together (Table 7). The results obtained are in good agreement with the relationship, established elsewhere, between the timing of formation of kimberlitic rocks, on the one hand, and their CDI_{Ti}^{Fe} and $DI_{Ti}^{(5)}$, on the other (see Fig. 23, Table 6).

Later ejections of the kimberlite magmas with high CDI and DI values and their further quenching premarily as massive rock varieties are not only thus explained, but should be considered as an inevitable consequence of the processes, the model of which was suggested earlier by the present author on the basis of comprehensive analysis of the physicochemical conditions of kimberlite formation (Milashev 1972a). According to the model, the chambers of a kimberlite magma appear in various parts of ascending convective currents of the upper mantle matter, as thermodynamic parameters vary greatly. The variability of the thermodynamic regime of initiation and subcrustal evolution of a kimberlite magma is caused by nonuniform heating of the upper mantle matter in the internal and peripheral zones of convection currents. In the internal zones where the temperature of substratum is maximal, melting starts at much higher pressure and, hence, at greater depth than in the peripheral parts of the ascending current.

The magmatic chambers rise to the base of the Earth's crust due not only to the ascending current of enclosing substratum, but also to autonomous motion by the mechanism of zone melting. The highest rate of flow is characteristic of the central parts of convection currents where, because of high temperatures, substratum rocks are the least viscous. The high temperature of substratum enclosing the parent magmatic chambers actuates the processes of zone melting which do not require a large consumption of energy under these conditions, and therefore proceed with a minimal contrast range with respect to iron, resulting in relatively low overall differentiation of the magma. The kimberlite magma from these chambers first rises to the base of the Earth's crust and, under favorable tectonic conditions, to hgiher levels, where it primarily forms large explosion pipes filled with tuffisite or eruption breccia.

Thus, certain relationships between the chemistry and relative age of kimberlite have been established and theoretically explained. The initiation of kimberlite volcanism in each field is marked by the least differentiated kimberlite having conspicuous negative contrast differentiation with respect to iron. Further, kimberlitic rock varieties appear whose CDI_{Ti}^{Fe} gradually varies from low negative values through near zero to high positive values. Of the rocks of the same petrochemical type having similar CDI_{Ti}^{Fe} values, the latest rocks are most likely to be represented by the most differentiated rock varieties, i.e., those having maximal $DI_{Ti}^{(5)}$ values.

CHAPTER 8

Structural Control of Kimberlite Occurrence and Factors of its Localization

Distribution and Tectonic Setting of Kimberlite

Most kimberlites are confined to ancient cratons. Manifestations of kimberlite volcanism are inferred to occur also on the cratons where primary sources of diamond have yet to be found. The firm data and theoretical concepts currently available suggest that kimberlite volcanism is typical only of the cratons and governed by the character of upper mantle evolution beneath vast craton areas of the Earth. The contention of some "kimberlite-like" ultrabasic rocks in the circum-cratonic fold belts reported in literature seems debatable and remains to be further investigated. Theoretically, it is just conceivable that kimberlite may occur within median masses; in this case their formation is related to the craton stage in the geological history of the regions.

The association of kimberlite volcanism with ancient cratons has gained general favor with many writers, but less agreement appears to exist on major factors of kimberlite localization. The concepts of the leading (and even exclusive) role of fundamental fractures in the spatial distribution of kimberlite, conceived at the beginning of the century, have gained proponents (Du Toit 1906; Harger 1906; Arseniev 1961; Sarsadskikh 1968). Regional fundamental fracture zones associated with kimberlite localization are generally drawn as relatively straight lines connecting two or more kimberlite fields. However, most zones owe their pattern to one or other concept, and are supported by neither geological nor geophysical data.

Of this group of hypotheses the most well-judged, rigorous, and orthodox seems to be that developed by Bardet (1964), who relates the distribution of kimberlites of West and Southern Africa and Siberia with the fractures which, in spite of significant difference in age, are arranged in the form of a regular geometric grid at an angle of 45° to the present north-south direction. Bardet (1964) ascribes the emplacement of such fractures to certain dynamic stress due to the Earth's rotation. This attractive hypothesis, however, cannot be reconciled with pole migration and continental drift.

Some authors who analyze kimberlite localization on a craton scale arrive at the conclusion of the association of kimberlite volcanism with major craton structures and consider fractures as subordinate factors of localization. The Paleozoic (Atlasov 1960) and Mesozoic (Trofimov 1967) syneclises, marginal parts of anteclises (Odintsov 1958), as well as sites of small (Spizharsky 1960) or great (Kovalsky 1963) thickness of the stratified sedimentary cover of the cratons are considered as structures favorable for kimberlite occurrence.

Sharp (1974) holds a very original viewpoint relative to the tendencies and regularities of the spatial distribution of kimberlite. He believes that the origin of diamond-bearing kimberlite is a direct consequence of plate tectonics and associates it with activity at the deepest levels of mobile plates. Kimberlite pipes which, according to Sharp (1974), generated in some regions due to plate collision, might have been either eroded or are not yet exposed. The zones of kimberlite pipes and dykes would be parallel to the mountain fold belts. Sharp states that such a pattern is confirmed by the South African kimberlite associated with the Cape Fold Mountains, and by the North American kimberlites occurring near the Appalachian fold belt. Sharp believes that young kimberlite, providing it is discovered in abyssal basins parallel to the present mobile island arc zones, would lend additional credence to his hypothesis.

Along with these opinions, often poles apart, there are statements showing that manifestations of kimberlite volcanism may take place within differing structural framework under persistent geotectonic conditions (Erlikh 1958; Alekseev and Diakov 1961; Milashev et al. 1963 and others). The most common geotectonic regularity in the distribution of kimberlite volcanism is its association with major structures subject to prolonged, slow, and persistent uplift. The vast majority of kimberlite fields are situated within anteclises. The failure to find kimberlite diatremes within syneclises is probably related to the suggestion that they may have been buried beneath thick younger deposits. Since kimberlites belong to extremely deep-seated rocks erupted on the Earth's surface, the regional patterns of distribution of kimberlite volcanism ought to be considered not only in the context of the structural framework of the sedimentary cover and basement of the craton, but with allowance for the structure of deeper parts of the lithosphere.

The study of the thermodynamic conditions of kimberlite formation and its relationship with the evolution of the upper mantle leads to the conclusion that the major factors relevant to the distribution of kimberlite volcanism over the cratons are governed by the most abyssal processes originating in the upper mantle at a depth of 300–400 km and then continuing into the lower levels of the Earth's crust. The processes are probably due to convection currents which impel immense masses of hot and lower-density substratum to float to the base of the Earth's crust. On reaching the upper levels of the mantle, plutonic material, impelled by convection energy, appears not to spread on horizontal planes, but to be guided by planes of structural weakness in the Earth's crust. Therefore, the upper surface of a resultant gigantic lens of plutonic rocks may be either situated in the immediate vicinity of the crustal base, or separated from it by thick sequences of relatively cold "local" substratum.

The driving mechanism of autonomous movement of the kimberlite magma in the upper mantle is likely to be zone (partial) melting (Milashev 1972a, 1974b). Because of the relatively small size of the chambers, the kimberlite magma progresses successfully only together with the uprising hot substratum. On reaching relatively cold rocks, composing the upper mantle, the kimberlite magma releases heat intensely and the upward movement "goes out slowly" over a length of a few kilometers. Hence, although the kimberlite magmas generate in various parts of the gigantic substratum block, captured by convection currents, they could reach the base of the Earth's crust (and its surface, under favorable conditions) only in the parts where substratum withdrawn from the depth was brought very close to the base of the Earth's crust.

Mobilization and radial transport of enormous masses of highly viscous sub-stratum are inevitably accompanied by deformation of the overlying mantle zones and result in the uplift of extensive portions of the Earth's crust; as plutonic material spreads tangentially, the uplift continues from the central into the peripheral zones of kimberlite provinces. The uplifted sites of the area undergo intense denudation, the amplitude and age of which in most cases can be reliably determined by geological methods. Analysis of the data on the Siberian platform indicates convincingly that suggestions of a genetic relation between kimberlite volcanism and convection currents within the upper mantle are consistent with the observations on the preferential distribution of kimberlite over the areas which suffered long, slow, and persistent uplift during certain periods of their evolution.

The ascent of enormous masses of hot lower-density deep-seated material to the base of the Earth's crust could not apparently be confined solely to effects of tectogenesis; most probably it must have been accompanied by intensification of magmatic activity, firstly, due to partial melting of rocks of the basaltic layer of cratons. For example, on the Siberian platform vigorous flood basalt magmatism following kimberlite magmatism can hardly be thought of as fortuitous. Fields of extensive development of flood basalt embrace almost the whole of the central and western parts of the Central Siberian kimberlite province; they are also distributed in the north-east of the platform where the other (North Siberian) kimberlite province is inferred to occur (Milashev 1974a and others).

Fissuring in Country Rocks and Structural Contacts of Kimberlite Fields

Kimberlite pipes, dykes, and sills are not distributed evenly over vast areas of kimberlite provinces, but concentrate in separate rather small sites which, on a complex ground, are recognized as kimberlite fields. The net area of all the fields in any province account for only a small proportion of the total area of the province. One of the possible reasons for the far from ubiquitous distribution of kimberlite over the territory of any kimberlite province is related to the variation in distance from the upper surface of the lens of substratum withdrawn from depth to the base of the Earth's crust in various parts of the platform.

Attempts are often made to relate kimberlite localization in some sites of kimberlite provinces to fundamental fracture zones, major folds, and present structures in the crystalline basement. However, a thorough analysis shows that the spatial association of kimberlite fields with the above structures is occasional and could hardly be genetic in nature. The distribution of the vast majority of kimberlite fields in Siberia and South Africa is not associated with deep fractures and is independent of the depth to the crystalline basement of the craton.

The assumptions that even deep fundamental fractures bear no relation to the localization of kimberlite volcanism is further supported by the results of detailed investigations. The well-known region of the Siberian platform embracing the Malaya and Bolshaya Kuonamka river basins provides the best examples. All the data obtained there indicate conclusively that the kimberlite magma in its upward movement did not follow the deep fault zone, several hundred meters in amplitude and above 300 km in extent — a zone of fundamental fractures of various displacement. The kimberlite bodies are not localized within this zone, but distributed within the western, uplifted, block (Milashev 1971). Thus, the original data do not support the widespread opinion of the localization of kimberlite fields in deep fundamental fracture zones. The results of joint geological and geophysical studies lead to the conclusion that such fractures were not feeders during the events of kimberlite volcanism; they separate large basement blocks differing in physical properties of rocks.

When analyzing possible causes of the preferential, but not ubiquitous distribution of kimberlite over the areas whose deep structure is favorable for penetration of a kimberlite magma to the base of the Earth's crust, it should be borne in mind that various parts of the Earth's crust substantially differ in rock composition, tectonic evolution, and other features; hence, they are shattered to a various extent and could not be identically permeable to magmas. It is quite obvious that sites of high permeability (as well as those exhibiting any extreme properties) are very limited as compared not only to the areas of manifestations of kimberlite magmatism, but to the regions characterized by small crustal thickness.

Purposeful geophysical and geological studies can reveal sites of high crustal permeability, probably related to a considerable shattering of the rocks. The results of earlier geophysical studies can be used to compile medium-scale maps of deep structure and to roughly outline probable zones of high permeability. Then deep seismic sounding can be used in combination with the exchange earthquake wave method and large-scale gravity observations. The fact that kimberlite emplacement within any field was virtually synchronous (on the geological time scale) somewhat alleviates the problem. This eliminates the necessity for correlation of the observed (or inferred) kimberlite distribution with a diversity of geological and tectonic features of the region, because it would be sufficient to restore the paleotectonic environment in certain sites of the territory for the period prior to the beginning of kimberlite volcanism in the region. It is true, of course, that the resolution of paleotectonic methods is not always sufficient to recognize sites and zones of high crustal permeability.

An integrated approach to the study of fissuring in country rocks is also based on geological methods. Since the processes of formation of zones of high crustal permeability and fractures associated with kimberlite localization could not but be accompanied by fissuring in sedimentary rocks, the sites of kimberlite occurrence should be characterized by anomalous development of fractures. Field, experimental, and theoretical investigations of the relationship between the fissuring in sedimentary cover and the fundamental fractures show that they include regional systems of fissures and fissures pertinent to the emplacement of deep fractures.

Much importance in the formation of regional systems of fissures has always been attached to the effect of pre-existing fissures on the orientation of newly forming fissures. Thus, Belousov (1962) notes that pre-existing fissures affect the distribution of strains involved and, hence, the position of later fissures which tend to associate with the earlier and already dead fissures, in some cases the former greatly depart from the position dictated by the strains applied.

A close study of the mechanism responsible for the formation of regional systems of fissures was made by Hodgson (1965). Field observations allowed him to conclude that fissures are formed at the onset of sedimentation and progressively develop into each new sedimentary unit. A fissure pattern of earlier rocks is reflected in young deposits, not yet cracked and being at the stage of consolidation, and thus controls the orientation of faults. Hodgson (1965) believes that such minor, but constant factors as microseismic and tidal motions, changes in the rate of Earth's rotation, and the like may be conducive to the development of faults and fissures from layer to layer. In the absence of other movements, the above factors revive pre-existing fissures and mark the planes of structural weakness in newly deposited sediments; new fissures are being embedded along the planes after lithification of the sediments.

Fissures genetically related to the formation of fundamental fractures probably predate faulting. Belousov (1962) emphasizes the gradual development of most faults. At first separate small faults appear throughout the zones, then they grow and merge; the process is believed to have continued through geological time. This suggests that both regional systems of fissures and fissuring genetically related to deep fractures should be correlative in orientation with the strike of deep fundamental faults and fractures within a region. Such correlations are suggested by field observations, unfortunately meagre so far.

It is difficult to carry out systematic areal study of fissuring in the country rocks because of the lack of outcrop in the areas of kimberlite occurrence; therefore the use of this very important factor in deciphering the interior structure of kimberlite fields is limited. This gap can be partially filled by the application of the morphostructural approach to the study of the structure of closed regions and primarily methods based on a genetic relation between crustal features and a drainage network of a region.

Faults and fissures are reflected in the topography owing to a number of exogenic processes. These questions are discussed in detail by Goldbraikh et al. (1968). They emphasize that all fissure and fracture zones, even those which had been buried beneath thick sediments, if they had not been healed by further processes, are reflected in the topography to a greater or lesser degree. The writers note a conservative character of faults which can be inherited for much longer periods than folds; this fact, along with high efficiency of exogenic processes, is beneficial for exhibiting all fissure and fault networks in the basement at the present exposure surface.

The most prominent and thus readily observed linear features of landforms are channels whose orientation is commonly correlative with the strike of structures in the basement. An almost complete or adequate coincidence in the orientation of drainage systems and fractures has been recorded in most regions under study. Some writers (Blanchet 1957) are so convinced of a close relationship between the drainage networks and faults that they use the observations in practice, not even trying to prove the relationship.

If the orientation of drainage networks reflects the strike of major fracture systems in the basement, then certain features of channels must also reflect a different extent of fracturing. An integrated approach to the study of major fracture systems in areas of kimberlite volcanism involves the development of areal schemes of rose diagrams, isopycnic sketch maps, showing isotropy of orientation and specificity of a drainage network (Milashev 1979). In view of the plotting on these maps, the direct measurements of the total extent of the drainage network per unit area and the extent in different directions of all streams shown on topographic maps of 1:100,000 scale are classified as original data. The author (Milashev 1979) presents theoretical

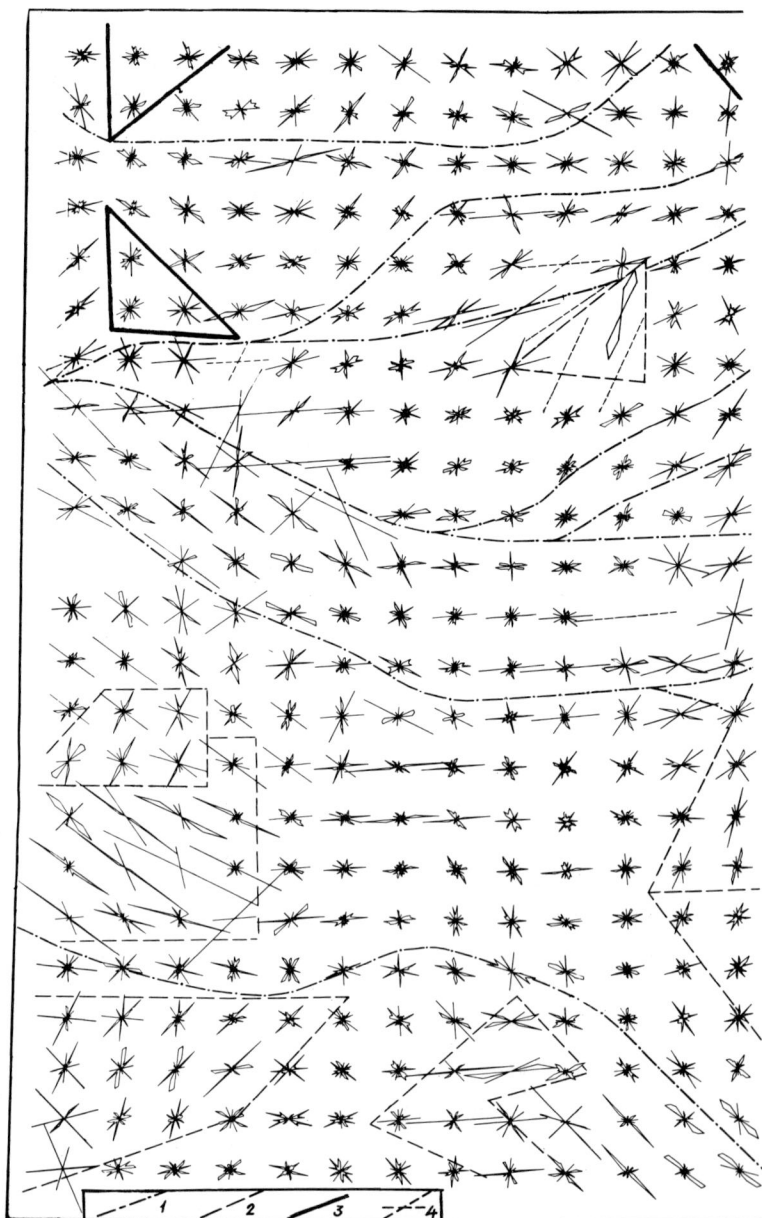

Fig. 24. Medium-scale sketch map of rose diagrams for major fracture system and zonation of the Malaya Botuobiya basin and adjacent areas. The construction is based on a "sliding window" technique at 10° intervals of recording; the area of an elementary unit is 100 km²; sliding pitch in longitudinal and latitudinal direction is 4.5 km and 5 km, respectively. *1* Main strikes of major

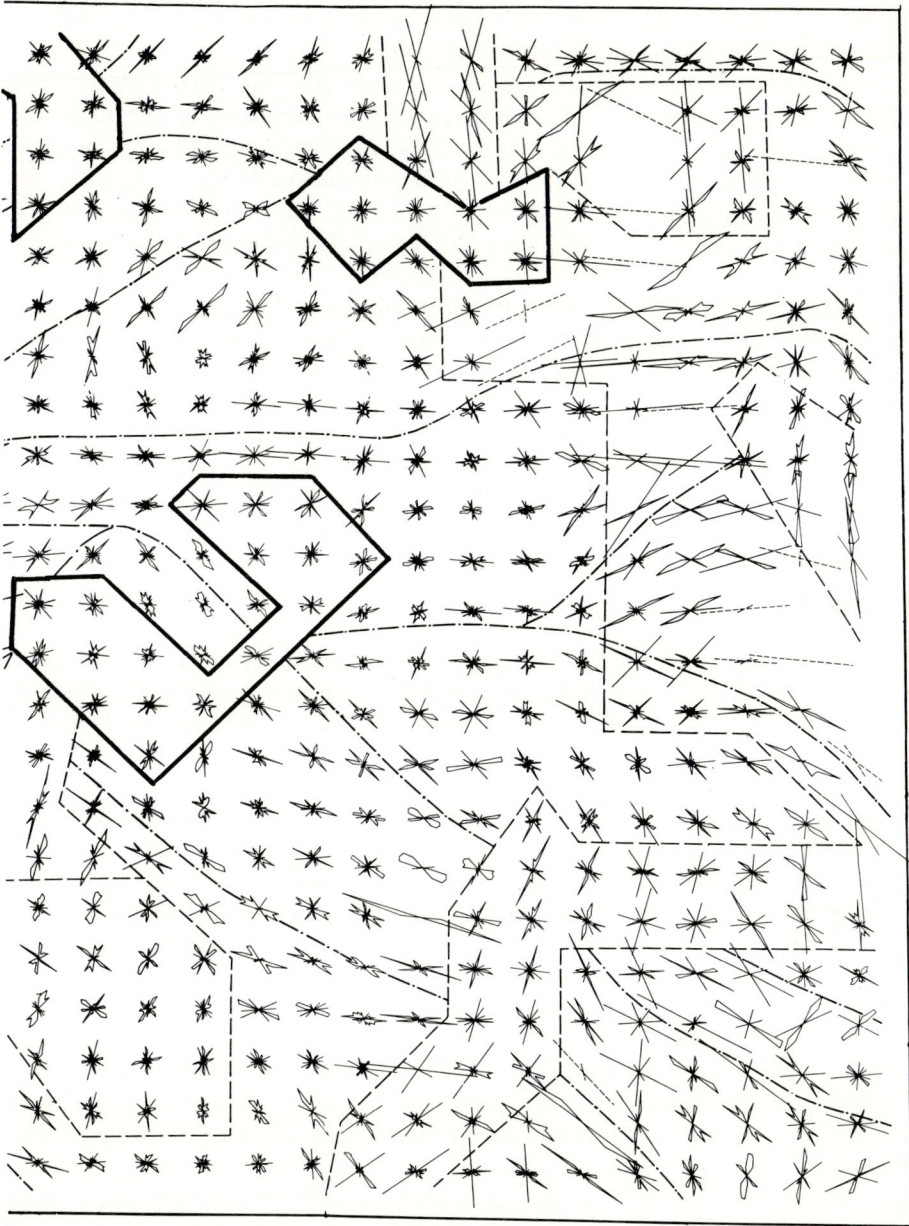

fractures; limits of sites: *2* differing in structural pattern of major fracture system, *3* essentially differing in character of major fracture systems; *4* rose diagrams whose scale is half reduced as compared to the others

statements, and techniques used for compile areal sketch maps, as well as implications for some kimberlite areas on the Siberian platform.

The Malaya Botuobiya basin provides an example of the usage of such an integrated approach to the study of major fracturing. An areal scheme of rose diagrams of the major fracture system of the territory (Fig. 24) enables the following important conclusions. Within the confines of the Malaya Botuobiya basin and adjacent areas the grain of major fracture systems runs sublatitudinally; this complies well with geological and geophysical data, suggesting en echelon structure of basement with sublatitudinal trend of blocks (Babayan et al. 1976).

Against the background of a general sublatitudinal trend of major fracture systems, two main types of sites (blocks) may be recognized. Those of the first type, although similar in the character of major fracture systems to the adjacent regions, differ in structural pattern and/or preferred orientation. Blocks of this kind are most common in the southern part and in the east of the region. Sites of the second type differ sharply in the character of their major fracture systems and, first of all, in their degree of isotropy of orientation from the surrounding areas. These blocks are situated in the northern part of the region. It should be noted that the construction of areal schemes of rose diagrams was made with due regard for the recognition of a block only in the case of no less than three isotropic roses not lying on one straight line (Milashev 1979).

All kimberlite bodies of the Malaya Botuobiya field occur within the site situated in the central part of the territory studied. It has been established that the remainder of the fields of the Central Siberian province are also confined to the blocks with highly isotropic orientation of major fracture systems (Milashev and Sokolova 1984); this is easily explained because these crustal sections are permeable to magma when tangential tectonic stresses of any direction are applied. This suggests that outlines of the blocks in which kimberlite has been found may be considered as structural boundaries of kimberlite fields. If no kimberlite has been found as yet within the blocks characterized by isotropic major fractures, they may be considered as sites favorable for kimberlite emplacement and therefore recommended for prospecting.

Until recently all the boundaries of kimberlite fields were in fact recorded geometric ("morphological") boundaries drawn at some distance from the terminal outcrops. These boundaries neither have an independent geological record nor do they explain the observed distribution of kimberlite. Therefore they neither determine natural geological (first of all, structural) outlines of the sites where kimberlite is known to occur nor the outlines of new promising sites where kimberlite has not yet been discovered.

The peculiar features of the geological structure of the boundary zones separating the fault blocks differing in density, isotropy of orientation, and structural pattern of major fracture systems have been recognized on the basis of field observations. There are three main types of boundary zones: (i) shatter zones with small displacement of blocks in contact; (ii) zones represented by a series of step faults; (iii) tilted zones which suffered folding of the hanging wall. In terms of tectonophysics the boundary zones are considered as linear manifestations of heavily disintegrated rocks of the sedimentary cover of the platform. They may be ascribed to a permanent pulsation in basement and deeper layers of the Earth's crust ("planet breathing"). In this context the zones separating the blocks, which greatly differ in character (first of all, in the degree of isotropy of orientation) of major fracture systems, seem to reflect deeper fracturing as compared to the zones bordering the blocks displaying a varied pattern of major fracture systems. The relation of zones of the first group to deep fractures is suggested by the occurrence of magmatic and, in particular, trachyte volcanic rocks in separate sites of the zones (see Chap. 14).

Special field studies reveal that major fracture systems are identical in crustal blocks of the same type and different in blocks varying in regard to the pattern of major fracturing. This means that this principle may be used for tectonic zoning of vast, primarily closed territories. The width of the boundary zones separating heterogeneous blocks does not exceed several tens of meters, therefore they can be shown as a line even on large-scale maps.

Consequently, the study of major fracture systems in diamond-bearing regions has provided a tool for the recognition of structural boundaries of kimberlite fields. This is, however, not the only advantage of the method. It contributes to our knowledge about the structure of the upper parts of the platform basement and probably about deeper layers of the Earth's crust. The tectonic relief, as well as the presence, size, and configuration of heterogeneous blocks of the platform basement are reflected at the present exposure surface in the regional pattern of dominant orientations of major fracture systems and in local distortion of its structural pattern.

Interesting information on some features of the regional geological structure, which are probably associated with deeper layers of the Earth's crust, can be gained from such parameters as density σ, isotropy of orientation q and specificity ρ of a drainage network (Fig. 25). The estimation of parameters and methods of construction of schemes have been discussed earlier (Milashev 1979); here we present only the formula for calculation of a dimensionless index of isotropy q for rose diagrams:

$$q = \sqrt{(l \sqrt{n})/L}, \tag{10}$$

where L and l are, respectively, the long axis and the short axis, perpendicular to the long one, of a circle diagram; n is the number of rays in a rose spaced at $10°$ (intervals with the length of streams less than 1 km are omitted as poorly representative).

As a first approximation, the index of specificity ρ of a drainage network can be expressed as the product

$$\rho = \sigma q, \tag{11}$$

where σ is the measure of density of streams found by the division of their net length by the area of unit cell in a given grid of observation and averaging.

Even a cursory inspection of density schemes and sketch maps showing the degree of isotropy of orientation of deep fracture systems reveals a distinct regular pattern in the distribution of maxima and minima of the indexes: against the background of a major north-west direction ($310°$) a subordinate north-east trend ($40–50°$) can be observed. These structures are easily discernible on the sketch maps showing the specificity of a drainage network which reflects the general variability of the two first indexes.

The dominant north-west trend of linear crustal structures inferred from analysis of major fracture systems have no analogs in a wealth of current geological and geo-

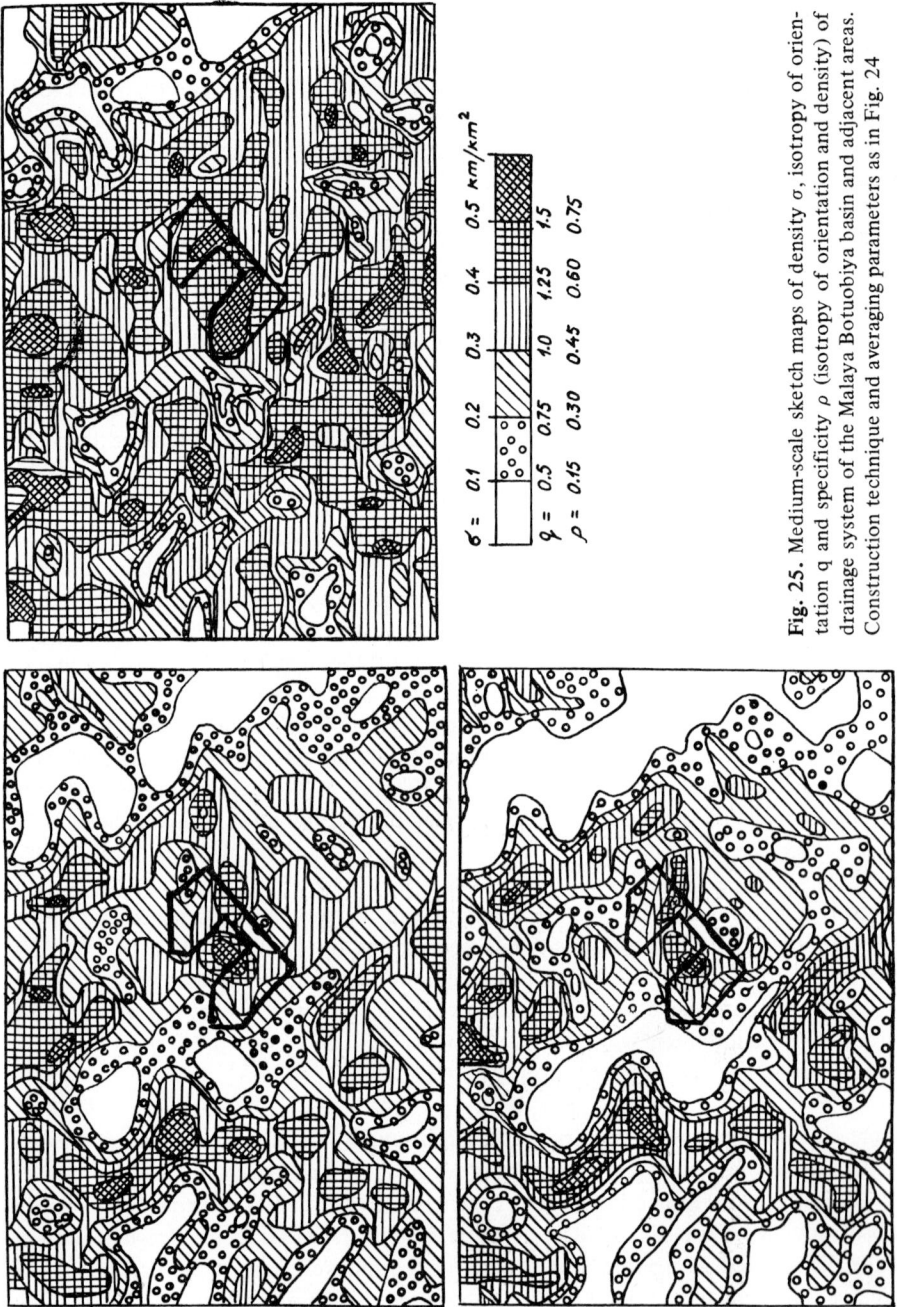

Fig. 25. Medium-scale sketch maps of density σ, isotropy of orientation q and specificity ρ (isotropy of orientation and density) of drainage system of the Malaya Botuobiya basin and adjacent areas. Construction technique and averaging parameters as in Fig. 24

physical data on the structure of the region. Within the structural framework they are presumed to be correlative with some features inherent in the deeper layers of the Earth's crust relevant, to a certain degree, to the formation of the Botuobiya Saddle, despite a slightly different trend. Linear minima of the indexes discussed are close in their character (as sites composed of the lowest-density rocks) and in orientation to north-west-trending systems of gravity highs recorded in the western and eastern parts of the region.

Thus, the involvement of data on major fracture systems and their interpretation with the aid of special techniques not only enables us to recognize crustal blocks differing in density, character, and structural pattern of fissures, but offers means to determine structural boundaries of the kimberlite fields, to locate sites favorable for kimberlite emplacement, and to gain new information on the structure of the upper parts of the basement and, probably, on the deeper layers of the Earth's crust. Of the results obtained from the study of the relationship between kimberlite distribution and fissuring in country rocks, of particular interest are the inferences about the concentration of kimberlite bodies in the crustal blocks pierced by a dense network of isometrically oriented fissures. These problems are dealt with in detail in Part III of the book.

Tectonic Factors Responsible for the Localization of Diatremes

Regular patterns of distribution of diatremes that are apparently confined to the faults within the kimberlite fields have attracted attention since the very beginning of study of primary diamond deposits. As early as the end of the 19th century, investigators revealed the tendency of kimberlites to concentrate in isolated sites and to form elongate linear groups of bodies. Linear groups ("chains") of pipes and dykes are generally subparallel or at an angle to each other. In some cases detailed studies have resulted in the discovery of one or more diatremes along the line connecting several known bodies. The validity of interpretation in terms of general geology, and scarce, but conclusive data on the confinement of some kimberlite bodies to the faults in country rocks have led to the situation that in analyzing kimberlite fields all the investigators attempted to find that the diatremes are structurally controlled by known or, more commonly, inferred faults.

Even in the first years of study of the Daldyn field, attention was drawn to a pronounced tendency of pipes and dykes to form some linear groups (Zveder and Shchukin 1960). The most extensive chains of kimberlite bodies run ENE. The northern group, which consists of ten pipes spaced at a 0.5–6.5 km interval, is the longest, totaling 23.5 km. There are two linear groups in the central part of the field. One of them, 2.5 km long, consists of three pipes, the other, 17.5 km long, includes five pipes. Finally, there is a linear group about 12 km long, which also consists of five pipes on the southern margin of the field. All the major chains are subparallel to each other and spaced at 10–15 km intervals. The northern group has apophyses oriented at an angle of 45° to the strike of the axial line of the chain.

An alternative interpretation of regular kimberlite distribution in the Daldyn-Alakit field is given by Tursky (1969). He considers that the explosion pipes and dykes occur along a NE-trending linear zone, 20–30 km wide. Within the zone the strike of the kimberlite bodies is EW-NE at $30°-35°$ to each other in plan. Some pipes occur at the intersection of the EW and NE-trending fissures. Thus, Tursky (1969) acknowledges the confinement of the kimberlite bodies to a narrow elongate zone and ascribes it to the presence of a NE-trending deep-seated fault in basement. The spatial distribution, orientation, and morphology of the kimberlite bodies suggest that this is a sinistral fault with an attendant system of NE-trending shear joints and sublatitudinal feather joints, as well as rare NW-trending tension joints.

A team of workers (Kharkiv et al. 1972) who made a close study of the geological structure and kimberlite of the Malaya Botuobiya region emphasize that the explosion kimberlite and basalt pipes occur mainly at the intersections of sublongitudinal deep-seated faults and NW-trending feather faults. However, not all the contentions are equally valid and reliable. The sublongitudinal deep-seated faults are reflected in the magnetic field and the land magnetic survey records them as anomalous zones 12 km wide. Within the zones the intensity of the magnetic field is 50–120 nT or, less commonly, a few hundred nT.

The magnetic field pattern suggests that the disturbing masses (dykes) tracing each of the above zones are divided into separate parts (sections), 5 to 30 km long, showing asimuthal variations with $7-25°$; the masses dip ESE or, less commonly, WNW, at an angle of $75°-80°$. There is no surface expression of the faults. In some cases slickensides, as well as pyrite and magnetite mineralization, are recorded in the Lower Paleozoic terrigenous-carbonate rocks within the zones. Although magnetic survey does not record inferred NW-trending feather joints and faults, in sites of their inferred localization gravity gradients and structural lows are reflected in the pre-Jurassic topography.

During the first stage of study of the Srednyaya and Nizhnyaya Kuonamka fields, about half the kimberlite fields were believed to occur within the confines of the basement faults overlain by sedimentary rocks and reflected in the magnetic field. The localization of the second half of the bodies in these fields is tentatively attributed to the faults inferred but not reflected in the magnetic field. Later, the hypothesis of the relationship between the kimberlite fields and the crustal faults in the Srednyaya and Nizhnyaya Kuonamka fields, as well as in 11 other fields of the north-eastern Central Siberian province has been statistically tested (Tabunov 1971a). For this purpose all known kimberlite and picrite outcrops were plotted on the map showing basement faults and compiled using comprehensive interpretation of gravity and magnetic data. With a maximal thickness of the sedimentary cover in the region attaining 3 km (averaging about 1.5 km), the width of the zone "affected" by faulting was taken to be 2 km. Only 85 bodies of 313 (i.e., 27%) occur within the 2–3 km-wide zone, whose longitudinal axis is a fault projection on the surface.

In five of the 13 fields, all the pipes and dykes occur outside the zone "affected" by faulting; in four fields the number of bodies falling within the 2-km-wide zone ranges from 11% to 37%, and only in four fields do more than 37% of diatremes fall into the zone of the inferred effect of faulting. It should be borne in mind that the fields of the latter group (Srednyaya and Nizhnyaya Kuonamka, Luchakan and Kuranakh fields) occur in the area where quite a number of faults are inferred in basement from geophysical data. It is not improbable that this is not the structural control, but merely a spatial coincidence of some diatremes with isolated sections of a rather dense network of fossil faults. The fact that the kimberlite bodies in these fields occur in the areas least pierced by faults is strong evidence against the kimberlite-localizing role of faults inferred from geophysical data in this region (Milashev 1971).

There are also adequately studied regions where faults are known not only in deep-seated basement rocks, but also in exposed strata enclosing kimberlite bodies. Regardless of the reliability of the mapped faults, their influence on the localization of the kimberlite bodies is not obvious and is far from being unambiguous. Only in few cases are the diatremes undoubtedly controlled by the observed fractures, because (i) the diatremes occur at the intersection of fractures and locally even follow the outlines of the nodal section; (ii) the orientation of long axes of the diatremes in the chains of kimberlite bodies concentrated along a single fault completely follows its strike.

In summary, the spatial distribution and localization of the kimberlite bodies within any field are governed by the fractures generally not identified geologically, and not always inferred geophysically. Some idea of the nature of the kimberlite-controlling fractures may be gained from the features discussed above.

It is well known that both geological and magnetic surveys result in an easy and distinct identification of fractures which are infilled by dykes of igneous (in particular, high magnetic) rocks, as well as of fissures which border the strata differing in habit, composition, and magnetic properties. Geophysical and geological data suggest that the great majority of the kimberlite-localizing faults are fractures which display neither large displacement nor emplacement of large portions of nonkimberlitic magmatic melts. The fact that these fractures show, as a rule, low gravity gradients and high conductivity suggests that they are represented by shatter zones subject to infiltration. It is quite probable that the formation of shatter zones was accompanied by fissuring in consolidated strata both in sites adjacent to the zones and those far away from them. Hence, the peculiar features of fissures piercing appropriate parts of the territory to some extent affected the localization and morphology of the kimberlite bodies and may be used to indirectly recognize the paths of the upward movement of kimberlite melts.

Study of the hypothetical relationship between the morphology and distribution of kimberlite bodies, on the one hand, and fissuring in the country rocks on the other was started soon after the discovery of kimberlite in Yakutia. The study was mainly accomplished by means of correlation of the orientation of long axes of pipes, the strikes of dykes, and those of chains of bodies with the trend of fissure systems in the country rocks. Correlations were performed with due regard not only for the number, but for a "measure": vectors on orientation diagrams were constructed on the basis of the frequency (the number) of kimberlite bodies having this or that strike, and the absolute extent of the long axes of pipes and dykes in various directions. As a rule, the diagrams of the second type are more demonstrative.

In various areas of the Central Siberian region, the kimberlite bodies were shown to be aligned primarily along the trend of major fissure systems. Within each field the explosion pipes and dykes are lined up in two or three, or, less commonly, in one or four or five directions. A point worth mention is that when some fields are lined up, being probably confined to a single zone of high crustal permeability, the long axes of most kimberlite bodies are aligned along the structural grain of the zone and quite often along one or two further diagonal trends. In these regions the fissures in the country rocks are also oriented in a similar way.

As mentioned above, in almost each field a rather high proportion of explosion pipes and dykes are grouped to form chains which include some (as a rule, three to five) lined-up bodies. In this case the orientation of long (in plan) axes of diatremes and the strike of dykes to each other, as well as the general trend of the chain, are not considered. This approach cannot be assumed to be advantageous, since some important data on the direction of tension during the formation of the bodies studied are lost. Partially this may be compensated for by the recognition of two terminal and at least one intervening type of the linear groups of kimberlite bodies, namely groups of (i) concordant type in which the orientation of the long axes of all the kimberlite bodies is roughly coincident with the strike of the group they form; (ii) discordant type in which the long axes of all the kimberlite bodies are oriented roughly across the strike of the group they form; (iii) intervening type in which the long axes of some kimberlite bodies are aligned along the trend of the group as a whole, while the long axes of other bodies differ greatly from the general trend of the group.

Most probably the kimberlite bodies of each concordant group occur within the confines of a single tectonic structure and belong to one and the same epoch of ultrabasic volcanism. Each intervening and, in part, discordant group represents a complex assemblage of kimberlite bodies which formed in varying tectonic conditions over a long period of time (a few millions of years). The axial lines of these groups trace not only the trend of faults controlling the distribution of concordant bodies, but also the sites where they are intersected by cutting fractures and where the localization of discordant bodies took place. This suggests a rather close relationship between the orientation of the long axes of kimberlite bodies and the trend of the major fracture systems, and explains some causes of this relationship.

All known kimberlites of the Siberian platform, and of the provinces of Africa pierce ancient sedimentary rocks. Based on the above mechanism of fissure emplacement, we may conclude that most fissure systems in the country rocks had been formed long before the onset of kimberlite volcanism. When rising into the higher crustal layers and emplacing pipes and dykes, the kimberlite magma had to use those systems of fissures which had been subject to tension during a certain tectono-volcanic stage. From these viewpoints the places where concordant linear groups are located must be considered as sites where country rocks had suffered maximal tensile stresses, and the strikes of the concordant groups as normal to the direction of tensile stresses. When interpreting the mechanism responsible for the origin and tectonic setting of discordant linear groups, it should be assumed that the bodies which form each group appear to be sited at the intersection of fissure planes with a pre-existing zone of structural weakness, whose strike roughly parallels the direction of tension.

If concordant and discordant linear groups of kimberlite bodies were formed during one or more phases, the formation of the intervening groups took at least two phases. Comprehensive study of the kimberlitic rocks in linear groups of each field allows the number of phases of their formation to be established. This, along with the results of comparative analysis of the orientation of linear groups, long axes of the kimberlite bodies, and the trends of fissures in the country rocks, gives the possibility of restoring the spatial distribution and orientation of tensile stresses for all the major phases of kimberlite field formation.

CHAPTER 9

Temporal and Spatial Rules in the Magmatic History of Kimberlite Fields

The formation of any diatreme cannot be divorced from the complex processes operative within the kimberlite field where it occurs. In this context the hypotheses of the origin of kimberlite pipes cannot ignore the major structural features of kimberlite fields, but, on the contrary, should incorporate and reconcile them with main rules pertinent to the magmatic history of the field. Taking into consideration the importance of spatial and temporal rules of volcanic activity within kimberlite fields for the problems discussed here, particularly for a better understanding of the mechanism and setting of the formation of cone-shaped bodies, we shall consider the problems in more detail.

Trends of contrast differentiation indexes (CDI) and those of differentiation indexes (DI) of the kimberlite magma may be used to reconstruct a general sequence of events of ultrabasic volcanism both in isolated sites and over the entire area of each field. The relationship between the petrochemistry of the kimberlite and its relative age is discussed in Chapter 7. It will only be recalled here that CDI_{Ti}^{Fe} increases in a successive series of kimberlites from earlier to later ones, thus allowing the division of the whole epoch of the formation of the field studied into stages. Based on $DI_{Ti}^{(5)}$ values, each stage is further subdivided into phases. Since the rocks which emplaced at the final phase of the early stage in the general case show larger $DI_{Ti}^{(5)}$ values than those of the initial phase of the later stage, the principal sequence of formation (temporal development) of any field can be inferred from the trend analysis of CDI_{Ti}^{Fe} and $DI_{Ti}^{(5)}$ values.

The validity of statements has been checked and supported by the construction of trends of these petrochemical indexes for three kimberlite fields used as standards (Milashev 1979). The CDI distribution shows a much higher degree of ordering as compared to DI. The areal distribution of the CDI shows one main feature in common for all the kimberlite fields: its low values occur in the middle of each field and increase toward the periphery of the field. The CDI may increase from the central part of the field either radially or into only two sides. The situations are characteristic of the temporal pattern (zoning) of central- or linear-type fields, respectively. The assumption that the trends obtained represent the development of the process of formation of kimberlite fields in time infers that in the central-type fields volcanism starts mainly in the center of the field, fanning and waning toward the periphery. In the linear-type fields, kimberlite volcanism starts predominantly in a narrow linear zone across the field and spreads on both sides of the zone toward the opposite boundaries of the field.

The spatial distribution of DI which is not correlative with the CDI trend can display not general, but individual and not so obvious tendencies. For instance, if in one of the three well-known fields the DI decreases radially, in the second field it increases also radially, while in the third, on the contrary, the trend surface of the index studied shows the prominent tendency to increase westward and south-eastward with the minimal values being recorded in the north, east, and south-west of the field.

Methods used in studies of spatial rules of magmatism differ in depth of penetration, and the results obtained permit deciphering the internal structure of kimberlite fields with a varying degree of thoroughness. An integrated approach provides not only a generalized pattern of the field, but reveals many important structural features, thus contributing greatly to a better understanding of the internal structure of kimberlite fields.

The method which ensures the deepest penetration but yields only a general picture with no detail is based on the possibility of determining a relative degree of variability of the thermodynamic regime during the plutonic and hypabyssal phases of evolution of kimberlite magmas and of establishing the permeability of deeper and near-surface crustal layers from the content of phenocrysts of olivine II and III (see Chap. 3). This approach offers strong possibilities in revealing major featues of intra-crustal evolution of the kimberlite magma, as well as in constructing permeability diagrams for different crustal layers which can be further correlated with the distribution of kimberlite bodies and structural features of enclosing rocks, recognized as the result of study of the major fracture pattern.

Calculations of permeability ν, ν_p and ν_h [Eqs. (4) and (6)] for all petrochemical groups and single-type rock complexes in four standard kimberlite fields show that, in spite of substantial variability of permeability coefficients for different petrochemical rock groups and single-type rock complexes within a given field, these variations are not common for these rocks even in two adjacent fields formed in similar environmental conditions (Milashev 1979). Differences in the chemistry of even extreme kimberlite varieties are insignificant and, hence, cannot cause the scatter observed in the content of phenocrysts of olivine in them. Therefore the absence of a certain relationship between the permeability indexes ν, ν_p and ν_h, calculated from olivine content and petrochemical rock features, seems quite natural.

In our opinion, variations of ν, ν_p and ν_h of identical rocks in various fields are best interpreted as caused by variable crustal permeability (and tectonic regime) in certain sites of the territory during the intracrustal evolution and the upward movement of adequate batches of kimberlite melts. The assumption that varying crustal permeability is the main reason for variability of olivine content in kimberlite suggests that the petrography of the rocks allows the estimation of averaged crustal permeability at great and small depth both within the whole of the kimberlite field and in sites of development of isolated groups of diatremes and single diatremes.

It should be noted that visual analysis not only of values for each pipe and dyke, but also of averaged ν_p and ν_h values is not efficient for compact groups of kimberlite bodies. In view of the considerable number of bodies, their uneven distribution in space and considerable variations of both coefficients, we usually fail to establish the regularity in variation of these coefficients within the fields studied. It is advisable to use trend analysis to obtain general tendencies in the areal distribution of coefficients ν_p and ν_h within a given field and within several adjacent fields. However, it is quite evident that the integral picture is obtained on account of the leveling of local features. Certain rules may be revealed to be the result of relatively slight generalization even with the 25% confidence level. In the course of further generalization, for instance with the 50%, 75%, or 100% confidence level, more common tendencies are displayed, but many important features of the internal structure of kimberlite fields become obscured and lost.

In the plutonic phase of magmatic history, the trend surface of the crustal permeability coefficient ν_p suggests the presence of one or two isometric or slightly elongate (in plan) highly permeable sites in each of well-known fields. The trend surface of crustal permeability coefficient is quite different at the hypabyssal phase of magmatic evolution. Trend analysis of the coefficient ν_h suggests the presence of three to five very elongate highly permeable sites in the near-surface layers of each kimberlite field. It should be stressed that not only trend surfaces of coefficients ν_p and ν_h, but their general outlines in plan are different. For instance, if the general trend outline of the coefficient ν_p from the isopleth of minimal positive values is relatively compact and even close to the isometric one, analogous trend sections of the coefficient ν_h have sinuous contours and rather frequent intracontour zero "windows".

Thus, the results of trend surface analysis of coefficients ν_p and ν_h in a given kimberlite field suggest that at the plutonic phase, i.e., at a depth of several tens of kilometers, the melts moved through one or two isometric or elongate highly permeable sites (zones), while in the hypabyssal phase at a depth of a few kilometers these paths were permeability zones, relatively small in area, linked or isolated in space. This suggests the branching of highly permeable zones as the surface is approached. The suggestion is in good agreement with the ideas of other writers, who studied the relationship between deep-seated and surface fractures (Radkevich 1960). It is noteworthy that isopleths of minimal positive values of the coefficient ν_h delineate the territory which is very similar in outline and size to zones differing from the surrounding areas in character and pattern of major fracture systems, as well as in the presence of kimberlite (Milashev 1979).

The coincidence in shape and size of the zones delineated from the trend of coefficient ν_h and the major fracture pattern can hardly be considered as being random. Moreover, the factor of randomness need not be involved at all, as the trend surface of the ν_h indirectly depicts some peculiar features of fracturing in the hypabyssal crustal layers within the confines of certain kimberlite fields. Hence, the trend surface analysis of the ν_h and the study of the fracture pattern largely complement and, if they coincide, support each other. The coincidence is especially important to obtain mutual control, because the methods are based on quite different sources.

The differences in crustal permeability to the kimberlite melts in some or other sites of the area are governed by the variability of the density, vertical and lateral persistency of faults, as well as by the difference in degree of opening or closing of fractures at particular phases of kimberlite volcanism. This means that the average crustal permeability at a scale of the whole field or a group of kimberlite bodies can be used only as a very general statement. In detailed studies permeability must be estimated separately for each phase of magmatism confined to certain zones of structural weakness within the kimberlite field studied.

These studies are intimately associated with an investigation into ore-controlling faults and spatial distribution of kimberlite pipes, dykes, and sills. In discussion of local factors of kimberlite distribution, it was shown that folding of any type and scale has no pronounced effect on the spatial distribution as a whole and on diatreme siting. In this context faults play a leading role. The lack of outcrop of almost all diamond-bearing rocks makes it difficult to obtain sufficiently comprehensive data which are evenly distributed over the area about the sites of occurrence, density, and orientation of the faults. The study of a drainage network on a compound ground permits medium-scale zoning of kimberlite fields and adjacent areas and recognizing sites differing in density, degree of isotropy of orientation, and general major fracture pattern (see Chap. 8). The study of the structure of country rocks in kimberlite fields based on the geology and petrology of the kimberlite bodies makes it possible to apply it to the recognition of ore-controlling and ore-enclosing faults. Let us examine these questions.

In fact, all kimberlite fields display a pronounced tendency in the distribution of certain pipes and dykes in the form of linear groups. The relationship between some linear groups of diatremes and crustal faults is emphasized in all the accounts of this topic, although field observations of kimberlite-controlling faults are few. Kimberlite fields where all bodies or most of them form distinct linear groups, thus ruling out any controversy in interpretation, are relatively rare. In most cases the distinct linear groups (chains) include the small number of pipes and dykes. The remainder of the bodies in the fields are believed to be associated with shear joints or are by no means confined to faults controlling the distribution of linear groups.

Since the structural control of kimberlite bodies not incorporated in chains is usually a point of controversy, the following simple approach may be used to estimate quantitatively and display the spatial relationships between the bodies to analyze the distribution of pipes and dykes within a field. One should visually plot all straight combinations of four or more bodies within the field studied and link them with straight lines, whose graphic presentation displays certain quantitative parameters.

Among these parameters the most subjective are probably the number n and the area S of the bodies calculated per unit length l of inferred ore-controlling faults. Both coefficients are somewhat interrelated, but the relation is not strictly proportional, because the size of kimberlite bodies even within one field varies over a wide range (see Chap. 5). The geological significance of the coefficients is approximately the same, and their product, which shows their variability alone and in combination, can be referred to as the coefficient of kimberlite impregnation of both observed and inferred ore-controlling faults:

$$k = nS/l^2. \tag{12}$$

Denoting the kimberlite impregnation levels of the faults enables us to graphically represent and analyze the spatial distribution of fractures distinguished with respect to this characteristic. Experience gained in the construction of such diagrams for three types of kimberlite fields of the Central Siberian province shows that within one field the projections of ore-controlling fractures encompass all bodies, while each of the two others contains 96% of pipes and dykes. Seventy percent of bodies of the first field, 85% and 80% of the other two fields occur at the intersections of inferred ore-controlling fractures. It should be also emphasized that most (73–81%) of the elongate diatremes and dykes in the study fields are aligned along inferred ore-controlling fractures (Milashev 1979).

The results obtained, with allowance made for rare, but reliable data on diatreme localization at the intersections of the fractures, as well as the coincidence of long axes of the diatremes with the trend of ore-enclosing faults, suggest that inferred faults obtained with the aid of the above techniques correspond, at a first approximation, to virtually existing fault systems controlling the distribution of kimberlite within certain fields. This is also confirmed by the fact that on the continuation of discordant diatreme axes one can find pipes occurring on the fractures, recognized from tracing chains of bodies, or at their intersections. The construction of appropriate rose diagrams facilitates the analysis of major trends in the orientation of inferred fractures according to size, extent, and impregnation with kimberlite.

The schemes of ore-controlling fractures give an insight into the internal structure of kimberlite fields by providing information on the sequence of formation or postdating, as well as that on the relatively active life of fractures and faults controlling the distribution of pipes and dykes; the schemes also help to locate sites of inferred occurrence of bodies which are not yet discovered.

The formation of kimberlite within any field continues for a rather long interval of time (see Chap. 7). With long duration of volcanism over the areas embracing several hundred to several thousand square kilometers, tectonic stresses cannot be constant, but vary in intensity and direction. As a result, some faults control the distribution of kimberlite bodies of only this or that phase, while others suffer multiple deformations and directly affect the localization of explosion pipes and dykes emplaced during the different phases of formation of appropriate kimberlite fields.

No general geological indications and features provide information on the succession of formation or renewal, as well as on the duration of active life of fractures and faults controlling the distribution of pipes and dykes. The petrochemistry of kimberlite associated with the faults may be used as an indirect criterion for the subdivision of faults by means of the relative time of their formation and relative duration of their active life during the epochs of kimberlite volcanism, because a certain relationship exists between the composition and relative age of kimberlites (see Chap. 7, Fig. 23). The technique of construction is simple: the location of kimberlite bodies, the orientation of their long axes, as well as the confinement to certain volcanic stages and phases [from CDI_{Ti}^{Fe} and $DI_{Ti}^{(5)}$ values] are plotted on the scheme of inferred ore-controlling faults.

If at the beginning of studies the division of magmatic history can be confined to stages, later studies require further subdivision into phases. Some fields contain, along with multiple bodies formed during different phases within some stages, single bodies made up of rocks which are similar in CDI_{Ti}^{Fe}, but that differ in $DI_{Ti}^{(5)}$ values and thus, strictly speaking, ought to be assigned to different phases of one stage. However, in the case of the relatively large number of pipes and dykes in the field studied, the recognition of individual phases from single pipes and dykes makes the analysis unwieldy. Better results can be obtained if a "sliding" degree of age subdivision is used: volcanic stages and phases can be recognized for petrochemical groups represented by numerous exposures of rocks, but the subdivision of not very abundant varieties can be confined to stages.

The relative orientation of inferred ore-controlling fractures and elongate kimberlite bodies varies from strictly parallel to mutually perpendicular. Because of the complex configuration of most diatremes it is difficult to precisely determine the orientation of elongation of many of them, the uncertain interval ("play") being about 20–30°. Therefore it is advisable to recognize two extreme cases: the orientation of the long axis of a kimberlite body is roughly concordant ($<\pm15°$) or discordant ($>\pm15°$) to the trend of inferred faults. In the case of concordant orientation, an inferred fault can be recognized as an ore-controlling one for kimberlite of certain phase and/or stage.

There are no special criteria permitting the determination of the extent of a fault section which effectively controls the occurrence of kimberlite of an appropriate age group. At a first approximation, the intersections with other faults and the sites of occurrence of discordant kimberlite bodies can serve as markers for this fault section. The obtained extent of faults which were active during the epochs of kimberlite volcanism and effectively controlled kimberlite distribution accounts for 40–50% of the total length of ore-controlling faults inferred from the "geometric" approach.

The activity of some faults along the strike was not constant with time, as a result of which different sections of each of them were localizing for kimberlite of either some earlier or relatively later phase of this or a subsequent stage. The occurrence on one fault section of two or more concordant bodies composed of rocks of different volcanic stages and phases suggests multiple activation of the fault in the process of formation of the field studied. It is worth mention that concordant kimberlite bodies associated with the faults which suffered two or three activations belong, as a rule, to strictly successive phases with only one "jump over" per phase.

CHAPTER 10

Kimberlite Magma Chambers

The Hypotheses

Questions dealing with the depth of generation and differentiation of the kimberlite magmas characteristic for magma chambers and with the genesis of the diamond have long been a subject of controversy and still attract widespread attention. Current ideas are divided into two main groups which are further subdivided into subgroups.

The workers who subscribe to the first group of the concepts consider that the kimberlite magmas originate in subcrustal depth within upper mantle rocks, where the thermodynamic conditions foster crystallization of diamond and pyrope. The magmas rise to the base of the crust, then rapidly reach the surface and solidify, thus conserving the diamond, which is metastable at low pressure. Proponents of the second group of ideas think that ultrabasic magmas uplifted from deep within the crust or from subcrustal layers acquire, in the process of evolution, properties which contribute to the generation of kimberlite in some intermittent vents at a depth of a few kilometers from the surface. As a result, in these vents explosion of hydrocarbons sucked from the enclosing strata or "the gas phase of magma induced by pulsations of the crust" give rise to the conditions required and sufficient for crystallization of the diamond, which formed from carbon of oil and gas (Vasiliev et al. 1961) or from graphite of metamorphic rocks underlying the platform basement (Trofimov 1967). A critical account of the above hypotheses is beyond the scope of the book; moreover, they have been discussed elsewhere (Milashev 1972a; Sobolev 1974 and others).

The hypotheses relating kimberlite origin and evolution to the upper mantle seem to be more theoretically devised, substantiated and supported by observations and, hence, more convincing. The place and time of crystallization of diamond, pyrope, pyroxene, and other minerals of "cognate inclusions" are widely disputed, hence the subdivision of the hypotheses of this group into two subgroups.

The proponents of the hypotheses of the first subgroup consider the cognate inclusions and their separate mineral constituents as fragments of upper mantle rocks; in this context, the kimberlite magma formed by partial melting of substratum would be a "transporter" withdrawing crushed, not completely melted material from the depth to the surface (Sobolev 1974). However, no mechanism responsible for the uplift of the kimberlite magmas from depths of 100–200 km in the upper mantle to the base of the crust is discussed, although the PT conditions dominating there give rise to a high plasticity of substratum which rules out any faulting.

The hypotheses of the second subgroup also attribute the generation of the kimberlite magmas to partial melting of mantle rocks, whereas they move radially,

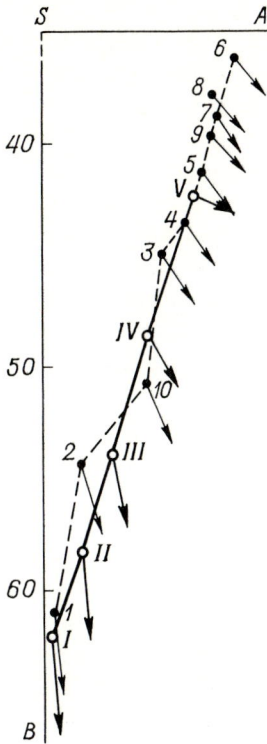

Fig. 26. Variation petrochemical diagram constructed for effusive ultrabasic and alkali ultrabasic rocks of the Maimecha-Kotui basin and for a probable composition of picrite-facies melts formed due to zone melting of variable volumes of upper mantle material. (Milashev 1974a). *1* Meimecha Formation, meymechite (5 analyses); Arydzhan Formation: *2* picritic porphyry (2); *3* olivine melilitite (2); *4* augitite (3); *5* olivine melanephelinite (10); *6* limburgite (3); Delkan Formation: *7* limburgite (2); *8* olivine analcime (4); *9* olivine melanephelinite (2); *10* mean weighted composition of rocks in the district. Melt composition with remolten mantle material to magma chamber volume ratio: *I* 1 (parent magma); *II* 10; *III* 20; *IV* 30; *V* 40

following the mechanism of zone melting. The progressive radial motion of the molten magma in this case is accomplished by melting-through the roof and sinking of almost the same amount of newly crystallized mineral phases in lower regions of the chamber. Both thermodynamic estimates and experimental data suggest that as the result of zone melting the newly crystallized mineral phases become more refractory than the phases which suffered melting; therefore, less refractory compounds and volatiles are accumulated in the melt. Hence, the radial propagation of magma chambers through the mechanism of zone melting is inevitably accompanied by gradual variations in magma composition.

The chemical balance and energetics of kinematic differentiation of the kimberlite and picrite magmas during their upward movement in upper mantle rocks are discussed in detail elsewhere (Milashev 1972a, 1974a). The truth of the theoretical assumptions and the validity of the equations used to calculate kinematic differentiation of the alkali-ultrabasic magmas were tested and supported by the study of a wealth of picrite-facies porphyritic rocks of the Maimecha-Kotui region of the Siberian platform. Effusive ultrabasic and alkali-ultrabasic rocks of the region and theoretically calculated compositions of the picrite-facies magmas resulting from melting-through of various quantitites of substratum are plotted on the petrochemical diagram constructed after Zavaritskii (Fig. 26). It shows that plots denoting melts, whose compositions were calculated theoretically, form a band which is, in fact, averaged relative to the variation curve of the effusive rocks of the Maimecha-Kotui region. The coincidence observed is not fortuitous and undoubtedly suggests the validity of the application of the principles of zone melting to genetic statements, the possibility of quantitative estimates of the extent of differentiation of the ultrabasic melts, as well as that of magma-derived kimberlite and picrite-facies rocks.

Without going into detail, we should only stress that the extent of differentiation of the kimberlite and picrite magmas is controlled by the difference in composition of melting and crystallizing phases, as well as by the volume of repeatedly melting substratum rocks; the volume is a function of distance passed by the magma chamber within the upper mantle through the mechanism of zone melting. The mineral phases which crystallized from the kimberlite magmas are unlikely to sink completely in the rear (bottom) part of the chamber, but may be partially caught up by convection currents to form segregations ("cognate inclusions").

On reaching the base of the crust, the kimberlite magmas rise rapidly along the zones of structural weakness; they either pierce their way to the surface to form diatremes or solidify near the surface as fissure and sheet intrusions. The assumptions of the presence of some intermittent chambers at a depth of 1—4 km (Trifimov 1967) or at the boundary of the platform cover and the crystalline basement (Vasiliev et al. 1961) are inconsistent with the geological observations and invalid in terms of energetics.

The numerous occurrences of kimberlite pipes, including Mwadui, the largest diatreme in the world, within granite and gneiss provide strong evidence against the presence of intermittent chambers of kimberlite magmas at the cover/basement boundary. On the request of Kovalsky, one of the authors of this hypotheses, drilling was performed on a pipe of the Srednyaya Kuonamka field where the depth to a basement is 100—200 m. The borehole penetrated 300 m of the kimberlite breccia, that is 100 m below the cover/basement boundary. Drill cores show no evidence of either vents or substantial variations in composition, texture, and structure of the kimberlite breccia.

Another important factor bearing witness against the presence of intermittent chambers of kimberlite magma is the occurrence of diamond in the rocks. Actually, if any intermittent chambers were present at a depth of 1—4 km, kimberlite magmas with diamond in them ought to stay in the chambers for some time (at least during infilling). However, thermodynamic estimates and experimental data suggest that such a detention would have resulted in the complete replacement of diamond by graphite, which is a stable modification of carbon under hypabyssal conditions.

The statements of the advocates of the hypotheses that high pressures set up and maintained in intermittent chambers through "the gas phase of magma induced by pulsations of the crust" or because of explosions of hydrocarbons sucked from the enclosing rocks, appear to be unsound and more open to criticism than the concept of intermittent chambers. The "alliance" between the gas phase and crustal pulsations, as the result of which the intermittent chambers set up pressures sufficient for diamond crystallization, seems invalid even if only because of the fact that an n-fold increase in gas pressure at a constant temperature requires the same-fold reduction in its volume. Any important regime-dependent pulsating reduction in the volume of an intermittent chamber can take place only if the rocks surrounding the vent are characterized by high elasticity and compressibility or, at least, high plasticity.

However, even in this case (quite inadequate for the depth of 1—4 km invoked by Trofimov) pressures inside, and, particularly, at contacts of, rigid blocks involved in pulsations would be much higher than those in intermittent kimberlite chambers which must have been more "compliant" owing to the presence of the gas phase. If pulsations or some other regional processes had really given rise to such stresses, then tectonic deformations, variable in type and large in scale, ought to be reflected in the highest crustal layers. Moreover, intermittent chambers themselves should have locally been brought to the present exposure surface, because in the kimberlite provinces of the world there are numerous areas where denudation penetrates to a depth above 1 km.

The contention about extremely high pressures in intermittent chambers retained due to explosions of self-detonating hydrocarbon mixtures sucked from the country rocks appears not only speculative, but can be grouped with the "perpetuum mobile" models if for no other reason than that it is based on the assumption that oil and gas are sucked (!) from the sedimentary rocks into the intermittent chamber in which the pressure is several tens of times higher.

It is also difficult to explain the presence of intermittent kimberlite chambers under hypabyssal conditions in terms of energetics. Even in the largest provinces the volume of kimberlitic rocks is known to total only several cubic kilometers (Milashev 1965), and within any field their volume does not exceed a few tenths of a cubic kilometer. The volume of one or several intermittent vents, assumed to be responsible for the formation of kimberlite bodies of any field, is unlikely to differ considerably from the above estimates. With such a small volume of a melt, some lengthy existence of an intracrustal vent would no doubt lead to the exhaustion of heat resources and hence to crystallization of magma.

All this suggests the presence of only mantle magma chambers rather than crustal (and, moreover) hypabyssal ones. The short-term journey of the magma from the base of the crust to its higher levels and its rapid solidification in small pipes and dykes warrant the preservation of the diamond, which is metastable under these conditions. Any prolonged detention of the kimberlite magmas on their way from subcrustal depth to the land surface would inevitably lead to complete graphitization of the diamond, since geological evidence implies that there are no local sites within the crust where pressures of 4000—6000 MPa might have been set up and maintained for a long time.

The Character and Number of Magma Chambers in Kimberlite Fields

Kimberlite rocks infilling pipes and dykes of even one field differ substantially in chemistry and relative time of intrusion (see Chaps. 2 and 7); this may be considered evidence for the presence of numerous magma chambers from which the rocks derived. Although the question is crucial for our study, it is difficult, both theoretically and methodologically, to estimate the probable number of magma chambers that took part in the formation of kimberlite bodies of any field.

Naturally, the first step in this direction is analysis of spatial distribution of chemically varied kimberlites in the well-known fields. In fact, results of comparative analysis of the petrochemistry of kimberlites occurring in different parts of the field should be used to estimate the number, approximate location, and peculiar features of magma chambers responsible for the formation of the rocks in the field studied. The discovery of zoning in the distribution of major representative elements within the field or in the isolated sites would suggest the presence of a single chamber evolving in the whole field or some local scattered chambers.

Visual analysis of the data reveals some vague trend in such a zonal distribution of kimberlites containing various amounts of representative elements. However, trend surface analysis gives no evidence for any zoning in the distribution of iron, titanium, and aluminum in the adequately studied kimberlite fields. With the 25% confidence level, maximal and minimal values in the content of the above elements occur along the periphery of all fields. Moreover, the character of trend surfaces shows that no further generalization of initial data (i.e., calculations with the 50% and 100% confidence levels) would reveal zoning in kimberlite distribution. Strong evidence for the validity of the results obtained is an almost complete analogy in the trend of iron and titanium, which are similar in geochemistry and exhibit a quite different, almost opposite character to the trend of aluminum, which differs sharply in geochemistry from the first two elements. The distribution of trends for major representative elements in kimberlite rules out their application in the estimation of the probable number of magma chambers even within the well-known fields. Hence, we should seek for other ways of estimation.

In theoretical calculations of the probable number of magma chambers responsible for the formation of the kimberlite bodies of any field, the quantity sought for may be restricted by two extreme assumptions: (1) all pipes and dykes in a given field were formed by magma generated in a single chamber; (2) each simple-structure body and each rock generation in complex bodies was formed from magma generated in separate magma chambers. This may be expressed by the following equation:

$$1 \leqslant N \leqslant B + G, \tag{13}$$

where N is the probable number of magma chambers in a given kimberlite field; B is the total number of simple kimberlite bodies; G is the total number of kimberlitic rock generations in complex bodies.

However, a high degree of reliability thus obtained is depreciated by a very wide range of equipermissible values of the quantity sought. Although it is quite obvious that the real number of magma chambers generating the melts in the course of formation of any kimberlite field lies within the range determined by Eq. (13), nevertheless, no universally known and adopted geological and petrographic approaches can provide more accurate data.

The problem can be approached in another way, namely, by using the data considered in the preceding chapters of the book, first of all, the relationship between the regime and the extent of differentiation of magma and that between the petrochemical features (types) and the sequence of formation of kimberlitic rocks. Important differences in contrast of differentiation with respect to iron (CDI_{Ti}^{Fe}) which reflects the energetics of processes operating in subcrustal depth allow the recognition of age groups of rocks (see Chap. 7, Fig. 23) whose time intervals in each field can be related to separate stages of its formation.

Each stage may have resulted in the formation of several (up to four) petrochemical types which, with similar CDI_{Ti}^{Fe} values, have similar or substantially different $DI_{Ti}^{(5)}$ values. According to the above principles and data, in the case of different $DI_{Ti}^{(5)}$ values, the rocks of even one petrochemical type should be assigned to various phases of the same stage. Since in the course of each phase rocks of several petrochemical types were formed, phases are further subdivided into steps. This suggests that a step is a time span during which a complex of rocks of the same petrochemical type was formed; the rocks, as evidenced by $DI_{Ti}^{(5)}$ and CDI_{Ti}^{Fe} values, relate to the same phase of the appropriate stage. It is not possible, at the present state of knowledge, to establish the order of steps within a phase. The total number of steps, S, of the kimberlite magma intrusions in any field can be expressed by the equation

$$S = a_{n_1} + b_{n_{1-2}} + \ldots + v_{n_8}, \tag{14}$$

where n_1, n_{1-2} ... n_8 are the observed petrochemical rock types (see Fig. 21); a, b ... v is the number of complexes strongly differing in $DI_{Ti}^{(5)}$ in the rocks of each petrochemical type.

The study of petrochemical conditions of kimberlite formation indicates that differentiation of the melts takes place almost exclusively at abyssal depths in the subcrust by the mechanism of zone melting. The outburst of the magma into deep-seated fracture zones and the rapid ascent up the fractures actually rule out further variations in bulk compositions of the magma and crystallized phenocrysts, whereas the intrusion into diatremes and near-surface fissures results in solidification of the magma, thus completing the conservation of deep-seated material with the exception of volatiles.

In a single magma chamber evolving through the mechanism of zone melting, differentiation with respect to contrast of the process for all major elements is governed by the thermodynamic regime and, as a whole, is fairly persistent. On the contrary, the extent of differentiation of the magma cannot be identical in upper and lower regions of the chambers: a relative decrease in the extent of magma differentiation in the apical part and increase in the near-floor part of the chamber, as compared to average values for the whole of the system, is an obligatory condition and, at the same time, a consequence of radial displacement of magma chambers resulted from the melting-through of the roof and the settling of crystals on the chamber floor, as temperature and pressure decrease steadily. Calculations of mass budget in the case of kinematic differentiation of kimberlite magmas suggest that the amplitude of inevitable variability of the composition of the melt at the chamber top and floor is relatively small at any given time. The amplitude is believed to be much lower than variations in differentiation indexes of the rocks, which, according to the above, should be assigned to a single phase in the formation of the field.

The above statements have enabled some constraints to be placed on the equipermissible values of the probable number of magma chambers, N, responsible for the formation of kimberlite of any field. The quantity N sought for is much greater than the number of stages St, but it is only slightly greater than, or equal to, the number of phases P and equal to the number of steps S of kimberlite magma intrusions, i.e.,

$$St < P \leqslant N \leqslant S. \tag{15}$$

Table 8. Quantitative distribution (vol.%) of petrochemical groups and types of kimberlitic rocks within major fields of the Central Siberian province

Petrochemical groups and types of kimberlitic rocks		Diamond-subfacies kimberlite fields					Diamond- and pyrope-subfacies	
		Malaya Botuo-biya	Markha-Alakit	Daldyn	Ver-khnyaya Muna	Chomur-dakh	Omonos-Ukukit	Omonos-Sukhan
Rock groups recognized with respect to relative CDI_{Ti}^{Fe} values:	$+(50 + 100)$	–	< 1	1	2	1	12	1
	$+(10 + 50)$	68	31	3	84	78	62	57
	$(-10) + (+10)$	18	53	54	4	2	15	42
	$-(10 + 50)$	10	10	40	7	19	5	–
	$-(50 + 100)$	4	6	2	3	–	6	–
Rock types: distribution in accordance with $DI_{Ti}^{(5)}$ values: $<25; 25-62$ / $63-100; >100$	1	28; – / –; –	–; 6 / 1; –	<1; 1 / 1	– –	1; 46 / 18; 1	–; 4 / 36; 27	–; – / 35; –
	1–2	6; 5 / –; –	45; 1 / –; –	–; 43 / –; –	–; <1 / –; –	–; 1 / –; –	–; 5 / 6; –	–; 40 / –; –
	1–3	–; – / 1; –	1; – / –; –	1; 2 / –; –	–; 17 / –; –	–; 2 / –; –	–; 5 / 1; –	–; 2 / 23; –
	1–5	–; 34 / –; –	–; <1 / –; –	– –	<1; 3 / –; –	– –	–; 2 / –; –	– –
	2	9; – / –; –	7; 1 / –; –	9; 13 / –; –	5; – / –; –	15; – / 4; –	5; 1 / –; –	– –
	2–6	4; 1 / –; –	–; 1 / 1; –	12; 1 / –; –	– –	– –	– –	– –
	2–8	7; – / –; –	– –	3; <1 / –; –	–; 6 / –; –	– –	4; 4 / –; –	– –
	3	– –	2; 9 / –; –	–; 1 / –; –	–; 12 / –; –	– –	–; – / 1; –	– –
	3–4	– –	–; 2 / –; –	–; <1 / –; –	–; 25 / –; –	– –	– –	– –
	3–8	– –	– –	–; <1 / –; –	– –	– –	– –	– –
	4	– –	–; 1 / –; –	– –	30; – / –; –	– –	– –	– –
	5	– –	–; 10 / –; –	– –	–; 1 / –; –	12; – / –; –	– –	– –

| kimberlite fields | | | | | Pyrope-subfacies kimberlite and picrite fields | | | | | |
Lucha-kan	Kura-nakh	Sred-nyaya Kuonam-ka	Nizh-nyaya Kuonam-ka	Motor-chuna	Dzhy-uken	Nizhnii Ukukit	Merchim-den	Verkh-nyaya Molodo	Kuoika-Been-chime	Orto-Yrygakh
6	3	26	5	–	42	–	27	–	49	–
85	97	63	70	83	44	99	67	24	44	80
9	–	11	18	17	5	1	6	75	7	20
–	–	<1	7	–	9	–	<1	1	–	–
–	–	–	–	–	–	–	–	–	–	–
–;10 / 27; 2	–;52 / <1;–	–;34 / 40; 1	8;38 / 4;–	–;19 / 64	–;26 / 5;–	–;53 / 26;–	20;24 / 37; 2	–;16 / 6;–	20;23 / –;–	– –
–; 4 / –;–	– –	–; 9 / –;–	5; 4 / –;–	–;17 / –;–	–; 5 / –;–	–; 1 / –;–	–; 4 / 1;–	–;73 / –;–	2;– / –;–	26;– / –;–
–;19 / 21; 1	–;48 / 1;–	2;– / 11; 2	–;7 / 6;–	– –	–;11 / –;–	–;14 / 6;–	– –	– –	–;18 / –;–	– –
– –	– –	– –	–; 4 / –;–	– –	–;45 / –;–	– –	12;– / –;–	1; 1 / –;–	– –	– –
– –	– –	– –	–; 8 / –;–	– –	3; 5 / –;–	– –	–;– / <1;–	–; 1 / –;–	– –	–;63 / –;–
– –	– –	– –	– –	– –	– –	– –	– –	– –	– –	17;– / –;–
– –	– –	–; 1 / –;–	– –	– –	– –	– –	– –	– –	– –	– –
–;10 / 7;–	– –	– –	–; 6 / –;–	– –	– –	– –	– –	– –	– –	– –
– –	– –	– –	– –	– –	– –	– –	– –	– –	– –	– –
– –	– –	– –	– –	– –	– –	– –	– –	– –	– –	– –
– –	– –	– –	– –	– –	– –	– –	– –	– –	– –	– –
– –	– –	– –	– –	– –	– –	– –	– –	– –	20;– / –;–	– –

Table 8 (continued)

Petrochemical groups and types of kimberlitic rocks	Diamond-subfacies kimberlite fields					Diamond- and pyrope-subfacies	
	Malaya Botuobiya	Markha-Alakit	Daldyn	Ver-khnyaya Muna	Chomur-dakh	Omonos-Ukukit	Omonos-Sukhan
5–6	--	3; 1 / -; -	-; 9 / -; -	--	--	--	--
6	--	-; 1 / -; -	-; 1 / -; -	--	--	--	--
6–7	--	1; 1 / -; -	-; 1 / -; -	--	--	--	--
7	--	-; 1 / -; -	--	--	--	--	--
7–8	--	2; - / -; -	--	--	--	--	--
8	--	1; 1 / -; -	-; 2 / -; -	--	--	--	--

It is not easy to prescribe boundary $DI_{Ti}^{(5)}$ values for rock complexes of one petrochemical type which can be related to separate magma chambers. With correct estimates of the discrepancy between the extreme and mean $DI_{Ti}^{(5)}$ values of the rock complexes of a single type the number of magma chambers in a given field can be calculated from the equation

$$N \approx a_{n_1} + b_{n_{1-2}} + \ldots + v_{n_8}. \tag{16}$$

According to linear dimensions *in plan,* most kimberlite fields are less than, or approximately equal to, the crustal thickness in appropriate sites of platforms. Had we not ruled out the possibility that the kimberlite magma during its ascent moved up the vertical and subvertical, also steeply inclined fractures within the crust, then all bodies made up of the single-type rock complexes, similar in DI, could have derived from a single magma chamber. Only for the fields whose linear dimensions substantially exceed the crustal thickness is it possible that there had been two or more chambers generating petrochemically similar magmas. For these fields the values calculated from Eq. (16) should be considered as a minimal value of the quantity sought. Introducing correction into the value, the distribution of the single-type rocks should be taken into account: the presence of such rocks in two isolated and widely spaced sites of a vast field may be attributed to their origin from two separate chambers.

kimberlite fields					Pyrope-subfacies kimberlite and picrite fields					
Lucha-kan	Kura-nakh	Sred-nyaya Kuonam-ka	Nizh-nyaya Kuonam-ka	Motor-chuna	Dzhy-uken	Nizhnii Ukukit	Merchim-den	Verkh-Molodo chime	Kuoika-Been-	Orto-Yrygakh
--	--	--	$\frac{10; -}{-; -}$	--	--	--	$\frac{-; <1}{-; -}$	$\frac{2; -}{-; -}$	$\frac{7; -}{-; -}$	--
--	--	--	--	--	--	--	$\frac{<1; -}{-; -}$	--	--	--
--	--	--	--	--	--	--	--	--	--	--
--	--	--	--	--	--	--	--	--	--	--
--	--	--	--	--	--	--	--	--	--	--
--	--	--	--	--	--	--	--	--	--	--

The range of magma differentiation and the quantitative distribution of rocks of various types display an essential variability in kimberlite fields of not only an entire province, but even a single facies zone of a province (Table 8). Data presented in Table 8 allow a number of important conclusions to be arrived at. From CDI values three or four groups are recognized in rocks of most (ten) fields, two groups in rocks of four fields, and five groups in rocks of four fields. It is worthwhile to note that most kimberlitic rocks of each field fall into one group, and only in three fields are the amounts of rocks falling into two adjacent groups roughly equal (Omonos-Sukhan, Dzhyuken, and Kuoika-Beenchime fields). The abundance ratios of rocks of various petrochemical types suggest similar tendencies: kimberlitic rocks of any one (more commonly, first) petrochemical type dominate most fields. In three fields the approximately equal amounts of rocks fall into two petrochemical types, and in one field they form three types.

Rocks in the diamond-subfacies fields are usually characterized by an essentially low extent of differentiation; moderate and high extents of differentiation are less common or exceptional, respectively. In fields of diamond- and pyrope-subfacies kimberlites most rocks display the moderate and high extent of differentiation with subordinate amounts of slightly and highly differentiated varieties. According to the general extent of differentiation and the abundance ratio of the rocks differentiated

Table 9. Some factors controlling the formation of major fields of the Central Siberian kimberlite province

Field	Number of			Probable number of magmatic		
	Bodies studied B	Intrusions I^a	Petrochemical types P	Stages St	Phases Ph^b	Steps $\approx$ chambers N^b
Malaya Botuobiya	9	12	7	4	7	9
Markha-Alakit	38	47	16	4	11	23
Daldyn	52	63	13	4	10	18
Verkhnyaya Muna	14	15	9	5	7	10
Chomurdakh	17	19	5	4	8	9
Omonos-Ukukit	35	37	7	5	10	13
Omonos-Sukhan	6	6	3	3	3	4
Luchakan	13	26	4	3	6	8
Kuranakh	10	15	2	2	5	4
Srednyaya Kuonamka	36	40	4	4	11	8
Nizhnyaya Kuonamka	41	45	7	4	8	11
Motorchuna	3	5	2	2	3	3
Dzhyuken	17	19	5	4	6	7
Nizhnii Ukukit	8	8	3	2	4	5
Merchimden	19	22	6	4	9	10
Verkhnyaya Molodo	13	13	6	3	3	8
Kuoika-Beenchime	10	10	5	3	5	6
Orto-Yrygakh	3	3	3	2	3	3

[a] The number of intrusions is taken to be equal to the sum of simple kimberlite bodies and kimberlitic rock generations in complex bodies [Eq. (13)].

[b] Numbers N and Ph were found under the assumption that all rocks of the Central Siberian province may be divided into four complexes according to $DI_{Ti}^{(5)}$ values: <25; $25-62$; $62-100$; >100 (see Table 8).

to varying extent, fields of pyrope-subfacies kimberlites and those of picrites intervene between these two field groups.

Interpretation of the petrochemical data presented in Table 8 on the basis of the above principles allows estimation of the number of volcanic stages and phases, as well as a rough determination of the possible number of magma chambers in 18 major fields of the Central Siberian kimberlite province (Table 9). In analyzing the results obtained, it should be taken into account that because of the subdivision of single-type rocks into four groups from $DI_{Ti}^{(5)}$ values, the real number of phases of kimberlite volcanism reaches the values shown in Table 9 only in some of the above-listed fields. To increase the accuracy and reliability of estimates for each field, the single-type rock complexes should be classified with due regard for their natural boundary $DI_{Ti}^{(5)}$ values. As a result, the number of these ("natural") rock complexes in the rocks of the

Table 10. Relations between factors controlling the formation of some fields of the Central Siberian kimberlite province (Table 9)

Facies	Subfacies	Field	Ph/St	N/St	B/St	I/St	P/St	B/Ph	I/Ph	P/Ph	N/Ph	B/N	I/N	P/N
Kimberlite	Diamond	Daldyn	2.5	4.5	13.0	15.8	3.2	5.2	6.3	1.3	1.8	2.9	3.5	0.7
		Markha-Alakit	2.8	5.8	9.5	11.8	4.0	3.4	4.7	1.5	2.1	1.7	2.0	0.7
		Mean value	2.6	5.2	11.2	13.8	3.6	4.3	5.5	1.4	2.0	2.3	2.8	0.7
	Diamond and pyrope	Chomurdakh Omonos-	2.0	2.2	4.2	4.8	1.2	2.1	2.4	0.6	1.1	1.9	2.1	0.6
		Ukukit	2.0	2.6	7.0	7.4	1.4	3.5	3.7	0.7	1.3	2.7	2.8	0.5
		Luchakan	2.0	2.7	7.7	8.7	1.3	3.8	4.3	0.7	1.3	2.9	3.2	0.5
		Srednyaya Kuonamka	2.8	2.0	8.0	8.8	1.0	3.3	3.6	0.4	0.7	4.5	5.0	0.5
		Nizhnyaya Kuonamka	2.0	2.8	10.2	11.2	1.8	5.1	5.6	0.9	1.4	3.7	4.1	0.6
		Mean value	2.2	2.5	7.4	8.2	1.3	3.6	3.9	0.7	1.2	3.1	3.4	0.5
	Pyrope	Merchimden	2.2	2.5	4.8	5.5	1.5	2.1	2.4	0.7	1.1	1.9	2.2	0.6
Picrite		Dzhyuken	1.5	1.8	4.2	4.8	1.2	2.8	3.2	0.8	1.2	2.4	2.7	0.7

same petrochemical type in each field is no larger than two or three. This suggests that the most probable number of phases of kimberlite volcanism taken, as a first approximation, as N value [Eq. (15)], in most fields is somewhat lower, as compared to the vaues of Table 9.

Assuming that the number of volcanic steps roughly corresponds to that of magma chambers, let us analyze values of Table 9 which reflect some peculiar features of the formation of major kimberlite fields of the Central Siberian province. Since the definition and reliability of any statistically recognized rules depend to a large measure on the number of objects studied, it seems expedient to take into consideration results obtained from kimberlite fields where the number of bodies studied is over 15. They are nine, including two fields of diamond-subfacies kimberlite, one field of pyrope-subfacies rocks, and one field of picrite-facies rocks of ultrabasic volcanism (Table 10).

The abundance ratios of characteristics involved in the formation of these fields suggest that an average number of magmatic rock types falling on one volcanic stage is maximal within the fields of diamond-subfacies kimberlites, progressively decreases in the fields of diamond- and pyrope-subfacies kimberlites and in fields of pyrope-subfacies kimberlites, and reaches minimal values in picrite fields. Ratios of values to the number of volcanic phases in any field also steadily decrease from the diamond- to pyrope subfacies kimberlites, grow again in picrites and even somewhat exceed analogous data in the pyrope subfacies.

The most "productive" chambers with respect to facies rock groups as a whole and within separate fields can be determined by ascribing the formation of each single-type rock complex within one field to a certain magma chamber and with due regard for considerable differences in the volume of various complexes. The data presented in Table 8 show that the most productive chambers within the fields of diamond- and pyrope-subfacies kimberlites have generated moderately and strongly differentiated magmas of only the first petrochemical type. In the diamond-subfacies fields the most productive magma chambers "supplied" weakly and moderately differentiated magmas placed into three petrochemical types: Type 1—2 (Markha-Alakit and Daldyn fields); types 1—5 (Malaya Botuobiya field); and type 4 (Verkhnyaya Muna field). In three of four pyrope-subfacies fields the most productive chambers have generated weakly, moderately, and highly differentiated magmas of petrochemical type 1. In the fourth (Verkhnyaya Molodo) field the most productive chamber was the source of moderately differentiated magma of intermediate type, type 1—2. The most productive chambers in two picrite fields have generated moderately differentiated magmas assigned to different petrochemical types: type 1—5 (Dzhyuken) and type 2 (Orto-Yrygakh).

CHAPTER 11

Diatremes of Diamondiferous Lamproites of Australia

Terminology

The term lamproite was first defined by Niggli (1923) and refers to potassium-rich mafic and ultramafic alkaline igneous rocks. Relative to all other igneous rock types, they possess high K/Na and K/Al ratios and in mineralogy they are unparallelled in other rock types as well. Rock types originally included in the lamproite clan were orendite, madupite, myomingite from the Leucite Hills (USA), jamillite, verite, and fortunite from South-Eastern Spain. Troger (1935) redefined the term lamproite as the extrusive equivalent of lamprophyres which are rich in K_2O and MgO. Rittman (1951), in his nomenclature scheme for volcanic rocks, placed lamproites in the "lamproitic phonolite", "lamproitic leucitite", or "lamproite trachyte" groups. Under the Streckeisen scheme (1967) lamproites would fall in the compositional fields of alkali trachyte, phonolite, and leucitite.

Sahama (1974) classified K-rich alkaline rocks into two groups (lamproitic or orenditic and kamafugitic) on the basis of chemistry and petrography. He placed orendite, lamproite, madupite, wyomingite, wolgidite, and others into the orenditic group. Prider (1960) used the term lamproite to embrace fitzroyite, mamilite, sedricite, wolgidite, and other rock types of Australia. He classified them on the basis of chemistry and mineralogy, whereas Borley (1967) included in this group jumillite, cancalite, fortunite, and verite from Spain.

The absence of a clear-cut definition of the term lamproite, and the description of similar varieties of these rocks under different names make the terminology of lamproites extremely confusing and inconsistent, as compared to that of other magmatites. For example, olivine-pyroxene-phlogopite-leucite lamproites in Spain are known under the name jumillites (Osann 1906) and the identical rocks in Vietnam are called cocites (Lacroix 1933). The fitzroite of Australia (Wade and Prider 1940) is an equivalent of the wyomingite of the USA (Cross 1897).

The nomenclature became even more problematic after the discovery in 1978 of diamondiferous rocks in Western Australia, the rocks being intimately associated with typical kimberlites and leucite-bearing mafic rocks. Most exposures of lamproites have been known here since the early 20th century and their petrography, petrochemistry, and geochemistry have been studied in detail. Owing to the occurrence of diamond and pyrope in some varieties of these rocks, they were ranked with kimberlites as economically important. They were also considered to be petrogenetically similar, as the upper mantle derivatives emplaced at PT conditions pertinent to diamond stability and ascending rapidly to the surface, where conservation of diamond took place.

Australian and American geologists tried to solve the problems of lamproite nomenclature; for this purpose they specified the definition of the term lamproite and made the systematics and nomenclature of these rocks more efficient. Jaques et al. (1984), following Wade and Prider (1940), defined lamproites as:

- an ultrapotassic, magnesian igneous rock. Lamproite contains, as primary pheno-crystal and/or groundmass constituents, variable amounts of leucite and/or glass, and usually one or more of the following minerals is present: phlogopite (typi-cally titaniferous), clinopyroxene (typically diopside), amphibole (typically titani-ferous, potassic richterite), olivine, and sanidine. Other primary minerals may include priderite, perovskite, apatite, wadeite, spinel and nepheline.
- Upper mantle-derived xenocrysts or xenoliths (including olivine, pyroxene, garnet and spinel) and diamond as a rare accessory mineral may or may not be present.
- Lamproites can be basic and ultrabasic and are characterized by a high K_2O/Na_2O ratio, typically greater than five, trace element compositions are extreme with high concentrations of Rb, Sr, Ba, Zr, Nb, Pl, Th, U, and light REE.

A similar, but more detailed definition of the term lamproite was given by Berg-man (1984).

Jaques et al. (1984) proposed subdividing lamproites and naming them on the basis of primary mineral composition. In particular, to substitute traditional nonin-formative terms fitzroyite and sedricite for names revealing the rock composition: phlogopite-leucite, diopside-leucite, and similar lamproites.

Lamproite Geology

Over 100 pipes, dykes, and sills composed of lamproites and kimberlites have been found in north-western Australia. They are concentrated in four regions, referred to as provinces by the Australian geologists (Atkinson et al. 1984). Three of these pro-vinces (North, East, and West provinces) are marginal to the Kimberley craton, and one (Wandagee) lies about 1000 km to the southwest, near the north-western boundary of the Pilbara craton. The distribution of kimberlites and lamproites in Western Australia is as follows. Only classical kimberlites are concentrated in the North Kim-berley province; kimberlites and minor olivine lamproites (including the richly dia-mondiferous Argyle Pipe) have been found in the East Kimberley province; only olivine, olivine-leucite, and leucite lamproites occur in the West Kimberley province; kimberlites have also been found in the Wandagee province.

The King Leopold and Halls Creek mobile zones border the Kimberley craton on the south-west and south-east, respectively. The mobile zones contain folded, meta-morphosed rocks, and Precambrian volcanics (1940 m.y.) intruded by basic and ultrabasic rocks and by granites. The youngest granite intrusions have been dated at about 1800 m.y. The mobile zones are bounded by ancient major transcurrent fault systems, and the movements on them appear to have continued at least until the Permian, and probably until the Triassic or even Early Jurassic.

At the north end of the Kimberley craton, pipes and dykes virtually barren of diamond are situated close to the north-west-trending major downwarp, but their trend is north-north-east, parallel to the structural trend of the Halls Creek mobile zone. They intrude Proterozoic terrigenous deposits of Carpentarian age. The kimberlitic rocks of the East Kimberley province form two fields. One of them, which includes the Argyle lamproite pipe and two kimberlite dykes, is situated within the Halls Creek mobile zone. The second field, composed of some subdiamondiferous kimberlite diatremes, is situated on the eastern margin of the craton. The age of the Argyle Pipe and accompanying dykes is uncertain; they are known to intrude Proterozoic (1057 ± 80 m.y. and 1158 ± 123 m.y.) deposits.

In the West (or, to be more precise, south-west) Kimberley region more than 100 separate pipes and dykes made up of olivine and leucite-bearing lamproites have been found. They occur in three main fields (Ellendale, Calwynyardah, and Noonkanbah fields) within a broad belt which extends from the King Leopold mobile zone across the Lennard Shelf to the Fitzroy Trough. The majority of the intrusions show evidence of strong structural control, notable west-north-westerly-trending faults at the margin of the Fitzroy Trough, northerly-trending fractures in the Kimberley Block, Lennard Shelf, and Fitzroy Trough, and east-west-trending en-echelon faults and folds within the trough (Jaques et al. 1984). An Early Mocene age for the West Kimberley lamproites has been established from the results of palynological study of overlying epiclastic rocks, as well as from K-Ar and U-Pb datings on both mineral separates and whole rocks. The isotopic data show a range in age from 20–25 m.y. in the Ellendale field to 18–20 m.y. in the Noonkanbah field, suggesting a southward migration of magmatic activity with time (Jaques et al. 1984). Diamonds have been discovered in 30 bodies of the Northern Ellendale field adjacent to the craton. Substantial, but subeconomic diamond content has been established only in olivine lamproites (Ellendale 4, 7, 9, 11), and its minerals have been recorded in both olivine-leucite and leucite lamproites.

At Wandagee 16 kimberlite bodies virtually barren of diamond form a north-south elongate cluster along the fault-bounded eastern edge of the Wandagee Ridge. The basement is covered by 2000 m of Permian and Cretaceous sediments. The kimberlites intrude Permian, and are buried beneath Cretaceous sediments of Aptian age. One of the Wandagee diatremes has been dated at 160 ± 10 m.y. by the U-Pb method on macrocrystal zircon, hence geological and radiometric dating evidence supports a Jurassic age of emplacement.

The Australian geologists emphasize that the lamproites differ in form and structure from the classical kimberlite pipes and from the kimberlite pipe model of Hawthorne (1975). Actually, the relatively large diameter of the crater zone and extremely narrow feeder pipes, resembling a champagne glass in shape, are typical of the Miocene olivine and leucite lamproites which intrude poorly cemented Paleozoic rocks in the Ellendale, Calwynyardah, and Noonkanbah fields. However, weakly eroded diatremes occurring in loose Mesozoic sandstones of the Bakwanga field of the Congo province, Africa, are known to have much larger diameters of craters (up to T-shaped in cross-section). Their (vertical) feeders consist of typical kimberlite breccia, while funnels are made up of altered kimberlite tuff (?) with abundant mixtures and terrigenous bands. The kimberlites are believed to have resulted from submarine eruption, parallel with the accumulation of arenaceous sediment of layer M-4 of the Lualaba Series (Meyer de Stadelhofen 1963). On the contrary, the markedly eroded Argyle lamproite pipe intruding Proterozoic massive quartz sandstone is very elongate in plan and characterized by steeply dipping to vertical contacts.

The presence in a crater of massive (magmatic) varieties of lamproites which intrude and locally overlie tuff, sandy tuff, and other deposits which infill the crater is believed to be a main typical feature of lamproite diatremes which distinguishes them from kimberlite bodies (Atkinson et al. 1984). However, if such a structure is exemplified by Ellendale 6 and some other diatremes, the geological cross-section of the Argyle lamproite pipe (Atkinson et al. 1984), studied in detail down to 300 m from the present surface, shows no magmatic lamproites. With due regard for a probable depth of erosion, it is still evident that no substantial amount of magmatic lamproites are present, not only in the crater removed by denudation, but in the feeder pipe down to a depth of a few hundred meters from the original crater of the Argyle pipe. Hence, this feature, as well as the morphology of the bodies, is not persistent in all lamproites and therefore cannot be used by the lamproite diatreme type model differing from the model of a kimberlite pipe.

Lamproite diatremes do not greatly differ from kimberlite diatremes in sizes, as even in the least eroded fields of the West Kimberley province the areas of only three pipes exceed 100 ha, but the majority of diatremes (in Ellendale field, 2/3) are under 10 ha. The largest in area is the Mwadui kimberlite pipe (142 ha), while the area of the largest lamproite pipe (Calwynyardah) measures 128 ha.

Composition of Diamondiferous Lamproites

Economic diamond content has been recorded only in the olivine lamproites of the Argyle Pipe, but lower content and minerals of cubic modification of carbon have been found in 29 of the 33 Ellendale lamproites tested for diamond, including most of the leucite lamproites. The most diamondiferous lamproites have proved to be olivine lamproites, which are also richer in chromite, pyrope, and chrome diopside. Most of the larger diatremes are olivine lamproites. It is noteworthy that in all Ellendale lamproites tested for diamond content, tuffaceous phases are more diamondiferous by at least one order of magnitude than the accompanying massive magmatic phases.

Most occurrences are either leucite lamproite with an olivine content usually less than 10% or are olivine lamproite with no leucite. Of the 45 occurrences at Ellendale, 14 are olivine lamproite, 27 are leucite lamproite, and only 4 are leucite-olivine lamproite (Atkinson et al. 1984). Some of the olivine lamproite bodies have minor leucite lamproite phases developed. At Ellendale 13, both types of lamproite occur as distinct magmatic phases. Ellendale 7 and 33 contain tuffs with predominant clasts of olivine lamproite and minor leucite lamproite.

The olivine lamproites of the Ellendale field contain large (up to several millimeters) rounded xenocrysts of olivine and smaller euhedral olivines, averaging 0.3 mm, set in a fine-grained, often glassy groundmass. Olivine content is usually more than 30%. The olivine is mainly altered to talc and sometimes to serpentine and carbonate. Phlogopite and diopside are almost constant. The coarser-grained varieties have ragged phlogopite plates up to 2 mm across, but the grain size decreases with depth and outward toward the pipe margins. Diopside occurs in variable proportions as needles from 0.05 to 0.2 mm long, and in pyroxene-olivine varieties of lamproites diopside is as abundant as olivine. The groundmass is frequently studded with perovskite granules and minute euhedra of chrome-rich spinel. Other common accessory minerals are apatite, barite, potassic richterite, and a rutile-like mineral, probably priderite. Macrocrysts of debatable origin present in the olivine lamproites include common chrome-rich spinel, and smaller amounts of pyrope and almandine-pyrope garnet, chrome-diopside, orthopyroxene, and diamonds, as well as minerals probably derived from crystalline basement metamorphic rocks.

The tuffs at Ellendale contain abundant clasts of magmatic lamproite and crystals of olivine, together with a wide range of xenoliths derived from country rocks and from rocks pierced at depth. At deeper levels and toward the pipe margins, the tuffs incorporate grains of detrital quartz presumed to have been derived from the wallrock. These sandy tuffs are frequently bedded. Horizons of quartz sandstone with very little lamproite content occur intercalated within the tuffs. At the junctions of tuff with magmatic lamproite there is frequently an apparent transition zone. Approaching the contact the magmatic lamproite becomes vesicular and then slightly brecciated. The transitional zone itself carries closely packed clasts of vesicular magmatic lamproite, whereas country rock clasts are absent.

The Argyle lamproite pipe, which is the only one within the East Kimberley province, consists of olivine lamproite tuffs intruded by rare thin (1–2 m) dykes of massive rock varieties. Dyke lamproites carry predominantly euhedral phenocrysts of olivine (10–25%) now replaced by talc and minor carbonate, as well as ragged-edged microphenocrysts of titaniferous phlogopite showing reversed pleochroism. The very fine-grained groundmass consists mainly of phlogopite, minor anatase, sphene, perovskite, and accessory apatite. Prominent opaques include aggregates of manganiferous ilmenite and titaniferous magnesio-chromite.

The Argyle diatreme is infilled predominantly with sandy lapilli-tuff including abundant quartz grains, and only in the center of the northern end of the diatreme is a large body of nonsandy tuff devoid of detrital quartz. The sandy tuffs are cut by occasional dyke-like bodies of "nonsandy" tuffisite thought to be the sutobrecciated equivalent to the magmatic kimberlite. The nonsandy tuff carries very fine-grained (0.5–4.0 mm in diameter) lamproite clasts containing abundant talc pseudomorphs after olivine, set in a matrix of broken altered olivine crystals and scaly micaceous minerals. Rounded, extensively altered nodules of peridotite are also present.

The sandy tuff consists of an assemblage of crystal-lithic tuffs, in places weakly bedded. It includes angular to rounded lamproite clasts set in a groundmass of olivine crystals, comminuted lamproitic material, and abundant (30–50%) rounded grains of quartz. The clasts contain large talcose pseudomorphs after olivine phenocrysts, which are often euhedral and glomeroporphyritic. Some clasts contain polygonal grains about 0.01 mm in size of potash feldspar, whose shape strongly suggests replacement of leucite. Xenocrystal minerals include magnesiochromite, almandine, rare chrome-diopside and orthopyroxene. The garnets are titanium and chrome pyropes, as well as diamond. Diamond at Argyle is most abundant in the sandy tuff. Diamond grades are considerably lower in the nonsandy tuff, and preliminary sampling of the magmatic lamproite intrusives suggests these may even be barren.

In the near-surface levels of the pipes, olivine macrocrysts and microphenocrysts set in a groundmass are usually replaced by secondary minerals, but is preserved in fresh material from drill core. The two generations of olivine present in the olivine lamproites overlap in composition; the most frequent composition is $Mg_{91=91.5}$ but both generations range up to $Mg_{93.5}$. Microphenocryst/groundmass olivine ranges to more Fe-rich compositions (Mg_{87}), as found in kimberlites. Olivine in the leucite lamproites ranges in composition from $Mg_{91.1}$ to $Mg_{77.3}$. Some of the larger, Mg-rich grains ($>Mg_{91}$) are probably of xenocrystal origin (Jaques et al. 1984).

Phlogopite is present to some extent in the majority of the rocks under study. Most of the olivine lamproites contain strongly pleochroic groundmass flaky phlogopite and some have microphenocrysts. Typically, they have pale pink-yellow cores relatively free of inclusions and more strongly pleochroic rims crowded with inclusions of diopside, perovskite, chromite, barite, and apatite. Microprobe analysis shows that the cores are compositionally distinct from most groundmass micas, being richer in Mg, Al, and Cr, and poor in Na, but many are zones to more Fe-rich, Al-poor rims with low Cr contents which overlap the groundmass mica compositions.

Mica is the dominant ferromagnesian phase in many leucite lamproites. Phlogopite phenocrysts have pale-colored cores free of inclusions surrounded by mantles of more strongly pleochroic mica with abundant inclusions of priderite, diopside, apatite, barite and, less commonly, leucite. These are mantled by mica similar to that in the groundmass.

Phlogopite in the rocks under study is similar and partially overlaps in composition, but phlogopite from the olivine lamproites is poorer in TiO_2 (1–8, average 4%) compared to that from the leucite lamproites (4–10%, average 7%) (Jaques et al. 1984). Thus, in both variability range and avarage TiO_2 content, phlogopite from olivine lamproites is closer to groundmass phlogopite from the kimberlites (0.6–6.0%, average 3% TiO_2) than to that from the leucite lamproites.

Clinopyroxene is present in the groundmass of almost all the rocks in question. In the olivine lamproites it typically occurs as small prisms and acicular crystals. The diopsides are of uniform composition, lying close to the diopside-hedenbergite join with Fe content varying within 9–15%. They are rich in Ti and poor in Al. Diopside occurs as a phenocryst, or, in glassy rocks, as a microphenocryst phase in many leucite lamproites. Phenocryst diopside is of similar composition to the groundmass pyroxene in the olivine lamproites, but exhibits greater solid-solution toward enstatite, implying higher crystallization temperatures. The groundmass and microphenocryst diopsides in the more leucite-rich rocks overlap the phenocryst compositions and extend to more Fe-rich compositions. In general, groundmass diopside is poorer in Al and Cr, and richer in Ti than the phenocrysts.

Amphibole is present in the groundmass of the coarser-grained olivine lamproites and forms phenocrysts in the leucite lamproites. These amphiboles are of similar composition; they are represented by potassic richterite rich in Mg, alkalies, and Ti, and poor in Al.

Leucite is present either as a phenocryst or groundmass phase in all except the most olivine-rich lamproites and glassy rocks. Fresh leucite is rare. Typically, it is pseudomorphed by K-feldspar, clay, chalcedony, or zeolite. The leucite is close to the ideal leucite stoichiometry.

Table 11. Analyses of diamondiferous olivine lamproites and average leucite lamproites from major fields of Western Australia

	Olivine lamproites from pipes					Leucite lamproites	
	Argyle (2 analyses)	Ellendale No. 4	No. 9	No. 7	No. 11	Ellendale field	Noonkanbah and Calwyny-ardah fields
		(2)	(2)	(2)	(1)	(16)	(10)
SiO_2	43.35	41.45	41.66	42.16	40.10	49.67	51.30
TiO_2	2.90	3.23	3.25	2.70	2.64	6.15	5.00
Al_2O_3	4.71	3.51	4.45	3.42	3.50	7.05	9.34
Fe_2O_3	2.90	6.00	4.55	4.74	3.80	5.68	5.61
FeO	4.64	1.49	4.27	3.71	4.98	1.49	1.43
MnO	0.12	0.11	0.14	0.12	0.14	0.08	0.17
BaO	0.09	1.41	n.d.	n.d.	n.d.	1.25	0.89
MgO	22.55	20.40	22.26	25.90	26.10	9.48	6.46
CaO	5.16	4.67	5.29	4.34	5.61	3.33	4.38
K_2O	4.62	4.49	4.27	4.16	3.46	8.20	10.01
Na_2O	0.28	0.32	0.42	0.44	0.45	0.39	0.30
H_2O	5.01	9.11	5.92	5.60	5.48	5.58	3.18
CO_2	1.28	0.48	2.72	2.14	2.34	0.10	0.98
P_2O_5	1.24	2.50	1.13	0.61	1.17	1.29	0.92
Total	98.79	99.17	100.34	100.04	99.77	99.74	99.97

n.d. = not determined

Chrome-rich spinel is present as a primary groundmass phase in olivine lamproite, where it typically occurs as small euhedra in the groundmass, and as inclusions in phlogopite and other later crystallizing minerals. The chromites range from rare titaniferous, magnesian, aluminous chromites through titaniferous magnesian chromite to rare titaniferous, chromian magnetite (Jaques et al. 1984).

Table 11 shows chemical analyses of olivine lamproites from five diatremes with weight content of diamond and analyses of leucite lamproites averaged for major fields, Western Australia. Analyses exhibit enrichment in K and Ti, typical of the rocks, and reflect a tendency to compositional variations in a series from olivine to extreme leucite lamproites. Diamondiferous olivine lamproites of the Argyle Pipe and four diatremes of the Ellendale field are similar in composition, but leucite lamproites in the Ellendale field and two fields of the Fitzroy graben differ markedly in the content of many components.

It should be emphasized that the leucite lamproites of the Fitzroy graben are richer in SiO_2, K_2O, TiO_2, Al_2O_3, and poorer in MgO as compared not only to olivine lamproites, but also to leucite lamproites of the Ellendale field. The remainder of the oxides in these rocks are similar in amount (see Table 11). Figure 27 displays the interrelationship of weight contents of major rock-forming oxides of the olivine and leucite lamproites. A steady, almost linear increase in SiO_2, K_2O, TiO_2, and Al_2O_3 content with a simultaneous decrease in MgO content allows the arrangement

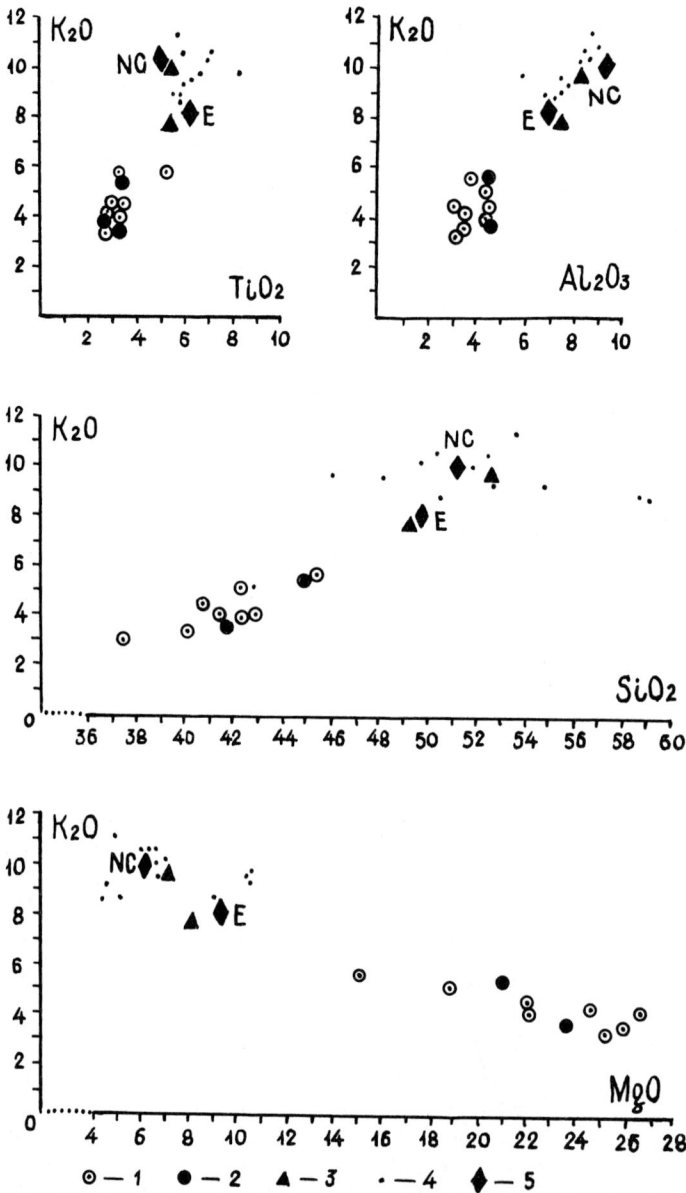

Fig. 27. Correlation of weight contents of rock-forming oxides in olivine and leucite lamproites of Western Australia. (Atkinson et al. 1984; Jaques et al. 1984). *1* Diamondiferous olivine lamproites of the Ellendale field; *2* olivine lamproite of the Argyle Pipe; *3* olivine-leucite lamproite; *4* leucite lamproite; *5* average composition of leucite lamproite in the Ellendale (*E*), Noonkanbah and Calwynyardah (*NC*) fields

of a series from diamondiferous olivine lamproites through leucite lamproites of the Ellendale field to barren lamproites of the Fitzroy graben, which are probable products of a single evolutionary process.

Inclusions of holocrystalline ultrabasic rocks, referred to as cognate inclusions in the kimberlites, are always present in the olivine lamproites, their tuffs and tuffisite breccia. These rounded, oval, or discoid euhedra reach 10 cm in diameter. Dunite xenoliths are common; sparse lherzolite and harzburgite nodules also occur.

In structure and texture the inclusions are identical to analogous xenoliths in the kimberlites. They differ in mineralogy: in olivine lamproites, plutonic inclusions are essentially represented by spinel dunite and peridotite, whereas pyrope mainly forms macrophenocrysts in the olivine lamproites and in the heavy mineral concentrates of bulk samples of the rocks.

Lamproites Versus Kimberlites

On the confirmation of an economic diamond content at Argyle, substantial, but subeconomic diamond content in some other olivine-lamproite diatremes in Western Australia, as well as uneconomic diamond content in the leucite lamproites of the same region, many papers were published on the subject of lamproites. The authors have proposed reclassifying many alkaline and alkaline-ultramafic rocks, previously referred to as kimberlites, to lamproites. Complete or even selective discussion of the proposals and considerations pro and contra the new classification are beyond the scope of our study, but it should be noted that two rock groups petrographically inequal and petrologically varied are under revision. One group includes leucite lavas of the Gaussberg Volcano, melanephelinites from the dykes at Mount Bayliss, Priestly Peak, Antarctica, etc.; the second group embraces diamondiferous kimberlites of Murfreesboro (USA), Chelima (India) and the like (Bergman 1984). Allowances have been mainly made for chemistry and, where possible, for mineralogy. However, if such an approach is quite expeditious, although not unquestionable, for the first group, the validity and expediency of reclassification of K- and Ti-rich kimberlites to lamproites should be questioned.

From petrographical and mineralogical definitions of the term kimberlite given by Lewis (1887), Williams (1932), Dawson (1971), Clement et al. (1977), and Mitchell (1979), porphyritic ultramafic alkaline rocks, diamond- and pyrope-bearing or devoid of these minerals, which differ from classical kimberlites in their very high K content with proper variations in mineralogy, should be termed lamproites.

Meanwhile, as early as the initial stage of study of the Yakutian kimberlites, it was proposed to divide a diversity of porphyritic mafic and ultramafic alkaline rocks composing pipes and dykes within platform regions into two large groups, which differ in the conditions of genesis and evolution of the parental magmas (Milashev 1963, 1965). The parental magmas placed in one group generated in the subcrust under PT conditions appropriate to a terrane of stable crystallization of diamond and pyrope, and rapidly reach the near-surface levels where conservation of barophilic minerals took place. This regime of rock formation is characterized by the theory of kimberlite facies of magmatism, the critical minerals of which being diamond and/or pyrope.

Depending on environmental conditions, rocks of analogous chemistry and mineralogy, but barren of diamond and pyrope, are placed into the picrite facies of platform ultramafic and ultramafic alkaline magmatism.

From this viewpoint, and because of the absence of constraint with respect to the content of secondary chemical elements in our definition of the term kimberlite, which may be applied to any porphyritic ultramafic and ultramafic alkaline rocks, the diamondiferous olivine lamproites of Western Australia may be classed with the kimberlite-facies rocks without any reserve. Petrochemical correlation of olivine lamproites which contain weight concentrations of diamond with the Yakutian kimberlites, intermediate or extremal in composition (Milashev 1965; Milashev et al. 1971; Ilupin et al. 1978) suggests that the differences observed in SiO_2, MgO, CaO, CO_2, and H_2O content may be due to regional(?) features of deuteric rock alterations, in particular, due to a greater extent of carbonatization of the kimberlites and by the lack of talc, which is, on the contrary, very abundant in the lamproites. The enrichment of diamondiferous olivine lamproites in K, partly in P, with similar Fe and Al content, as compared to intermediate kimberlites, appears to be petrologically more essential. The concentration of these representative elements is a function of a parental magma composition, and their compositional variations allow the regime and the extent of differentiation of the parental magmas to be elucidated (Milashev 1972a). It is noteworthy that in most diamondiferous olivine lamproites of Western Australia concentrations of Al, Ti, and total Fe lie beyond the highest contents of these elements in the kimberlites, and with respect to K content even the most alkaline Yakutian kimberlites are inferior to lamproites (Fig. 28).

This suggests that the diamondiferous olivine lamproites are anomalously K-rich porphyritic ultramafic alkaline rocks of the kimberlite facies which, unlike the classical kimberlites, can generally be named potassic kimberlites.

The boundary between the classical and the potassic kimberlites is essentially statistical in character. Figure 28 shows that neither eight contents of TiO_2, Al_2O_3 separately, nor one of the two petrochemical factors of the lamproite, K_2O/Al_2O_3 can be used as the basis for subdivision of these rocks. In eleven kimberlite pipes of Yakutia, the K_2O/Al_2O_3 value exceeds 0.7, recommended by Bergman (1984) as a lower limit for the lamproites. However, this value is apparently overestimated, since, according to data presented by Bergman himself, in leucite lamproites of Spain (Murcia and Almeria), some leucite-olivine lamproites of Australia, olivine lamproites of the USA, and olivine-phlogopite lamproites of England (Cornwall) K_2O/Al_2O_3 = 0.56–0.65. Typical kimberlites from some dozen diatremes of the Yakutian diamond province yield similar values.

The second important petrochemical factor of lamproites is K_2O/Na_2O, the lower limit of which, according to Bergman (1984), equals 4. However, in all micaceous, in most nonmicaceous and, hence, in "average" kimberlites of Yakutia, this ratio is above 4, sometimes reaches 10 and even 20.

Thus, the diamond-bearing olivine lamproites of Australia "overlap" the classical kimberlites not only in content of oxides of most representative elements, but in numerical values of geochemical factors used to distinguish the lamproites from other rocks. The coincidence observed reflects a certain similarity in composition and genesis of the rocks. It should be stressed that, according to the above factors, the olivine

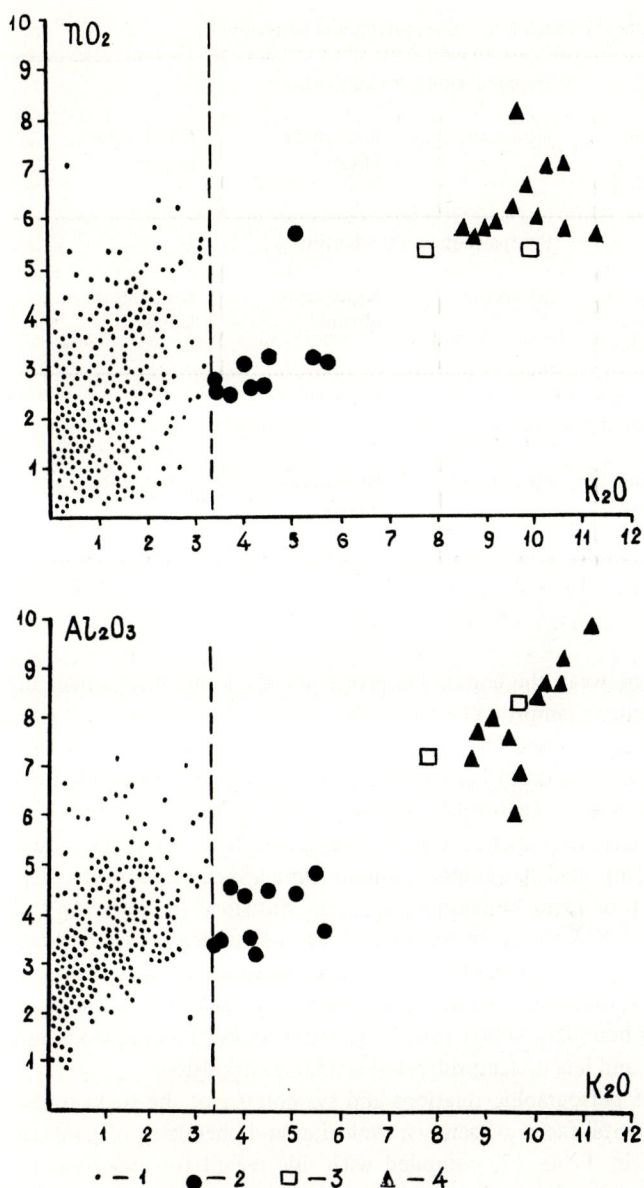

Fig. 28. Weight contents of oxides in kimberlites, comagmatic picrites of Yakutia, and lamproites of Western Australia. (Milashev 1965; Milashev et al. 1971; Ilupin et al. 1978; Atkinson et al. 1984; Jaques et al. 1984). *1* Kimberlite and picrite; *2* diamondiferous olivine lamproite; *3* olivine-leucite lamproite; *4* leucite lamproite

Table 12. Petrographical relations between kimberlite, picrite and lamproite

	P, MPa	Diamond-subfacies kimberlites			
		Non-mica-ceous	Micaceous	K-richterite olivine	K-richterite leucite
Kimberlite facies	4000				
		Pyrope-subfacies kimberlites			
		Non-mica-ceous	Micaceous	K-richterite olivine	K-richterite leucite
	2000				
		Picrites and picritic porphyries		Lamproites	
Picrite facies		Non-mica-ceous	Micaceous	K-richterite olivine	K-richterite leucite

```
        1    2    3    4    5    6    7    8    9% K₂O
```

lamproite may be placed between the leucite lamproite and the kimberlites which are consanguineous with the leucite lamproite (see Fig. 28).

An increase in alkalinity beyond the limits shown by micaceous kimberlite and accompanied by a common decrease in basicity led to the appearance first of K-richterite and then leucite. Hence to distinguish between common (non-micaceous and micaceous) kimberlites and K-rich varieties of kimberlite-facies rocks, the latter should be named K-richterite olivine and K-richterite leucite kimberlites, respectively. The boundaries between the four main kimberlite types are statistical in character and correspond roughly to: 1% K_2O between nonmicaceous and micaceous kimberlites; 3.3% K_2O between micaceous and K-richterite olivine kimberlites; 7% K_2O between K-richterite olivine and K-richterite leucite kimberlites (Fig. 28). The same K_2O contents are also taken as boundary values for picrite-facies rocks (nonmicaceous and micaceous picrites, olivine and leucite lamproites) of similar composition.

The above notions of petrographic relations and systematics of the rocks under study, based on environmental facies of genesis, evolution and chemistry of parental magmas, are summarized in Table 12, compiled with due regard for experimental data on differences in minimal pressures for stable crystallization of diamond (4000 MPa) and pyrope (2000 MPa) and the resultant suggestion on the subdivision of the kimberlite facies into two main subfacies, namely, diamond subfacies and pyrope subfacies (Milashev 1963, 1965, 1972a).

Such an approach to the solution of the problem under consideration raises at least two new questions: (1) Where should olivine lamproites, barren of diamond and pyrope, be placed and should they be renamed? (2) Should the diamond-bearing leucite lamproites be grouped with the kimberlite facies rocks?

Since the olivine lamproites barren of diamond and pyrope are the derivatives of the low baric picrite-facies magmas whose rocks preserve purely petrographic names, they should not be renamed.

The question of the possible placement of diamondiferous leucite lamproites to the high-pressure (kimberlite) facies should be answered positively. Actually, the presence of diamond in rocks unambiguously points to the genesis and evolution of parental magma under high PT-conditions, as well as its sufficiently rapid rise to the surface and cooling, thus providing means for the conservation of the mineral metastable at low pressure. This is suggestive of metamorphic terrains where rocks, irrespective of composition, metamorphosed under certain PT-conditions, form single metamorphic facies. Nobody finds it surprising that, for example, tremolite marbles belong with the greenschist facies, and kyanite-staurolite-almandine schists with no amphibole rank with the amphibole facies.

Part II Nonkimberlite Diatremes

CHAPTER 12

Alkali Basaltoid and Carbonatite Diatremes

Nonkimberlitic rocks composing explosion pipes within the confines of ancient and young platforms may be placed into three major groups on the basis of their composition. The first group contains alkali basaltoids and carbonatites closest in composition, mode of formation, and often regionally, associated with kimberlites. Groups 2 and 3 are represented by trap formation rocks and trachyte, respectively.

Cone-shaped bodies of alkali basaltoids and carbonatites are known from many regions of the world. They occur mainly along the periphery of kimberlite provinces. The diatremes show a wide age range, i.e., from Proterozoic to Recent. This is very important because the presence of ancient vents and active volcanoes extruding products of alkali-basaltoid and carbonatite composition gives insight into the processes of diatreme emplacement.

Anabar District

A large number of diatremes were reported from the Anabar district of the Siberian platform; they are filled with badly altered rocks having relict alkali minerals, and in petrology and petrochemistry may be easily differentiated from kimberlites and picrite porphyrites. The rocks are very common on the left side of the Anabar River mainly in three localities, where a linear north-north-east striking group extends for a distance of about 40 km (Fig. 29).

The southern, Nomokhtookh, site within the river basin of the same name joins the northern end of the Nizhnyaya Kuonamka kimberlite field. The number of explosive bodies totals 56, five of them being made up of kimberlite and picrite porphyrite tuffisite breccia. The Tundrovy site lies about 10 km north-east of the Nomokhtookh site at the upper Talakhtaakh Creek. Here, nine explosive bodies and nine not magnetically confirmed cone-shaped bodies occupy an area of about 7 km^2. It is noteworthy that known and inferred explosion pipes are circularly positioned, forming a sort of ring structure. Orto Yrygakh, the northernmost site, lies less than 10 km north of the Tundrovy site at the upper reaches of the creek of the same name. The site area is about 40 km^2. Of 18 known diatremes, two bodies are composed of picrite porphyrite tuffisite breccia, one pipe anomaly remains unproved. There is a clear tendency for a decrease in density of diatremes (the number of diatremes per unit area) from south to north, viz. 7, 2.5, and 0.5 in the Nomokhtookh, Tundrovy, and Orto Yrygakh sites, respectively.

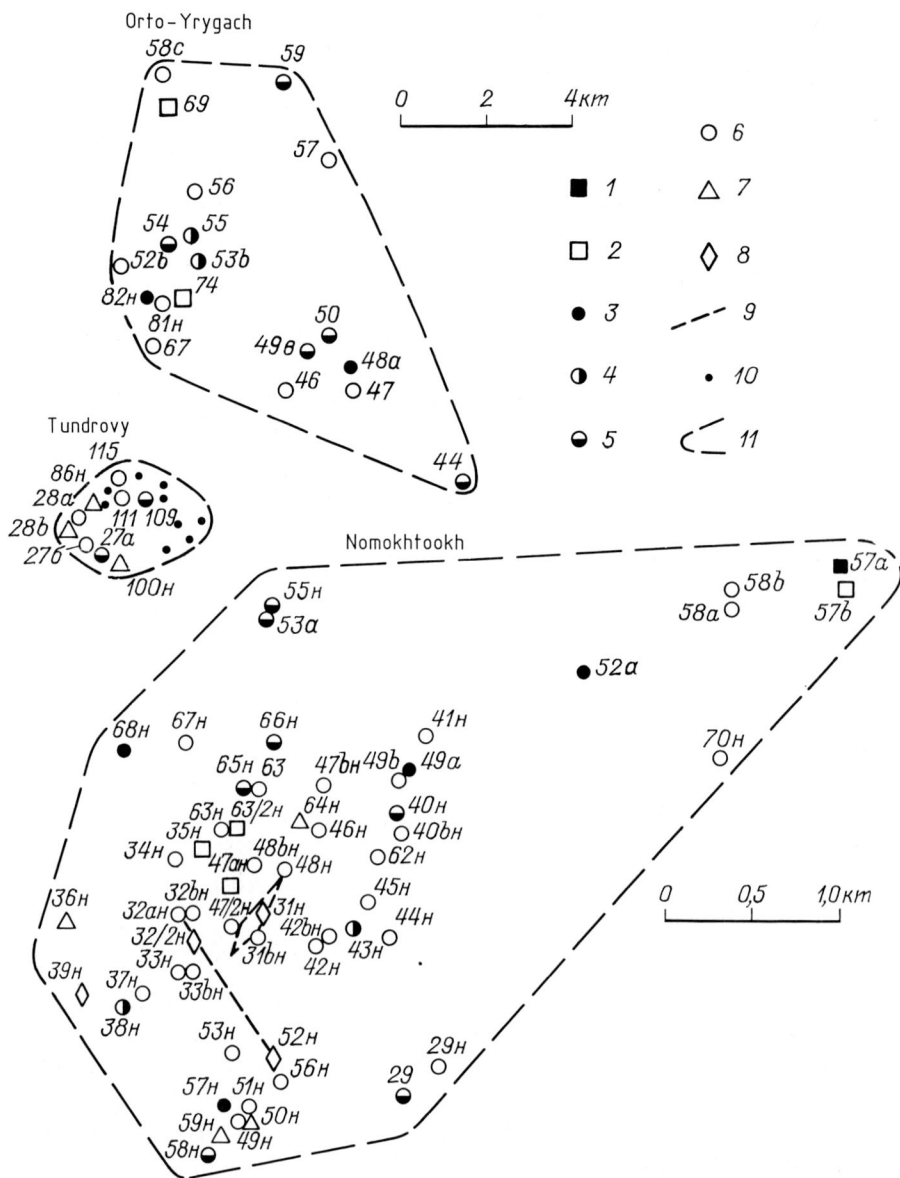

Fig. 29. Distribution of explosion pipes on the left side of the Anabar River. *1* Kimberlite tuffisite breccia; *2* tuffisite breccia of picritic porphyries; alkali basaltoids: *3* strongly carbonatized massive varieties; *4* eruption breccia; *5* tuffisite breccia; *6* xenotuffisite breccia; explosion shatter breccia of: *7* carbonate rocks, *8* carbonate rocks with iron and other mineralization; *9* shatter and mineralization zones in carbonate rocks; *10* nonproved anomalies of the pipe type; *11* inferred boundaries of sites

Explosion pipes of alkali basaltoids (Fig. 30) do not differ in morphology from kimberlite diatremes (see Figs. 1–6, 11, 12). However, their average area is much larger (0.91 ha) than that of kimberlite pipes, not only from the adjacent Kuonamka fields (0.54 ha), but from all the remaining fields of diamond- and pyrope-subfacies kimberlites (0.33 ha) and purely pyrope-subfacies kimberlites (0.27 ha) of the Siberian platform.

Explosion pipes of the Nomokhtookh site intrude carbonate and terrigenous-carbonate Upper Proterozoic, Lower and Middle Cambrian deposits. In two other sites the rocks enclosing pipes are Middle Cambrian carbonates. The age of the diatremes was determined on fragments of charred wood. According to I.A. Shilkina (USSR Academy of Sciences Botanical Institute), these plant remains belong to the genus *Arancariopitys* sp. ex formal group Protopinaceae. The representatives of the genus occur in the Rhaetian or Lower Jurassic to Upper Jurassic or Lower Cretaceous deposits inclusively (Milashev 1968a).

The rocks filling cone-shaped bodies in the Anabar district, except for the typical kimberlites and picrite porphyrites, are heavily altered and, by their characteristics, may be assigned to alkali basaltoids. Based on almost full carbonatization and the presence of REE, some investigators placed the rocks along with fully carbonatized kimberlites and picrite porphyrites to carbonatites not specifying the composition of unaltered rocks (Marshintsev 1974). This approach obviously led to a simple characterization of superposed mineralization and has not contributed to understanding the composition and genesis of source rocks and hence to the volcanic activity of the district.

It is difficult to ascertain the original mineral composition of the rocks and hence to precisely locate them in a general system of igneous rocks. The restoration of a primary composition is complicated because primary minerals are almost completely replaced by secondary minerals and because most of the rocks discussed are represented by xenotuffisitic breccia, having a large proportion of various minerals derived from beds intruded by diatremes. However, despite the difficulties, the rocks, even when studied in thin sections, can be easily distinguished both from heavily and fully carbonatized kimberlitic and picritic rocks and carbonatites proper on the basis of their texture and relict minerals. More likely they are to be placed into a separate family.

Three types of rocks, one of them with two varieties, are recognized by means of the peculiar structure, pseudomorph composition, and relict minerals within the family. Four groups, viz. massive rocks, eruption breccia, tuffisite breccia, and xenotuffisite breccia are classed on the basis of structural and textural features in each type. It is difficult to identify rocks of the types with any particular rock because of the poor knowledge of primary composition of porphyritic xenoliths and their groundmass. However, we are quite sure of their assignment to a group of alkali basalts.

Massive varieties of these peculiar rocks are discussed below. We do not provide the analysis for the remaining structural-genetic groups (eruption and tuffisite breccias) because typical features of each rock group can be found under the description of certain structural groups of kimberlitic rocks.

Type I. This type includes most of the rocks of the family discussed. They are represented mainly by xenotuffisite breccia and tuffisite breccia; eruption breccia and massive varieties are subordinate (see Fig. 29).

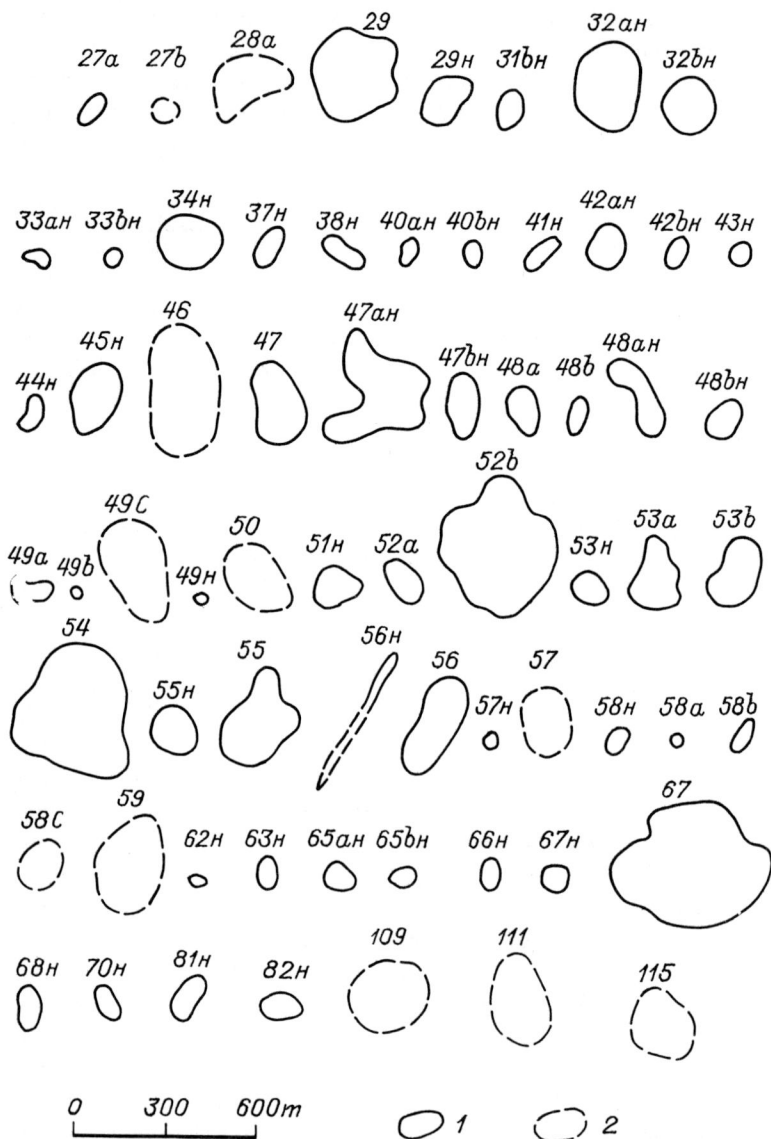

Fig. 30. Shape of explosion pipes of alkali basaltoid rocks in the Anabar district. (Milashev 1968a). Outlines of bodies in plan: *1* observed by magnetic survey; *2* inferred

Rock fabric is porphyritic relic with relict microlite groundmass. Porphyritic xenoliths are chiefly replaced by carbonate. Primary minerals are phlogopite, apatite, opaque mineral, and relict (xenogenic) grains of clinopyroxene and amphibole. Carbonate pseudomorphs after phenocrysts are oval-elongate and subisometric, rarely irregular in shape. Based on texture, amount, and distribution of subordinate elements, they may be subdivided into four groups.

Isometric and elongate hexagonal, often deformed, with fusion rinds on their surface, and embayed pseudomorphs always account for several percent to 10% of the rock. They are composed of brownish carbonate monocrystals with abundant dust inclusions. Pseudomorphs measure 1—5 mm, averaging 2—3 mm. Most pseudomorphs of the group have a dark rim owing to the concentration of dust inclusions (about 0.001 mm) in peripheral zones. Inclusions are opaque and light grey in reflected light. The width of the rims is 0.05—0.10 mm. The internal contact is vague, but discernible. Large and some average pseudomorphs are structurally complex: carbonate rims with numerous inclusions are polycrystalline, while nuclei are infilled with monocrystalline carbonate. There are intergrowths of apatite and amphibole. The hexagonal outlines of some pseudomorphs and especially the presence of dark rims imply that earlier phenocrysts were of the sodalite group, more probably hauyne and/or nosean.

Pseudomorphs (several percent to 10%), forming oval-elongate ribbon and prismatic-mosaic aggregates, 0.5—5.0 mm in size, are common, though not numerous. They seem to have been formed after olivine phenocrysts.

Pseudomorphs (1—15%) with fusion rinds on their surface, or rounded, isometric, and elongate are very common. They are chiefly composed of monocrystals or aggregates of two to five crystals of colorless carbonate with subordinate scattered black opaque minerals and dust particles, light grey in reflected light.

There occur also (up to 5%) angular-prismatic carbonate pseudomorphs in which ore dust and iron hydroxides form a pattern of intersecting straight and locally sinuous lines. Relics of almost colorless clinopyroxene (cNg = 44—46°; +2 V = 58—59°) occur in nuclei of psuedomorphs.

Isometric and irregular aggregates of acicular aegirine crystals are probably xenolith fragments derived from fenitized rocks.

Phlogopite occurs as oval, often deformed scales, 1—4 mm in size and idiomorphic crystals measuring less than 1 mm. Phlogopite content, as a rule, does not exceed 1—3% and only in varieties rich in idiomorphic crystals (of metasomatic genesis?) amounts to 10%. Mica is often fully replaced by chlorite. There are also polycrystalline phlogopite aggregates containing apatite, amphibole, aegirine, and locally feldspar. The growths seem to be fragments derived from the shattering of fenitized xenoliths common in rocks of the type discussed.

Apatite forming porphyritic phenocrysts is a part of the groundmass and occurs in xenoliths of fenitized rocks. Porphyritic phenocrysts have rod-like bipyramidal or prismatic habit. The largest crystals are lightly fused; they measure 0.3—1.5 mm and account for 3—5%. Apatite of the groundmass consists of occasional prismatic crystals, 0.02 × 0.1 mm in size.

Opaque minerals occur in the form of idiomorphic, oval, and irregular grains up to 88 mm in size. They amount to about 1—3%. Most opaque minerals are magnetite and subordinate pyrite with occasional grains of ilmenite and chromite. Some rocks are impregnated with iron hydroxides. Ferrugination intensity varies considerably even within a single body.

Amphibole was observed in all thin sections, although the amount does not exceed 1%. It occurs both in polycrystalline growths and groundmass often containing phlogopite, apatite, and magnetite, set in groundmass as small angular-prismatic grains.

Amphibole of occasional grains and polycrystalline growths is identical in optical parameters.

Amphiboles have high dispersion of optical axes r > V and strong pleochroism: Ng — straw-yellow, Np — dark blue-green; cNp — 16–21°; –2 V = 80–83°, allowing their placement in the arfvedsonite group. Occasional grains and growths of arfvedsonite seem to result from the shattering of fenitized xenoliths by the formation of the rocks discussed. Less numerous are fairly large (1–5 mm) oval and subidiomorphic grains of amphibole pleochroic: Ng — green, locally brown, Nm — pale yellow, Np — light yellow or brownish yellow; cNg — 17–21°; –2 V = 81–82°. Optical parameters suggest the relation of the mineral to hornblende with fairly high alkali content. The genesis of this variety of amphibole remains uncertain because occasional grains are subidiomorphic in outline; however, similar amphibole enters into the composition of amphibole-pyroxene and phlogopite-amphibole-pyroxene inclusions, aparently xenoliths.

The groundmass is chiefly carbonate, often ferruginous carbonate. Apart from fine dispersed iron hydroxides, the groundmass contains ore dust. Microcrystalline carbonate aggregates responsible for the microlitic relic fabric of the groundmass is easily discernible in the cryptocrystalline substratum. Rocks whose microlites are oval-elongate in outline resembling rice grains (0.02–0.1) × (0.1–0.3) mm in size are very common. They account for 50–70% of the groundmass, locally dropping to 10%. The rocks are classed as variety "a".

Variety "b" of type I includes rocks whose pseudomorphs after microlites have the shape of a prism with tapering ends, and measure (0.01–0.02) × (0.05–0.10) mm. The pseudomorphs are, as a rule, composed of monocrystals of transparent carbonate. Their content varies from several to a few tens percent, averaging 10–15% of the groundmass.

Some rocks contain pseudomorphs after microlites of both varieties, rod-like microlites being smaller than oval ones. Small (0.02 × 0.05 mm) carbonate aggregates, rectangular and oval-isometric in outline, amounting to several percent, occur in the groundmass of type I rocks.

The habit and size of pseudomorphs apparently owe their variety to the original composition of microlites. We may assume that they were represented by melilite (oval-elongate aggregates), clinopyroxene (rod-like), and nepheline (rectangular and oval-isometric).

Type II. Rocks of this type were found in only five diatremes of the Orto-Yrygakh site (49c, 50, 56, 57, 59). They incorporate two structural-genetic groups, viz., tuffisite breccia (four diatremes) and xenotuffisite breccia (56). Based on autoliths, massive rocks of type II have relic porphyritic texture with a relic microlitic groundmass. Phenocrysts are chlorite-micaceous pseudomorphs, amphibole, clinopyroxene, opaque minerals, and apatite.

Chlorite-micaceous pseudomorphs are oval-elongate and rounded-isometric in shape. Only the smallest are subidiomorphic short prismatic dipyramydal in outline. Pseudomorphs have a reticulate texture due to interlaced transverse-fibrous veinlets of fine scaly, almost colorless micaceous mineral. Cells between fenestrulas are filled with cryptocrystalline or chlorite-serpentine mass rarely with fine to medium-grained carbonate. Pseudomorphs measure 0.2 × 0.3 to 3 × 4 mm and amount to 10–15%.

Their habit resembles that of pseudomorphs after olivine; however, the chlorite-micaceous composition and rare small inclusions (relics?) of ortho- and clinopyroxene prevent drawing a final conclusion as to the original composition of replaced pheno-crysts.

Amphibole is short prismatic and rod-like, often with twin crystals without crystallographic bounds at ends. They measure 0.1×0.2 to 1×2 mm, averaging 0.3×0.7 mm. There are magnetite and apatite growths. Pleochroism: Ng — light green to light brown; Np — straw yellow or light brownish yellow (locally both varieties occur in one crystal, the brown one being in the nucleus, as a rule); cNg $17-22°$; -2 V = $81-85°$. Based on these properties, amphibole may be grouped into hornblende.

Phlogopite forms oval often deformed flakes up to 3 mm in size and idiomorphic crystals $0.1-0.3$ mm in diameter. Mica accounts for 1%, locally increasing to 5%. There occur growths with clinopyroxene and replacement of the latter by mica, and pleochroism from brown and light brown along Ng to colorless and yellowish along Np.

Clinopyroxene is idiomorphic or occurs as twin crystals, often with fusion rinds, amounting to about $0.5-1.0$%. Corroded phenocrysts are 2×4 mm in size, idiomorphic phenocrysts measure 0.5×1.0 mm. Optical constants of idiomorphic crystals with fusion rinds on the surface are the same: cNg = $43-46°$; $+2$ V = $57-59°$.

Aegirine radiate aggregates are undoubtedly xenolith fragments derived from fenitized rocks.

Opaque minerals often form crystals with distinct faces of octahedral; irregularly shaped grains and opaque dust also occur; they total $2-5$%. Aggregates measure 0.7 mm, averaging (without dust particles) 0.3 mm. Most opaque minerals are magnetite with subordinate ilmenite and chromite. Leucoxene rims on most magnetite grains suggest a considerable proportion of titanium.

Apatite occurs as short prismatic dipyramidal and rod-like crystals and irregular-shaped grains. They are usually not larger than 0.5 mm. Small crystals sometimes form inclusions in amphibole and phlogopite and growths with magnetite.

The groundmass is chiefly carbonate with subordinate ore dust. Pseudomorphs after acicular and prismatic microlites composed of colorless carbonate are easily discernible in cryptocrystalline carbonate, brownish in transmitted light. Microlites measure $(0.005-0.02)$ $(0.3-0.15)$ mm, amounting to $10-30$%.

Type III. Rocks of this type were found in three diatremes, viz., at two localities in the Nomokhtookh site (43n, 52a) and in one locacity of the Orto-Yrygakh site (48a). Two of them are composed of the massive variety and the third consists of explosion breccia. Thus, rocks of type III differ from the other types in the relationship or structural-genetic groups, represented mainly by tuffisite and xenotuffisite breccias.

The rocks are porphyritic, with crypto-microcrystalline texture of the fully altered groundmass. Porphyritic phenocrysts are phlogopite ($5-20$%), opaque minerals ($2-5$%), carbonate pseudomorphs after olivine and pyroxene ($1-3$%), and apatite ($2-7$%).

Phlogopite is most abundant in rocks, building up Pipe 52, the lowest content was found in Pipe 48a. It consists of thick tabular crystals, rounded and oval, $0.1-5.0$ mm in diameter. Despite a high content of phlogopite, polycrystalline growths are rare. Many large and average crystals display short columnar habit: their dimensions along the third axis are greater than along the jointing. Mica is brown or orange along Ng and yellowish or colorless along Np. Often phlogopite is partly or fully replaced by green strongly birefringent mica-like mineral, pleochroic dark green along Ng to yellowish or colorless along Np. Some large and some average crystals are permeated along jointing with carbonate lenses whose proportion is locally larger than that of phlogopite.

Opaque mineral consists of oval or irregular grains, partly idiomorphic, and some skeletal crystals. Apart from large grains (up to 5 mm), abundant ore dust is present as part of the groundmass. The octahedric habit of crystals with leucoxene rims allows them to be classed as titanomagnetite.

Based on shape and internal structure, carbonate pseudomorphs are divided into two types. The former includes the most abundant (1–2%) pseudomorphs marked by the combination of reticulate and mosaic texture due to the presence of inequigranular sites outlined by films of light grey opaque pelitic material. Large and medium-sized pseudomorphs are rounded-isometric and oval in outline and small ones look like idiomorphic crystals of short columnar bipyramidal habit. Texturally, the pseudomorphs are identical to serpentinized and carbonatized olivine phenocrysts. Dentate saw-like outlines between separate sites of pseudomorphs suggest the replacement of olivine in these rocks, like in kimberlites (Milashev et al. 1963), occurring mainly along 021.

Carbonate pseudomorphs of type II are oval-elongate in outline or look like slightly clastogenic grains with fusion rinds on their surface; they are 0.5–1.5 mm in size and account for about 1%. Pseudomorphs are composed of colorless carbonate impregnated with iron hydroxides and ore dust forming a pattern of intersecting straight and sinuous lines. There are no relics of original mineral. However, the presence of clinopyroxene relics in similar pseudomorphs from rocks of the above types suggests that these are fully replaced crystals of clinopyroxene.

Apatite forms idiomorphic rod-like and short prismatic dipyramidal crystals and rounded-isometric grains (0.05–2.0 mm). Some crystals have fusion rinds on the surface. Their maximal and minimal amount is in the rocks of Pipe 48a and 52a, respectively.

The groundmass is composed of micro- finely crystalline carbonate with subordinate ore dust (2–5%), iron hydroxides (2–7%), and perovskite almost fully replaced by rutile and leucoxene.

In summary, all the above recognized types of altered rocks are easily distinguished on the basis of their texture, composition of relict minerals, morphology, and texture of pseudomorphs. The rocks of type I are characterized by the presence of arfvedsonite, aegirine, xenoliths of fenitized rocks, and carbonate pseudomorphs after phenocrysts of minerals from the sodalite (?) group. The type II rocks are marked by abundant chloritemicaceous pseudomorphs after olivine (?) and by the presence of hornblende. Finally, the rocks of type III differ greatly from the above two types in their high content of phlogopite, in the absence of amphibole and fine-grained groundmass without relic microlitic structure.

Apart from the rocks discussed above, consisting mainly of altered alkali basaltoids, shatter breccia is present, and contains no igneous rocks. Breccia consisting of fragments derived from sedimentary country rocks occurs in linear fault zones and in cone-shaped bodies, isometric in plan. Breccia in the linear fault zones is undoubtedly of tectonic origin. As a rule, it is strongly ferruginated and displays REE mineralization.

Shatter breccia in cone-shaped bodies is known from the Nomokhtookh and Tundrovy sites. It is very similar to tectonic breccia in structure, texture, and in sediment composition. It differs in its lower proportion of ferruginated varieties and some peculiar features of REE mineralization. The explosion nature of the breccia is evidenced by its occurrence in the form of cone-shaped bodies. In our opinion, the fact that it is very common in the Anabar xenotuffisite breccia district with an extremely low proportion of effusive material is strong evidence for its explosion genesis. Therefore, it seems logical to assume the existence of extreme modifications of xenotuffisite breccia, i.e., explosion breccia, fully composed of shattered sedimentary rocks, deprived of effusives.

Therefore, it is reasonable to class breccia filling cone-shaped bodies and composed of fragments derived from country rocks as explosion shatter breccia. It looks like typical breccia composed mainly (80–95%) of fragments of limestone, dolomite, and their clay and silty varieties. Fragments range in size from some millimeters to 20 cm. It is difficult, as a rule, to tell "fragments proper" from cement because small fragments fill the space between larger ones. Some varieties of the breccia, apart from dominating shattered carbonate products, also contain a large proportion of terrigenous grains of quartz, feldspar, and the like. The minute fractions were the first to be replaced during ferrugination of breccia. When the process was intensive, iron hydroxydes replaced not only small- and medium-sized grains, but even large fragments, occurring along cracks and around the periphery.

Alkali basaltoid rocks filling diatremes of the three sites of the Anabar district show a similarity in chemical composition (Table 13). If relict structures, textures, and original minerals are ignored, they can be erroneously taken for carbonatites. However, chemical analyses of major rock-forming oxides suggest that this bulk composition is a result of secondary processes and its contents are far from original. Therefore, it is not reasonable to use all the rock-forming components, most of which were redistributed and lost in the course of post-magmatic processes, to elucidate the main features of original composition and to reliably establish a petrochemical analogy to so strongly altered rocks. Better results may be obtained by means of so-called representative elements, i.e., iron, titanium, chromium, aluminum, potassium, and phosphorus, whose contents have not changed greatly in the course of secondary processes.

The content of iron, aluminum, and potassium is much higher in scenotuffisite breccia than in the remaining structural-genetic rock groups of the same type, owing to the ferrugination characteristic of the breccia, and because technically it is impossible to completely extract xenoliths of carbonate and clay-carbonate rocks. There is no important difference in the contents of other elements, because their abundance in alkali basaltoids and that of terrigenous-carbonate rocks is fairly close. To estimate mean weight content of repesentative elements in alkali basaltoid rocks, we used analyses of massive varieties, eruption and tuffisite breccias only because analyses of xenotuffisite breccia introduce errors.

The variation of mean weight content of representative elements in alkali basaltoid rocks of different types is not important for absolute values; however, the data

Table 13. Variability and average chemical composition (wt.%) of highly carbonatized alkali basaltoids from diatremes of three main sites in the Anabar district. (Marshintsev 1974)

Oxide	Nomokhtookh Variability range	Average (38 analyses)	Tundrovy Variability range	Average (7 analyses)	Orto-Yrygakh Variability range	Average (27 analyses)
SiO_2	4.32–19.22	12.19	8.62–20.96	12.63	7.04–23.80	14.61
TiO_2	0.01– 0.99	0.29	0.03– 0.47	0.19	0.03– 1.71	0.40
Al_2O_3	1.03– 9.67	3.37	1.60– 5.90	2.99	1.26– 6.24	3.34
Fe_2O_3	0.16–17.09	6.90	2.13– 7.92	4.01	1.93–36.67	4.48
Cr_2O_3	0.001–0.12	0.02	0.01– 0.06	0.02	0.01– 0.05	0.02
FeO	0.64–18.31	3.74	0.36–17.50	4.37	0.65–12.73	4.25
MnO	0.04– 3.04	0.59	0.08– 0.53	0.32	0.04– 0.61	0.31
NiO	0.01– 0.10	0.02	– –	– –	0.001–0.12	0.02
CaO	4.38–36.11	27.51	16.34–34.28	27.22	14.19–40.12	26.89
MgO	2.39–13.36	9.66	8.65–17.40	11.70	4.85–16.34	10.37
K_2O	0.63– 4.05	1.33	0.54– 3.80	1.40	0.20– 2.98	1.04
Na_2O	0.16– 1.82	0.59	0.06– 0.46	0.30	0.25– 2.80	0.84
CO_2	19.80–33.21	27.79	22.88–33.44	29.19	13.50–34.10	20.08
P_2O_5	0.10– 6.28	1.88	0.69– 2.62	1.36	0.16– 5.08	1.24
H_2O^+	0.06– 5.48	1.51	0.47– 3.28	1.46	0.20– 2.92	1.10
S	0.02– 0.52	0.19	0.02– 0.94	0.23	0.02– 0.45	0.15
F	0.07– 0.62	0.28	– –	– –	– –	– –
SO_2	– –	– –	– –	– –	0.21– 2.31	1.02
Ignition losses	0.78–25.58	1.91	0.07– 4.44	2.18	1.09–39.44	9.91
Equiv. O for F and S	– –	–0.30	– –	–0.26	– –	–0.17
Total	– –	99.47	– –	99.31	– –	99.90

available suggest a similarity of important difference in chemical composition of certain rock types (Tables 14, 15). The rocks of both varieties of type I are almost identical in composition and differ slightly as to the mean weight content of all representative elements. The type II rocks display a more unique chemistry: they differ from the type I rocks in mean weight content of most elements and from those of type III in phosphorus content.

Fenitized xenoliths of crystalline rocks are found in many diatremes. Their content averages several percent. Macroscopically, the xenoliths are oval and rounded inclusions up to 10 cm in diameter. These are finely to medium-crystalline rocks with newly formed alkali amphibole, aegirine, epidoe, and subordinate albite, potash feldspar, and some other minerals.

Table 14. Weighted mean content of representative elements in carbonatized alkali basaltoids of the Anabar district

Structural rock group	Element	Rock types and varieties			
		I		II	III
		Ia	Ib		
Massive	Fe	10.84	8.94	–	5.81
	Ti	0.22	0.05	–	0.54
	Cr	0.023	0.018	–	0.048
	Al	2.87	1.52	–	1.89
	K	0.33	0.47	–	0.85
	P	2.12	1.46	–	1.21
Eruption breccia	Fe	5.55	–	–	12.35
	Ti	0.26	–	–	0.23
	Cr	0.012	–	–	0.006
	Al	1.92	–	–	2.41
	K	0.75	–	–	0.65
	P	0.78	–	–	1.78
Tuffisite breccia	Fe	5.60	4.84	5.54	–
	Ti	0.21	0.10	0.45	–
	Cr	0.022	0.007	0.063	–
	Al	1.07	1.35	2.32	–
	K	0.57	0.65	1.17	–
	P	0.63	1.10	0.36	–
Xenotuffisite breccia	Fe	6.94	12.92	–	–
	Ti	0.33	0.27	–	–
	Cr	0.015	0.018	–	–
	Al	1.67	2.02	–	–
	K	1.04	0.75	–	–
	P	0.58	0.52	–	–
Weighted mean for rock types (xeno- tuffisite breccia exclusive)	Fe	5.83	5.46	5.54	6.73
	Ti	0.22	0.09	0.45	0.50
	Cr	0.021	0.010	0.063	0.038
	Al	1.26	1.37	2.32	1.96
	K	0.58	0.62	1.17	0.82
	P	0.70	1.16	0.36	1.29

A cluster distribution pattern of diatremes at high concentration of many tens of pipes over a limited area of isometric sites suggests their relation to isolate hypabyssal laccolithlike (?) alkali or alkali basaltoid massifs having an area of about 10 km^2. This assumption is also supported by a very common REE mineralization trend found in alkali basaltoid rocks and that of kimberlites, and along shatter zones in sedimentary country rocks (Milashev 1968a).

Geophysical studies in diatreme areas of the region revealed a low gravity field reflected in variations of isanomals of regional anomalies. However, appropriate

Table 15. Discrepancies between weighted mean contents of representative elements in alkali basaltoids of the Anabar district

Correlative rock types and varieties	Fe	Ti	Cr	Al	K	P
Ia/Ib	U	U	U	U	U	U
	$\dfrac{0.34}{2.01}$	$\dfrac{1.22}{2.01}$	$\dfrac{1.30}{2.01}$	$\dfrac{0.35}{2.01}$	$\dfrac{0.26}{2.01}$	$\dfrac{1.68}{2.01}$
Ia/II	U	C	C	C	C	$C_{0.01}$
	$\dfrac{0.61}{2.00}$	$\dfrac{7.04}{3.45}$	$\dfrac{7.97}{3.50}$	$\dfrac{5.79}{3.45}$	$\dfrac{6.82}{3.45}$	$\dfrac{U_{0.001}}{2.92}$
Ia/III	U	$C_{0.02}$	U	U	U	U
	$\dfrac{0.70}{2.01}$	$\dfrac{U_{0.01}}{2.64}$	$\dfrac{1.92}{2.01}$	$\dfrac{1.88}{2.01}$	$\dfrac{1.38}{2.01}$	$\dfrac{1.85}{2.01}$
Ib/II	U	C	$C_{0.02}$	$C_{0.05}$	$C_{0.02}$	C
	$\dfrac{0.13}{2.06}$	$\dfrac{3.89}{3.72}$	$\dfrac{U_{0.01}}{2.77}$	$\dfrac{U_{0.02}}{2.11}$	$\dfrac{U_{0.01}}{2.61}$	$\dfrac{5.41}{3.71}$
Ib/III	U	U	U	U	U	U
	$\dfrac{0.67}{2.36}$	$\dfrac{1.03}{2.45}$	$\dfrac{1.56}{3.18}$	$\dfrac{1.29}{2.36}$	$\dfrac{1.14}{2.36}$	$\dfrac{0.27}{2.36}$
II/III	U	U	U	U	U	C
	$\dfrac{1.64}{2.06}$	$\dfrac{1.70}{2.06}$	$\dfrac{1.26}{2.16}$	$\dfrac{0.68}{2.06}$	$\dfrac{1.47}{2.06}$	$\dfrac{5.12}{3.72}$

Notes. Xenotuffisite breccia is omitted. Discrepancies between weighted mean content of elements: C – considerable, U – unimportant. Statistical data t: in numerator – calculated data, in denominator for the confidence level P = 0.001 with considerable discrepancy of mean values and for P = 0.05 with unimportant discrepancy. When $t_{P=0.05} < t_{cal} < t_{P=0.001}$ the confidence levels and calcualted t values are given.

anomalies have not been observed in the magnetic field. Remnant anomalies were established by averaging the regional gravity field. Two of the above sites of alkali basaltoid diatreme coincide fully and one site partly in plan with two negative gravity anomalies (Fig. 31). Negative residual anomalies are attributed to the tectonic structure of the basement. According to the estimates of G.D. Balakshin, local residual gravity anomalies over the alkali basaltoid diatreme sites may originate from stock-like intrusive bodies of the basement, they are 2–3 km in diameter and have a negative anomalous density of 0.1 g cm^{-3} (Marshintsev and Balakshin 1969).

The geophysical studies thus also support the existence of isometric (in plan) intrusions at a depth below the clusters of alkali basaltoid diatremes. However, the question of shape, size, and depth of an intrusion remains unsolved because of the

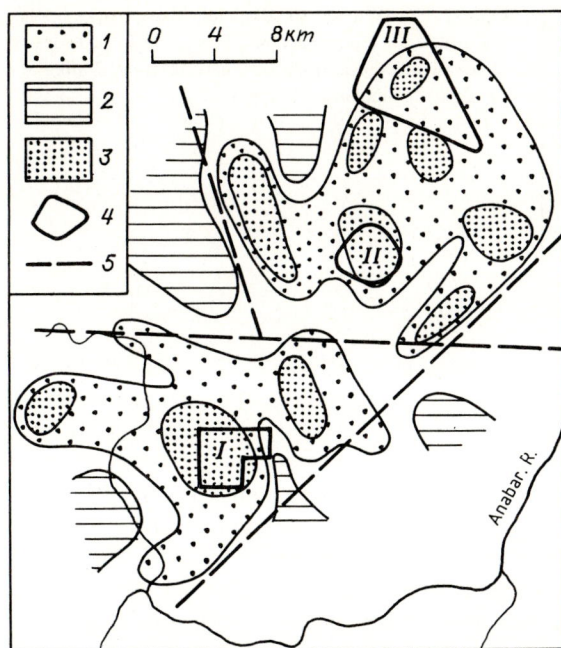

Fig. 31. Distribution of residual gravity anomalies in the districts of occurrence of alkali basaltoid diatremes in the Anabar district. (Marshintsev and Balakshin 1969). Residual gravity anomalies: *1* negative; *2* positive; *3* local lows; *4* distribution of alkali basaltoid diatremes (*I* Nomokhtookh, *II* Tundrovy; *III* Orto-Yrygakh); *5* inferred faults

paucity of data necessary to give preference to one of the assumptions (hypabyssal laccolith or a stock at a depth of above 2.5 km).

Ingili District

An extensive (about 300 km^2) massif of ultrabasic alkali rocks and carbonatite, and dyke swarms dispersing fan-like and associated ankaratrite pipes occupy the eastern slope of the Aldan shield and the Ingili River basin (Fig. 32). When in the Ingili district rocks filling explosion pipes and dykes were first discovered, they were called kimberlites (Kaminsky and Potapov 1968, 1969). However, detailed study showed that the rocks differ greatly from kimberlites, and Kaminsky (1969) named them ingilites. The term ingilite seems obsolete because its petrology allows it to be classed as ankaratrite.

The district is underlain by Late Proterozoic and Early Cambrian terrigenous and carbonate rocks. The generally steep north-east dip (1–3°) is complicated by small transverse flexures. Faults occur mainly as vertical fractures with small amplitude displacement.

Based on the mode of occurrence, one of the cone-shaped bodies of the district is classed as stock, the remaining being typical diatremes. They are irregular elongate in outline. Cone-shaped bodies are closely related to faults. For example, the Nina Pipe is confined to a north-east-striking fault (azimuth of 55°) and extends in the same direction. According to magnetometric data, the pipe shows a steep south-east dipping (80–85°). The Raschetnaya Pipe is associated with the same fault and has a steep

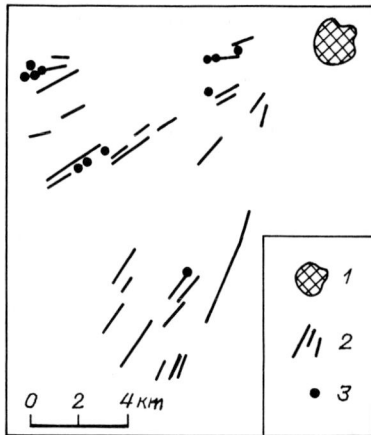

Fig. 32. Distribution of alkali ultrabasic rocks in the Ingili district. (Kaminsky and Potapov 1969). *1* Massif of alkali ultrabasic rocks and carbonatites; *2* ankaratrite dykes; *3* anakaratrite pipes

south-east dip. Three pipes, namely Nizhnyaya, Verkhnyaya, and Nakrytaya Pipes, are located on the north-east striking fault, the Finalnaya Pipe occurs 6 km away from the massif. The long axis of the diatreme follows the north-eastern direction. The part of the area with which the Finalnaya Pipe is associated is characterized by numerous faults, easily deciphered on aerial photographs, but not reflected in topography. To the east, the pipe is bounded by sublatitudinal faults more than 2 km long. Fairly extensive and morphologically complex limburgite dykes are related to faults west of the pipe.

The Lower Cambrian deposits overlying some cone-shaped bodies and radiological datings (K-Ar ages) suggest that these volcanic rocks were emplaced during the Late Proterozoic.

Country rocks have a slope of 20° at contacts with some diatremes (Raschetnaya). In the Izhevskaya Pipe the country rocks are intensely shattered and pierced by thin ankaratrite veinlets. The associated ankaratrite breccias are so crowded with fragments of country rocks that produce an impression of the gradual transition from the pipe to the country rock. There is a 5–6-m-thick skarn zone in dolomite at the contact with the stock-like Leksena Pipe.

Dykes vary in thickness from 0.5 to 5 m, locally reaching 12–15 m. They are hundreds of meters to several kilometers long. The strike is sublongitudinal (in the south) to sublatitudinal (in the west). The contacts with the country rocks are sharp, numerous apophyses and offshoots are present. In the contact zone the country rocks are strongly fissured. The traces of slip are observed along the cracks. The direction of microjointing often follows the major fractures in the vicinity of dykes.

Cone-shaped bodies are infilled with eruption breccia, while stock-like bodies and dykes are composed of massive ankaratrites. Eruption breccia consists of fragments derived from different (mainly, country) rocks (10–80%) set in a magmatic matrix. The matrix texture of eruption breccia is similar to that of massive rocks: porphyritic and relict porphyritic with microlitic groundmass. Porphyritic phenocrysts and groundmass average, respectively, 40% and 60%. Porphyritic phenocrysts are olivine (5–20%), clinopyroxene (0.25%), phlogopite (1–5%), picroilmenite (up to 2%). All the minerals are also present in the groundmass.

Table 16. Average chemical composition of rocks (wt.%) infilling explosion pipges and dykes of the Ingilit district. (Kaminsky 1969)

Oxide	Content	Oxide	Content
SiO_2	29.74	CaO	15.69
TiO_2	5.33	Na_2O	0.45
Al_2O_3	5.53	K_2O	1.04
Cr_2O_3	0.10	P_2O_5	0.72
Fe_2O_3	10.48	SO_3	0.17
FeO	5.46	Ignition losses	15.00
MnO	0.17		
MgO	10.06	Total	99.94

Olivine forms aggregates of two generations differing in size, grade of idiomorphism, and nature of secondary alterations. Olivine aggregates of earlier generation are 3 mm (less commonly 30 mm) in size, and are oval with traces of corrosion. The fayalite component accounts for 16–17%. Olivine aggregates of later generation are fully replaced by secondary minerals. Crystallography and composition of secondary products (serpentine, garnierite, magnetite) evidence a primary olivine composition of small (0.1–0.5 mm) relic idiomorphic pseudomorphs.

Clinopyroxene forms phenocrysts of two generations and microlite of the groundmass. Its content varies from 10% to 70% in the rocks (0–25% in phenocrysts). The first generation of clinopyroxene consists of large tabular crystals intergrown with darker pyroxene with a different extinction angle. Pyroxene of the second generation forms short prismatic crystals, polysynthetically twinned. Based on optical constants and chemical analysis, clinopyroxene may be classed as magnesium salite.

Phlogopite forms porphyritic phenocrysts and microlites of the groundmass. Its content in rocks totals 1–20%. Phenocrysts have no distinct crystallographic outlines, but opacite rim and traces of deformation are present. Secondary phlogopite is fairly common in rocks.

Picroilmenite consists of phenocrysts (5–20 mm) and small grains set in matrix. Its content varies from one to several percent. Based on chemical analyses, magnesium oxide in the mineral accounts for 5.7–9.0 wt.%.

The groundmass is composed of clinopyroxene and phlogopite microlites, cubic and rounded perovskite granules, and ore dust. In some dykes rocks contain short prismatic microlites fully composed of carbonate (pseudomorphs after melilite?).

The average chemical composition of the rocks discussed is listed in Table 16. Insense postmagmatic alteration of rocks and impurities in breccia (unavoidable in analysis) in the form of xenogenic products, mainly due to limestone and dolomite, change the primary composition considerably. As a result, the rocks studied display low contents of silica, aluminum, and alkali metals, as compared to the average limburgite of Zavaritskii. Rocks from the Ingili district differ from kimberlites and and associated picrite porphyrites in their high content of titanium and iron. However, the presence of clinopyroxene phenocrysts and microlites is the main petrological feature of the rocks. The above evidence, along with the absence of barophilic pyrope, obviates the necessity for placing ultrabasic porphyritic rocks of the Ingili district into

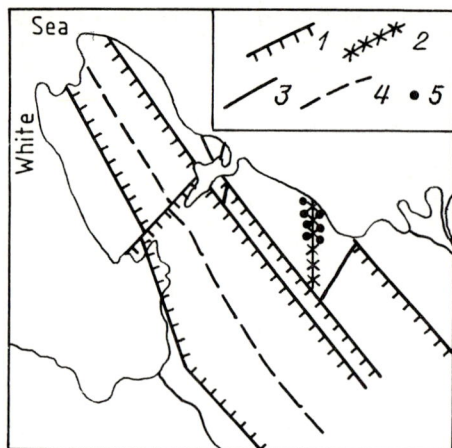

the kimberlite facies. These are rocks in the picrite facies according to our classification (Milashev 1963a, 1974a) and therefore, in terms of traditional petrology, they owe their name to the abundance ratio of major rock-forming minerals.

Onega Peninsula

In the later 1960's, explosion pipes composed of alkali basaltoid rocks and their breccias were discovered on the northern Russian platform and on the eastern slope of the Baltic shield. The number of cone-shaped bodies and local magnetic anomalies of cone-shaped type total 20 (Stankovsky et al. 1973; Kaminsky 1976). On the Onega Peninsula, a general gentle plunging of the crystalline basement under a transgressively onlapping sedimentary cover, typical of the north-western termination of the Russian platform, is complicated by a north-west-striking graben in the axial part of the peninsula (Fig. 33). Within the graben, the crystalline basement has suffered deep subsidence completely compensated by Vendian time. Thus, the Onega graben may be considered as part of the crystalline basement, i.e., a preplatform edifice emplaced and developed during the Late Proterozoic. The Onega graben is asymmetrical in the cross-section: the south-western flank is structurally simple, while the north-eastern flank has a complex step-wise structure. Pipes occur within the second (from the graben axis) step of the north-eastern flank and related to minor faults, feathering major fractures. Monchiquite explosion pipes are found on the projection of the Onega graben into the Kandalaksha Bay.

No relations between the pipes and the Paleozoic rocks have been established. Direct evidence for a Precambrian age of the diatremes is the fact that the Verkhovsky fault controlling their distribution cannot be traced within the Baltic Series (Cambrian) terrain (Stankovsky 1973). The shape (*in plan*) of most pipes is close to isometric, their diameters vary from 100 to 420 m (Fig. 34). The contacts with country rocks are sharp and "dry". Sedimentary rocks have altered to a clay-like mass, pierced by a dense network of cracks and thin (up to 5 mm) calcite veinlets at a distance of 4—5 m

Bolvantsy

Karakhta

Kurtyaevo

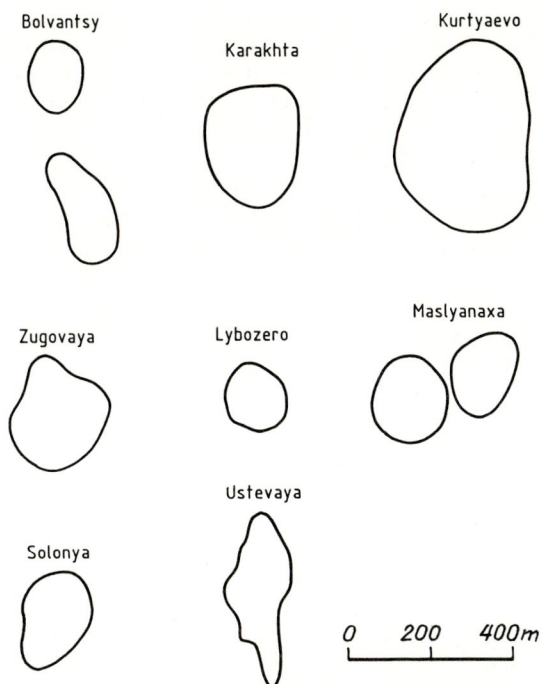

Fig. 34. Shape in plan of explosion pipes on Onega Peninsula. (Kaminsky 1976)

Zugovaya

Lybozero

Maslyanaxa

Solonya

Ustevaya

0   200   400m

from exocontacts. At a distance of 5–10 m from the contact, country rocks are strongly brecciated. A degree of brecciation decreases with distance from the contact and disappears completely at a distance of 35–40 m away. Small and large fragments of sedimentary country rocks form much of the volcanic breccia in the endocontact zones. The proportion of fine clastic sedimentary material in breccia drastically decreases from the contact toward the center of the pipes, and only relatively large fragments are present at a distance of 15–20 m from the contact. Apophyses, 4–8 m thick, are observed in some pipes.

The rocks infilling explosion pipes on the Onega Peninsula are classed as alkali basaltoid volcanic breccia (Kaminsky 1976). In the endocontact zones breccia contains very numerous fragments of the Upper Proterozoic country rocks along with xenoliths of the Archean gneiss and ultrabasic holocrystalline rocks. The proportion of xenogenic material decreases to 1–5% in breccia from the central parts of the diatremes.

Volcanic material proper forms a peculiar breccia of autolith type: oval and rounded aggregates of rocks of earlier generations (autoliths) set in a matrix of the same composition. Both autoliths and matrix have a breccia structure: small aggregates (chondrules), 0.5–0.7 mm in size, set in a matrix of the same composition. Chondrules consist of glassy, usually badly altered base which, together with plagioclase (also strongly altered) microlites, forms the groundmass. Porphyritic phenocrysts, acting as peculiar cores, are present in the center of most, especially large, chondrules. Owing to the microlitic texture of the base, typical of the rocks, chondrules acquire a concentric pattern, caused by a circular arrangement of microlites around the cores.

Table 17. Chemical composition (wt.%) of volcanic breccia from explosion pipes of the Onega Peninsula. (Kaminsky 1976)

Oxide	Bolvantsy (7 analyses)	Karakhta (19)	Kurtyaevo (9)	Lugovaya (9)	Lyvozero (38)	Maslyanaya (3)	Solonaya (11)	Ustievaya (12)	Average (108)
SiO_2	43.68	48.60	47.56	50.21	38.51	56.46	46.47	43.51	46.88
TiO_2	0.85	0.87	0.86	0.74	0.94	0.67	0.64	0.71	0.78
Al_2O_3	11.64	10.85	10.50	10.40	11.64	10.20	9.11	10.18	10.57
Cr_2O_3	0.01	0.018	0.027	—	0.016	—	—	—	0.018
Fe_2O_3	7.65	4.02	4.92	5.38	4.87	4.75	5.67	7.48	5.72
FeO	2.31	3.51	3.57	3.49	4.73	3.06	3.31	2.47	3.31
NiO	—	0.016	0.024	—	0.012	—	—	—	0.017
MnO	0.32	0.30	0.19	0.27	0.30	0.19	0.24	0.32	0.27
MgO	12.50	13.38	13.26	11.30	15.47	10.36	14.85	17.27	13.55
CaO	8.65	5.74	8.93	6.71	9.53	4.61	7.39	6.02	7.20
Na_2O	2.57	3.63	3.43	4.26	3.25	3.76	3.66	3.48	3.51
K_2O	1.50	1.54	1.42	1.55	0.37	1.90	1.11	0.56	1.24
P_2O_5	0.60	0.34	0.31	0.34	0.58	0.32	0.44	0.41	0.42
V_2O_5	—	tr.	—	—	0.04	—	—	—	0.02
SO_3	—	0.02	—	—	0.21	—	—	—	0.09
Ignition losses	8.60	5.72	4.67	5.18	9.38	3.36	6.57	7.37	6.36
Total	100.88	99.56	99.71	99.83	99.85	99.64	99.46	99.78	99.96
H_2O^-	4.18	3.51	3.15	1.65	3.73	1.25	—	—	2.91
CO_2	1.50	0.95	0.42	—	1.75	—	—	—	1.16
S_{total}	tr.	0.06	0.11	0.03	0.14	0.08	0.04	0.03	0.06

The chondrule matrix accounts for 10—90% of the rock volume. The matrix, like chondrules, consists of the base and phenocrysts of pyroxene, olivine, nepheline, garnet, mica, and hornblende.

Pyroxene — augite-diopside, locally with subordinate aegirine — forms phenocrysts of two generations. Large phenocrysts (up to 1.5 mm) are often zoned, resorped, and irregular. There are also small (0.1—0.3 mm) idiomorphic crystals of clinopyroxene.

Olivine, although not abundant, occurs in almost all diatremes. Commonly it is replaced by serpentine and carbonate. The optical properties of a few relics of the mineral suggests a considerable proportion (25—60%) of ferrous component.

No nepheline is present in the rocks discussed. Its original existence is assumed owing to peculiar square, tabular, and hexagonal pseudomorphs consisting of chlorite with cancrinite inclusions.

Garnet is not a typical mineral of alkali basaltoids, therefore its noticeable (up to 40% of a heavy fraction) content in volcanic breccia of the Onega Peninsula deserves special attention. Garnet represented by idiomorphic crystals up to 0.5 mm in size was observed in all thin sections. Most garnet is pink or pale pink, there are a few orange and even violet grains (Bolvantsy, Karakhta, and Lyvozero Pipes); 98% of garnet is pink and can be classified as almandine. Some garnet varieties are orange and violet and contain 43% (Karakhta Pipe) and even 63% (Bolvantsy) pyrope component (Kaminsky et al. 1975).

The chemical analyses of volcanic breccia from explosion pipes on the Onega Peninsula (Table 17) suggest a similarity of the rocks to monchiquite and alkali basaltoid known from explosion pipes in many districts. Volcanic breccia of the Onega Peninsula differs from similar rocks in the presence of rock-forming almandine and accessory pyrope. The presence of even minimal amounts of barophilic minerals implies peculiar environmental conditions at some stage (or stages) in the formation of the rocks.

According to the present author's concept of the typomorphism of pyrope and diamond for physical and chemical conditions in the kimberlite facies (Milashev et al. 1963; Milashev 1965, 1972a), the volcanic rocks infilling the explosion pipes on the Onega Peninsula should be assigned to the kimberlite-facies rocks. However, this does not mean that the Onega volcanic breccia is to be classed as kimberlite breccia. Kimberlite is a highly ultrabasic rock containing no porphyritic phenocrysts of pyroxene (Milashev et al. 1963). The classification of the Onega volcanic breccia as kimberlite-facies rocks of the alkali basaltoid volcanism allows them to be named kimberlites in a limited sense. This is also not inconsistent with principles of petrology. The petrology of metamorphic rocks provides an example of a similar approach to a facies name and rock diagnostics. Nobody would doubt the validity of the assignment of amphiboles proper, but also amphibole and micaceous gneisses to the amphibolite facies, and the fact that not only granulite (garnet-plagioclase rocks), but also bipyroxene and garnetiferous gneiss, as well as some other rocks, are placed into the granulite facies.

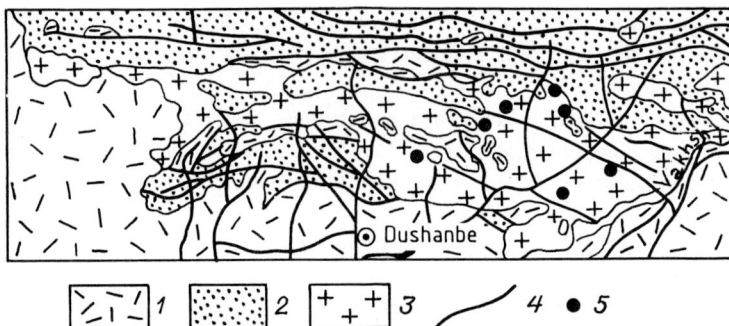

Fig. 35. Geological structure of the South Gissar and distribution of explosion pipes. (Baratov et al. 1970). Deposits: *1* Meso-Cenozoic; *2* Paleozoic; *3* intrusive bodies; *4* faults; *5* explosion pipes

South Gissar

The South Gissar encompasses the southern slope of the Gissar Range and the western part of its spur, viz., the Karatega Ridge. The fact that the Paleozoic fold area of the South Tien Shan is regionally underlain by Meso-Cenozoic deposits allows this district to be assigned to the southern margin of Hercynian systems of the Tien Shan. Sinitsyn (1957), who developed a tectonic zonation of the Tien Shan, has recognized the South Gissar zone owing to a peculiar geological structure of the district. Within this zone Lower and Middle Paleozoic deposits are absent from the section, while Upper Paleozoic rocks are very common; the major folding is late Hercynian in age. The tectonic regime of the southern slope of the Gissar Range became typically platform after the termination of the Hercynian orogeny. The composition and thickness of the Triassic, Jurassic, Cretaceous, and Paleozoic deposits support the above statements (Baratov et al. 1970).

Repeated magmatic activity and its considerable extent are peculiar geological features of the South Gissar. For example, more than half the district is underlain by granitoid batholiths. Volcanic formations and associated subvolcanic and hypabyssal intrusions are also fairly common. Seven magmatic complexes, different in age and composition, can be recognized. One is tentatively dated as Precambrian, five are Devonian to Late Permian, and the last is Early-Middle Triassic(?).

Of particular interest is the Early-Middle Triassic complex because it incorporates compositionally different explosion pipes and because its emplacement took place under subplatform tectonic conditions. The complex is widespread in the South Gissar zone and includes monzonite stocks, trachyandesite sills, explosion pipes of picrite porphyries, monchiquite limburgite, and analcime basalts, as well as dykes of trachytoid syenite porphyry, trachybasalt, alkali lamprophyre, and picrite. The age of the rocks was inferred on the basis of their active contact with red molasses of the Late Permian-Early Triassic Khanakin Formation and the absence of similar magmatic rocks in the Mesozoic deposits, unconformably overlying the Paleozoic rocks (the Mesozoic sequence starts with the Early Triassic deposits). The youngest formations of the complex are explosion pipes and dykes of camptonite (K-Ar age is 199–215 m.y.),

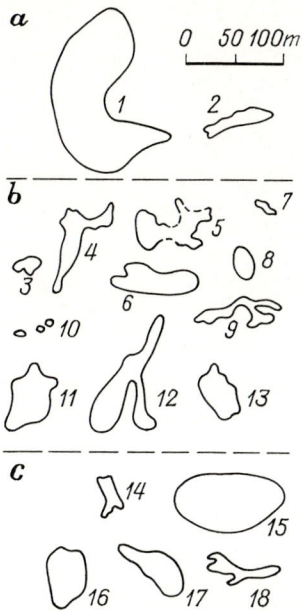

Fig. 36a–c. Shape in plan of explosion pipes of the South Gissar. (Baratov et al. 1970). **a** Pipes of picritic composition: *1* Tuvish; *2* Khelmagz; **b** pipes of monchiquite-limburgite composition: *3* Odzhuk; *4* Surkhob; *5* Chimkuidy; *6* Kaloch; *7–9* Dev-Dara I, II, and III; *10–13* Dashti-Mazar I, II, III, and IV; **c** pipes of analcime-basaltic composition: *14* Kadkob; *15* Sayed; *16* Pandema; *17, 18* Kurban I and II

monchiquite, and picrite (170–174 m.y.) of a probable Triassic-Jurassic age (Baratov et al. 1970).

About 30 explosion pipes within the Gissar granitoid pluton are known from the South Gissar. Some of them, especially the largest pipes, are controlled by fracture zones cutting the grain of the terrane; the zones were formed due to faulting of the basement in the rocks of the higher structural stages. Smaller pipes are locally related to "going through" fracture zones, concentrating in feather fractures.

At the present exposure surface the size of the pipes varies from 5 by 10 to 50 by 250 m. Morphologically, there are isometric and flattened bodies. Owing to a partly preserved top, some pipes are horseshoe in outline at the present erosion surface. Many diatremes composed of monchiquite-limburgite rocks and local analcime basalts (Fig. 36) exhibit an irregular, odd, amoeboid shape. Diatremes owe their peculiar morphology to a relatively slow volcanic activity, whose intensity was not enough to fully develop pipes occurring at the intersection of several faults. Being arranged in groups, diatremes even at the present exposure surface are linked by dykes of similar rocks, but containing no xenoliths, i.e., by massive varieties (Dev-Dara I, II and other pipes). Contacts of most diatremes have a steep centroclinal slope (60–85°).

In complex structure pipes, the earlier rock generations are represented by explosion breccia, while the later generations consist of massive porphyritic rocks having almost no xenoliths (Surkhob, Dev-Dara III and other pipes). Some dykes display a radial pattern with respect to the pipe center. The best example is provided by the Odzhuk Pipe, composed of monchiquite-camptonite eruption breccia. Six analcime-bearing olivine-pyroxene camptonite dykes, subradial with respect to the center, cut the diatreme. The dykes are 0.4–1.0 m thick and 6–20 m long (Baratov et al. 1970). Camptonite, picrite, and analcime basalt dykes occur in the enclosing rocks near the

Table 18. Chemical composition (wt.%) of volcanic rocks and deep-seated inclusions from explosion pipes of the South Gissar. (After Baratov et al. 1970; Kutonin et al. 1973)

Rocks	Pipes and dykes	Number of analyses	SiO_2	TiO_2	Al_2O_3	Fe_2O_3	FeO	MnO	MgO	CaO	Na_2O	K_2O	P_2O_5	Cr_2O_3	Ignition losses	Total
Picritic porphyry	Tuvish	3	38.71	1.66	12.57	4.55	7.20	0.21	10.97	13.19	1.13	1.79	0.61	–	7.86	100.45
Explosion breccia of picritic porphyry	Khelmagz	1	55.23	1.43	20.57	8.00	2.66	0.09	1.13	1.81	3.80	3.00	0.38	–	2.59	100.69
Monchiquite	Odzhuk	2	44.75	1.24	13.88	4.90	5.17	0.19	9.90	11.42	1.45	2.44	0.37	–	4.81	100.52
Eruption breccia of monchiquite	The same	3	49.63	0.99	14.93	6.44	3.36	0.14	5.61	6.76	1.60	3.58	0.40	–	6.71	100.51
	Dev-Dara II	1	43.61	2.15	13.35	5.46	6.02	0.23	16.96	11.84	1.59	2.41	0.80	–	4.78	98.50
	Kaloch	2	44.84	1.54	13.21	3.73	7.36	0.16	11.98	8.85	2.30	2.07	0.48	–	3.41	99.93
Camptonite-monchiquite	Dev-Dara II	2	44.70	1.92	15.52	4.54	5.75	0.18	7.80	8.03	1.92	2.98	0.60	–	5.89	99.83
	Dev-Dara IV	1	39.26	1.97	17.78	7.21	5.28	0.28	7.45	4.94	1.20	4.56	0.63	–	9.97	100.53
	Dashti-Mazar	5	45.88	0.89	14.48	6.84	4.95	0.15	6.40	9.93	2.36	2.95	0.38	–	5.12	100.33
Camptonite	Chumkuidy	2	42.92	1.60	13.68	3.29	6.91	0.12	10.02	11.37	2.42	2.72	0.31	–	5.04	100.40
Analcime basalt	Ranginau	2	50.82	0.98	17.01	8.00	4.50	0.10	2.76	5.32	3.44	4.20	0.41	–	3.22	100.76
	Kurban I	3	49.40	0.95	17.44	4.36	4.38	0.14	1.92	6.08	2.57	4.06	0.44	–	8.23	99.97
	Kurban II	1	44.80	1.07	17.89	1.73	4.83	0.09	2.54	9.54	2.37	3.12	0.47	–	11.66	100.11

Green clinopyroxenite																
Inclusion in picritic breccia	Tuvish	4	50.52	0.36	3.40	2.03	2.86	0.11	16.86	19.80	0.45	0.19	0.05	0.38	2.90	99.91
Inclusion in monchiquite-limburgite breccia	Surkhob	1	43.27	0.18	3.74	3.29	6.33	0.28	10.95	14.15	0.14	0.13	0.10	–	17.35	99.91
Black clinopyroxenite																
Inclusion in picritic breccia	Tuvish	20	43.48	0.84	12.23	3.08	5.23	0.11	13.40	17.69	0.59	0.21	0.03	0.07	2.60	99.56
Inclusion in monchiquite breccia	Odzhuk	1	45.11	0.62	11.00	4.30	3.42	0.22	13.48	18.55	0.54	0.16	0.07	–	2.62	100.09

pipes. The dykes are 200 m, less commonly 1000 m long and 5 m thick. They strike mainly north-easterly.

Picrite porthyries and eruption breccia of the rocks are known from two explosion pipes, one of them being the largest (Tuvish) and the other (Khelmagz) of medium size among the diatremes of the South Gissar. They are composed of eruption and tuffisite breccias of picrite porphyries, and only a small portion of the Tuvish Pipe contains massive picrite porphyries, probably emplaced during the second step of intrusion. Picrite porphyries are composed of carbonate, serpentine, and talc, forming pseudomorphs after olivine phenocrysts (10–15%) set in a matrix of microlite, diopside-augite with subordinate pleonaste, titaniferous phlogopite, analcime, and serpentine-chlorite material. There are a few large (up to 3 mm) clinopyroxene phenocrysts. Abundant clastic material (50–90%) and highly hydrothermally altered matrix from tuffisite breccia.

Xenogenic material should be extracted prior to the explosion breccia analysis (Table 18), otherwise the results are absolutely misleading.

Apart from the fragments of enclosing rocks, ultrabasic xenoliths are common in explosion breccia of picrite porphyries. Spinel peridotite, and black and green clino-pyroxene were reported (Kutolin et al. 1973). Fresh peridotite inclusions have not been found; they are virtually apoperidotitic reticulate and granoblastic serpentinites composed of serpentine and carbonate. Primary minerals are red-brown spinel and partly clinopyroxene. Similar inclusions are very common among ultrabasic nodules in basaltoids from different regions.

Green clinopyroxenite consists of diopside with sporadic amounts of olivine, orthopyroxene, and secondary phlogopite. Black clinopyroxenites contain augite and green spinel amounting to 5–20%. Occasional grains of olivine and a small proportion of secondary phlogopite are found in some samples. Black pyroxenites in the picritic breccia of the Tuvish Pipe differ considerably from green pyroxenites – judging from this average chemical composition – in their low content of silica, calcium, magnesium, and chromium and very high content of titanium, aluminum, and iron.

Those who have studied ultrabasic inclusions from picrite pipes of the South Gissar classify spinel peridotites and green clinopyroxenites as mantle formations. The inclusions of black clinopyroxenites are interpreted in a different way. The fact that clinopyroxene and splinelides of black clinopyroxenites are very similar in chemical composition and physical properties to high pressure phenocrysts present in alkali basaltoids in many regions, as well as the experimental data available (Green and Hibberson 1970), provide an opportunity to classify the inclusions as accumulative (segregation) formations (Kutolin et al. 1973).

Most explosion pipes in the South Gissar are infilled with monchiquite-limburgite and, in particular, with breccia of the rock. Massive varieties and matrix of eruption breccia of the group display a porphyritic texture, while the groundmass is microlitic. Phenocrysts (7–15%) are olivine, as a rule, fully replaced by carbonate, serpentine, and occasionally by talc. Pyroxene (15–20%) occurs as microlites (diopside-augite) and xenocrysts (diopside). Biotite (5–7%) occurs as strongly opacitized and corroded xenocrysts (?). Plagioclase is represented by occasional andesite grains set in a matrix, and abundant, more acid, clasts. All the minerals are set in a brownish volcanic glass

(40–60%). Chemical analyses of the rocks discussed (see Table 18) confirm that they form part of a large group of alkali basaltoids.

The content of xenoliths in eruption breccia of monchiquite-limburgite averages 30%, but some rocks are actually crowded with fragments of enclosing granitoids (Kaloch and other pipes). Based on composition and genesis, clastic material is divided into several groups: country rock xenoliths, xenoliths of the granulite-facies rocks, and eclogites; fragments derived from abyssal ultrabasic rocks. In this context the most interesting are inclusions of peridotite and pyroxenite. There are a few occurrences of peridotite, composed of serpentine pseudomorphs after olivine (50–60%), pyroxene (20–30%), phlogopite (10–15%), and spinelide (up to 5%).

Green and black varieties of pyroxenite inclusions occur in monchiquite-limburgite breccia, as in the above picrite breccia. The inclusions differ slightly from those in the picrite-porphyrite breccia in mineralogy, petrology, and chemical composition. It is noteworthy that the presence of two groups of clinopyroxene inclusions, viz., green containing diopside and chrome-diopside and black containing highly aluminous, titanium-rich augite, is probably a common feature of both alkali basaltoids from the South Gissar and many other regions of the world. Therefore, aggregates of black clinopyroxenite may be classed as segregations.

Occasional inclusions of eclogite composition are found in explosion breccia from the Odzhuk Pipe. They are composed of green diopside and pinkish-yellow garnet with a 42–55% pyrope component. The reaction nature of garnet history suggests that inclusions were formed as a result of eclogitization (Baratov et al. 1970).

Analcime basalt and eruption breccia of the rocks are not very common in explosion pipes of the South Gissar. These are dense rocks with a small proportion (5–15%) of andesite, diopside-augite, opacitized biotite, and fully replaced olivine phenocrysts set in brown volcanic glass containing microlites of plagioclase and clinopyroxene. Altered glass products contain a notable amount of analcime, present also in amygdules. Petrochemically, the rocks are equivalent to analcime-orthoclase basalt; however, orthoclase has not been found in these rocks, it is only a normative mineral component.

Xenogenic material amounts to 40% in eruption breccia of analcime basalt. A peculiar feature is the presence of different varieties of enclosing granitoids and granulite-facies rocks and the absence of fragments of eclogite and plutonic ultramafic rocks. This suggests less depth for the melts which gave rise to the formation of analcime basalt as compared to that of picrite and monchiquite-limburgite emplacement.

Minusinsk District

In south-western Central Siberia (some 150 km south-west of the twon of Krasnodar) within the North Minusinsk basin, several explosion pipes were found. They occupy the slopes of the Kopiev Dome, which inherited the anticlinal structure of the Lower Paleozoic folded basement of the Minusinsk trough. The core of the dome exposes Lower and Middle Devonian volcano-sedimentary rocks building up basal horizons of a flat-lying sedimentary cover of the basin. Above come continental-lagoonal and marine Upper Devonian and Lower Carboniferous deposits, whose total thickness is 1500–2500 m.

Table 19. Chemical composition of the rocks (wt.%) of the Kongarovskaya explosion pipes. (Kryukov 1964)

Oxide	Black fine-clastic breccia of second generation	Brown coarse-clastic breccia of first generation	Spinel lherzolite (inclusion)	Basalt with inclusions of ultrabasic rocks	
				Small proportion	Large proportion
SiO_2	48.16	38.03	42.99	42.70	39.88
TiO_2	1.91	1.35	0.23	1.96	2.19
Al_2O_3	12.96	10.40	3.75	12.75	10.48
Fe_2O_3	5.88	9.97	1.53	4.65	5.18
FeO	1.72	2.02	8.43	7.95	8.04
MnO	0.16	0.16	0.18	0.19	0.37
MgO	4.52	10.04	34.28	12.68	15.72
CaO	7.63	7.88	5.50	9.49	6.90
Na_2O	1.39	1.30	0.31	2.87	1.64
K_2O	3.49	2.19	0.14	1.50	0.46
P_2O_5	0.43	0.64	0.11	0.75	0.78
Ignition losses	11.94	16.14	2.54	1.85	7.70
Total	99.47	100.12	99.99	99.34	99.34

All in all, six explosion pipes were discovered in the district: Tergeshskaya, Belevskaya, Kongarovskaya, Baradzhulskaya, Marskaya, and Krasnenskie Ozera pipes. They are rounded and elongate in outline; their lateral dimensions vary from 100 to 540 m. The study of pipe contacts and the pattern of magnetic anomalies suggest vertical and steep-dipping diatremes, narrowing like a funnel with depth. The long axes often follow the strike of the tectonic structures to which they are related.

The pipes are infilled with eruption breccia, locally represented by several generations; viz., three and two generations in the Tergeshskaya and Kongarovskaya Pipes, respectively. Breccias of several generations differ in matrix to clastic material ratio, in size and dominating composition of fragments. Breccias of all diatremes are composed chiefly of angular fragments derived from the Devonian and Carboniferous sedimentary rocks and subordinate Devonian effusives. There are a few fragments of metamorphosed Cambrian rocks, schists, and igneous rocks. Of interest is the fact that breccias of all diatremes contain rounded inclusions of ultrabasic and eclogite-like rocks: garnet (pyrope) lherzolite, spinel websterite, monomineralic olivine, pyroxene (augite and diopside), and phlogopite nodules, eclogite-like scapolite-bearing rocks.

The matrix consists of tiny fragments set in chlorite-serpentine-carbonate groundmass; iron hydroxides and silicification are also present. It is relict porphyritic fabric with relict microlitic groundmass. The presence of olivine and pyroxene relics in pseudomorphs after microporphyritic phenocrysts, along with chlorite-serpentine leucoxenized groundmass, suggest the limburgite composition of the rock (Kryukov 1962, 1964). Chemical analyses (with allowances made for errors introduced by xenolith impurities) support this assumption (Table 19).

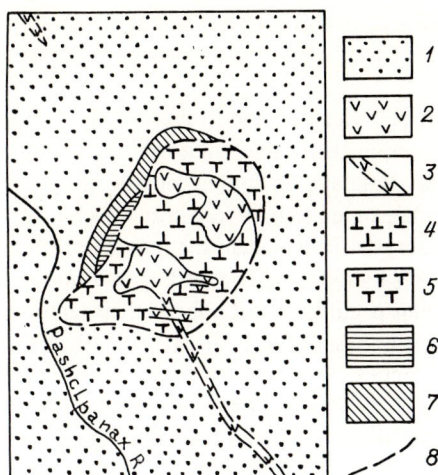

Fig. 37. Geological sketch map of the Kongarov-skaya Pipe. (Kryukov 1964). *1* Quaternary deposits; *2* basalt; *3* basalt dykes inferred from magnetic surveys; limburgite breccia: *4* fine-clastic breccia of the second generation; *5* coarse-clastic breccia of the first generation; *6* Carboniferous coal-bearing deposits; *7* Devonian redbeds; *8* outlines of the explosion pipe inferred from magnetic survey

All breccias are introduded by basalt necks and dykes continuing into the sedimentary rocks which enclose the pipes (Fig. 37). Basalts are black fine-grained fresh massive rocks. The extent of crystallization depends on the closeness to the contacts. The rocks are fully crystallized and often doleritic in the central parts of necks and dykes. Approaching the contacts, rocks become porphyritic with a microdoleritic matrix. Hyalobasalts are typical of the exocontact zones. Pyroxenite and peridotite inclusions are found among eruption breccias in basalts composing the necks, while they are absent from basalt dykes cutting sedimentary strata. Ultrabasic inclusions are presumed to have been entrained by magma during the eruption of breccia. Rock-forming minerals of the basalt in optical properties and composition are similar to those of the flood basalt formation; this is also true of the chemical composition of the rocks.

Mongolia

The Recent volcanism of Mongolia holds a unique position in the Cenozoic tectono-magmatic activity of Central Asia: it is comparable in extent, nature, and age with that of the East African province. Basalt extrusions in Mongolia reflect a peculiar Cenozoic phase in the geological history of the Central Asian province. The phase was marked by an overall uplift of the region and upwarping of the peneplened surface in Paleogene-Neogene time. The processes reached the climax in the Pliocene. Rifting began in different parts of the province in Pliocene-Pleistocene time. The basalts of Mongolia owe their emplacement to the Recent activation which interrupted the platform development of the region, which began at the Late Cretaceous-Early Paleogene.

Recent volcanic rocks of Mongolia are known from two major zones. The zone in Central Mongolia is sublongitudinal and runs from a rift basin of Lake Khubsugul (continuation of the Baikal rift) through the Khangai arch to the South Gobi desert. The other zone is a continuous basalt sheet covering the Deriganga plateau. Only inextensive basalt fields are known outside these zones. Of interest is that most basalt terranes were formed as the result of the activity of numerous rapidly extinguisting and migrating volcanic foci.

The Khangai volcanic area and the Dariganga plateau related to the Recent blocky arch are underlain by peculiar rocks whose equivalents are unknown from other regions of Cenozoic volcanism in Mongolia. Potassic, including leucitic, lavas of Khangai and sodium basalts of the Dariganga plateau are rich in deep-seated inclusions, the former also containing high pressure megacrysts. In Khangai, the most interesting and peculiar volcanic events took place in the Taryat basin. The outline and internal structure of the Taryat volcanic terrane owe their peculiarity not only to their tectonic history, but to a pre-volcanic topography of the basin. The basin is a sublatitudinal graben which forms part of the system of young structures of the Khangai uplift emplaced during the Late Neogene. The graben is 70–80 km long and 10–15 km wide (Kepezhinskas et al. 1975). The floor of the basin is the top of flat-lying lava sheets where the pre-Cenozoic basement is exposed. Six small volcanic edifices were found in the western part of the basin. The rivers flowing across the basin at places cut volcanic rocks and expose the pre-Cenozoic basement.

The subdivision of the Cenozoic volcanic rocks of the Taryat basin is as follows: (1) Pliocene lava composing relics of the highest terraces; (2) Pleistocene lava of 40–90-m terraces of the basin floor; (3) Holocene lava forming occasional flows at the level of flood-plain terraces, and volcanic edifices of the Pleistocene sheets.

A succession of the Pliocene lavas exhibits a twofold structure. Its lower two-thirds consists of thick (up to 20 m) potassium limburgite and augitite flows containing megacrysts of aluminum-rich augite, sanidine, and xenoliths of spinel lherzolite. A fine columnar (up to 10 cm in diameter) jointing is a peculiar feature. The third of the Pliocene succession is composed of several flows (vasicular at the top) of trachybasalt and trachytic andesite basalt, up to 3–6 m thick each, with thick columnar (50–70 cm) jointing.

Limburgite and augitite flows are intruded by explosion (?) and intrusive bodies. Isometric (in plan) bodies, up to 100 m in diameter, composed of orthoclase trachybasalt and trachyandesite dolerite may be tentatively classed as explosion ones. The actual identity of the rocks and lavas from the upper third of the Pliocene succession and their absence in the other parts of the Cenozoic section allow their interpretation as volcanic vents created during Pliocene extrusions. Intrusions are represented by small stocks (about 200 m in diameter), sills (up to 5–6 m thick and 300 m long), and dykes (1–3 m thick) of orthoclase trachybasalt, leucite tephrite, and leucite basanite, containing abundant deep-seated xenoliths. The intrusions were accompanied by deformation and brecciation of enclosing lava sheets, and the formation of eruption breccias forming lenticular beds up to 3 m thick at the base of the sills.

The Pleistocene lavas are the most common Cenozoic volcanites in the Taryat basin. They comprise an extensive complex, 70–90 m thick. All in all, there are ten distinct lava flows, each has a scoria crust and a massive center marked by thick columnar jointing. The lower and upper parts of the succession consist of potassic basanite and limburgite, respectively.

The Holocene lavas are located mainly in the western Taryat basin. Thin (3–5 m) sheets linking extrusion centers represented by six cones about 1000 m at the base,

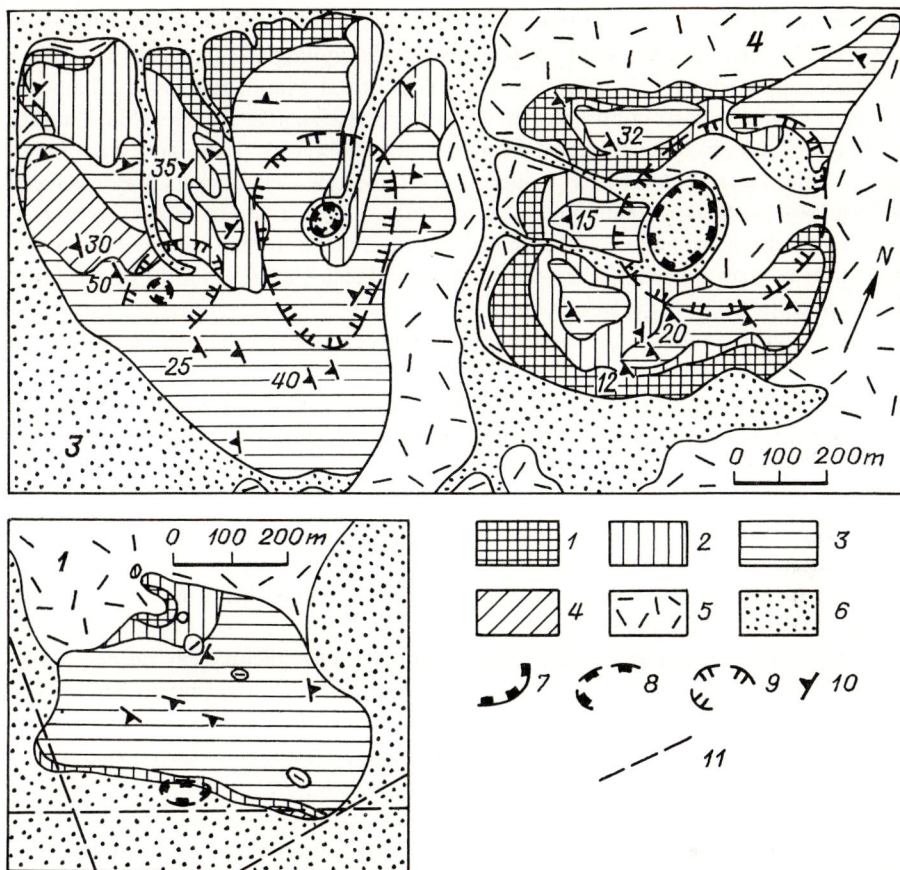

Fig. 38. Structure of volcanoes in the Shavaryn-Tsaram district. (Kaminsky 1980). Garnetiferous alkali basaltoids: *1* first step of the first eruption cycle (volcanic ash, scoria); *2* second step (ash); *3* first step of the second cycle (alteration of scoria and agglutinates); *4* second step (porous lavas, rare dense lavas); *5* Paleozoic sedimentary and igneous rocks; *6* Recent deluvial-proluvial deposits; volcanic vents: *7* established; *8* inferred; *9* craters; *10* attitude for volcanic rocks; *11* faults covered by drift

100–200 m high, with a crater diameter of 200–400 m. The best-preserved volcanic edifices have a slope of 45°. The cone is opened to the south toward the lava flow, which forms a sheet with an area of about 200 km^2. Apart from lava, the volcano has extruded a mixture of volcanic sand and ash with numerous lapilli and bombs up to 1 m in size. Leucite basanites and tephrites, potassium hawaiites, potassium limburgites, and augites occur among Holocene rocks. Of special interest is recently discovered eruption garnet basanite breccia. It occurs in the mountainous part of the Khangai Mountains and within the Terkhiin-Tsarannur Depression filled with Cenozoic alkali basalt sheets with locally preserved eruption edifices.

A group of vocanic edifices at the urochishche of Shavaryn Tsaram is the best known, and bears the same name. Edifice 1 (Fig. 38) is a relic of the upper part of a cone whose lower and crater parts are overlain by Pleistocene basalts; it measures 340–590 m. The volcanic rocks are stratified, beds dipping at an angle of $10°$ reflect the topography of the Paleozoic basement. The central part of the vent is composed of dense varieties of eruption breccia containing glassy shards. Rocks of the marginal parts are porous to scoria-like and resemble tuffisite breccia. There are fragments of breccia infilling the central part of the vent. Macroscopically, the dark grey matrix of the eruption breccia, is composed of fragments of megacrysts of garnet, pyroxene, sanidine, plagioclase, and of clasts of sedimentary, metamorphic, and igneous rocks. The latter contain bombs of glassy limburgite and deep-seated xenoliths of peridotite and pyroxenite up to 0.7 m in size.

Based on the minerals present, four volcanic rock groups can be recognized: basanite, alkali basalt, and leucite varieties of these rocks (Kaminsky 1980). All the minerals are identical or similar in composition. Olivine occurs as large (up to 2 mm), corroded (first generation), and small (about 0.01 mm) isometric (second generation) crystals with fusion rinds on the surface, containing 11–12% of the fayalite component. Titanium-augite forms megacrysts (up to 8 cm) and microlites set in the groundmass. Enstatine occurs as phenocrysts, one hundredth of a millimeter to 1 mm long. Sanadine is found in garnet basaltoids as megacrysts and rare phenocrysts. Their texture is marked by transitions to the triclinic system, resulting probably from crystallization at high (1000–2000 MPa) pressure.

Plagioclase forms laths in rocks of all types, while phenocrysts occur only in alkali basaltoids. Its composition is labradorite (An_{80-82}) in basanites to labradorite-bytownite (An_{70}), labradorite (An_{52-62}) in alkali basalts. Garnet forms mega- and phenocrysts. The former are 10 cm in diameter, they are irregular and orange to red. The pyrope component amounts to 52–56%, sometimes 62%. Next in abundance are almandine (19–31%), andradite (3–16%), and grossular (10–11%).

The K-Ar age on sanidine megacrysts in 2.0, 2.6, and 6.2 m.y. (Kaminsky 1980).

Volcanic edifices 3 and 4 are located 3.5 km north of edifice 1. Edifice 3 is 900 m in diameter; this is a two-crater volcano. The main (northern) crater measures 300–450 m and is clearly visible in the topography. Its floor, 70 m in diameter, is covered by Recent drift. The second crater (150–200 m in diameter), with a vent 30–40 m in diameter, was active only during the later phases of eruption and is recognized from the occurrence of lava flows. Edifice 4 is a one-crater volcano, 600–700 m in diameter; it occupies most of the edifice area. The craters of both volcanoes are surrounded by lava flows and well-stratified pyroclastics. The slope of the volcanic beds reaches $40°$. There were two cycles of the eruption, each beginning with the extrusion of pyroclastics, punctuated by effusion of gaseous lavas producing shikhlunite-type scoria basalt. The extrusion of lava rich in volatiles took place at the end of each cycle.

The rocks building up volcanic edifices 3 and 4 are almost identical to those of edifice 1. They differ mainly in the absence of large garnet megacrysts and in the few clinopyroxene megacrysts in rocks of edifices 3 and 4. Xenoliths represented exclusively by spinel lherzolite are very scarce and small in size.

The Cenozoic volcanic rocks of the Taryat basin are dominated by rocks of the alkali basalt group. There are a few more acid and less alkali differentiates. Petrochemically, the Cenozoic lavas form a highly saturated association of potassium-type alkali basalts (Kepezhinskas et al. 1975) (Table 20). Sodium basalts of the Dariganga plateau display low alkali with sodium content twice that of potassium.

Table 20. Chemical composition (wt.%) of Cenozoic volcanic rocks of the Taryat depression and Dariganga plateau. (Kepezhinskas et al. 1975; Filippov et al. 1976; Kaminsky 1980)

District, volcano	Rock	Number of analyses	SiO₂	TiO₂	Al₂O₃	Fe₂O₃	FeO	MnO	MgO	CaO	Na₂O	K₂O	P₂O₅	Ignition losses	S_total	Total
Taryat depression																
Narin-Gichegene V.	Leucite basanite, lava train	1	45.50	2.47	16.60	3.48	7.59	0.19	6.30	6.80	5.00	5.00	–	0.44	–	99.37
Khorog Volcano	Potassic limburgite, bomb	1	46.00	2.38	14.70	1.49	9.29	0.24	7.90	7.60	4.50	4.10	–	0.83	–	99.03
	Potassic basanite, flow	1	47.00	2.70	14.20	2.63	7.62	0.20	8.40	8.50	3.98	3.06	–	0.94	–	99.33
	Analcime-bearing potassic limburgite, flow	1	45.30	2.45	13.50	2.63	8.76	0.23	10.70	7.75	4.36	2.98	–	0.62	–	99.28
Shavaryn-Tsaram Volcano 1	Eruption breccia of potassic basaltoid															
	Breccia cement	1	51.25	2.07	17.20	2.75	7.25	0.13	5.74	5.54	3.73	4.12	–	0.34	–	100.12
	Glassy part of a matrix	1	45.60	2.65	17.66	3.84	7.12	0.15	4.44	6.93	5.85	4.56	–	0.78	–	99.58
	Leucite basanite	5	50.62	2.22	13.94	3.99	6.16	0.13	6.66	6.01	4.21	3.78	0.96	1.05	0.02	99.75
	Leucite limburgite	1	51.10	2.39	14.90	2.64	7.16	0.15	7.57	5.04	4.52	3.91	0.98	0.54	0.02	100.92
	Leucite alkali basalt	6	49.26	2.38	14.99	3.34	6.82	0.23	5.94	5.87	5.20	4.37	1.29	0.50	0.03	100.22
	Alkali basalt	8	51.15	2.39	14.75	3.48	6.70	0.15	6.53	5.38	4.42	3.78	0.93	0.79	0.02	100.47
Volcanoes 3 and 4	Leucite basanite	3	48.08	2.87	15.13	4.80	6.47	0.15	5.35	5.39	5.02	4.94	1.38	0.54	0.02	100.14
	Leucite limburgite	4	46.38	2.55	14.38	8.69	2.79	0.16	5.35	7.15	4.61	4.28	1.40	2.42	0.03	100.19
	Limburgite	3	47.08	2.35	14.57	10.45	0.65	0.15	5.88	6.79	4.84	4.12	1.32	1.48	0.02	99.70
	Leucite alkali basalt	4	47.21	2.57	14.97	7.61	3.71	0.16	6.10	6.48	4.73	4.24	1.36	1.04	0.03	100.21
	Alkali basalt	1	48.20	1.72	13.80	5.42	5.42	0.15	10.70	7.40	3.40	2.80	0.64	0.92	0.02	100.59
Dariganga plateau	Sodium alkali basaltoids	77	47.11	2.65	12.63	4.02	8.15	0.22	8.61	9.53	3.74	1.79	–	0.89	–	99.34

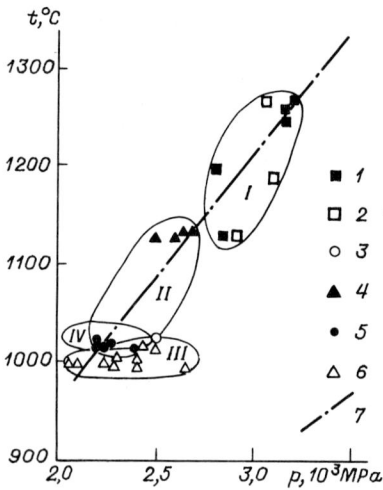

Fig. 39. PT-parameters responsible for the formation of deep-seated inclusions from basaltoids of Mongolia. (Kaminsky et al. 1979). *1* Garnetiferous lherzolite-harzburgite; *2* garnetiferous olivine websterite; *3* ilmenite lherzolite; *4–6* spinel ultrabasites; *1–3,5* Shavaryn-Tsaram; *4* Khangai Highland; *6* Dariganga Plateau; *7* inferred geotherm for the Khangai Highland. Fields of: *I* garnetiferous ultrabasites; *II, III* spinel ultrabasites; *IV* eclogitic rocks; *I, II, IV* Khangai Highland; *III* Dariganga Plateau

Large solitary crystals (megacrysts) of ferro-magnesium minerals and feldspar, along with xenoliths of deep-seated ultrabasic rocks, are present in the alkali-basalt series of Mongolia, as in similar volcanic rocks in other regions of the world. Megacrysts and deep-seated xenoliths are most diverse in potassium basalts. The megacrysts are clinopyroxene, garnet, phlogopite, sanidine, and xenoliths are spinel lherzolite with subordinate websterite, wehrlite, augitite, diopsiditite, aclogite, and eclogite-like rocks. In sodium basalts, megacrysts are clinopyroxene, olivine, sodium sanidine, while xenoliths are spinel lherzolite, websterite, wehrlite, augitite, and harzburgite.

It is noteworthy that megacrysts and xenoliths of ultrabasic rocks are not the same with respect to the number and quality even within a single petrochemical group and vary in eruption products derived from different volcanic centers. For example, all the diverse megacrysts (and xenoliths) characteristic of potassium basalts occur in eruption breccia of the Shavaryn Tsaram group, while ejectamenta from the Khorog Volcano contain only megacrysts of augitite, sanidine, and phlogopite. Rocks of the eruption center Salkhityn-ula yield most diverse megacrysts and xenoliths from sodium basalts of the Dariganga plateau.

The composition of paragenetic mineral associations from xenoliths studied using the thermobarometer technique shows the physical and chemical conditions under which these xenoliths were formed (Fig. 39). Filippov et al. (1976) estimated a pressure of 1500–1700 MPa and temperature of 900–1100°C for megacrysts from basanite of volcano 1 of the Shavaryn Tsaram group. Agafonov et al. (1975) obtained similar values of 1700 MPa and 940°C for the crystallization of garnet from the same rocks. Kaminsky (1980), using Hazen's diagram, concluded that sanidine megacrysts in basalts of volcano 1 formed at a pressure of 1600 MPa. Xenoliths of spinel ultrabasites from basalts of the Dariganga plateau display a wider range with respect to pressure and smaller temperature range for crystallization parameters: 2050–2650 MPa and 1000–1020°C (Kaminsky et al. 1979). The data obtained suggest that a depth at which the alkali basalt magma generated in the Cenozoic in Mongolia was no greater than 100 km.

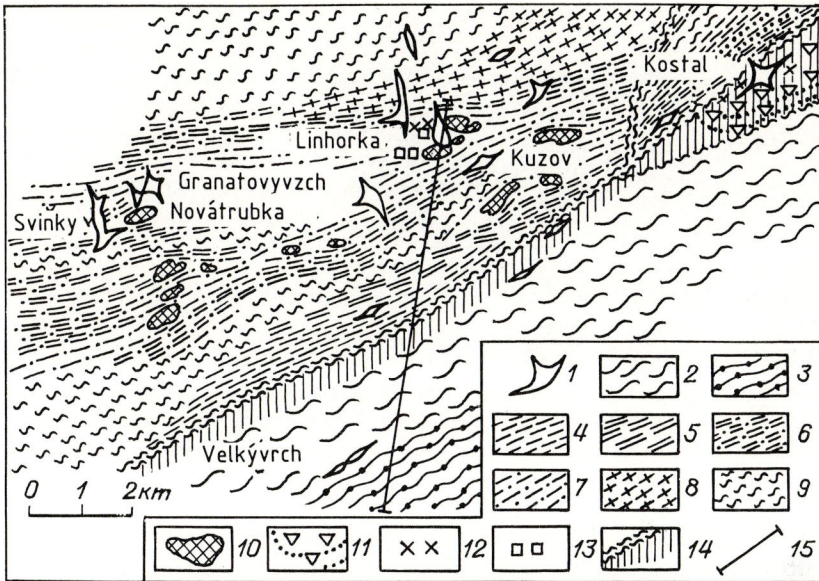

Fig. 40. Structure and composition of the crystalline basement of the pyrope-bearing diatremes of the České Středohoři Mountains. (Kopecky and Sattran 1966). *1* Paleogene-Neogene diatremes with xenoliths derived from basement (shape and size of diatremes at the level of the basement surface); *2* phyllite complex; *3* garnetiferous micaceous shale and gneiss; *4* muscovite and binary shale and gneiss; *5* paragneiss; *6* garnet-biotite gneiss containing kyanite; *7* granulite; *8* muscovite and binary orthogneiss; *9* migmatite; *10* pyrope peridotite; *11* nepheline syenite with contact zones of alkali metasomatites; *12* alkali hornblendite and pyroxenite; *13* diorite and gabbro-diorite; *14* fault zones; *15* geological section (see Fig. 41)

České Středohoři Mountains

In north-western Bohemia, the mining of the famous Bohemian garnet (pyrope) goes back to the 18th century. Even then pyrope was mined not only from loose Quaternary deposits but from garnet-bearing igneous rocks, i.e., garnet peridotite and vent breccia. Several tens of cone-shaped orebodies composed of massive and breccioid varieties of rocks ranging in composition from trachyandesite to limburgite and nephelinite are known from the pyrope-bearing diatremes.

The geology of the area is fairly complex, although only Cretaceous and Paleogene-Neogene sediments intruded by subvolcanic formations (Kopecky et al. 1967) are exposed. The area is divided into almost equal parts by the north-east-striking Litomerian fault zone (Figs. 40, 41). The submergence of the south-eastern block along the major fault took place in the Proterozoic, Late Paleozoic, Mesozoic, and Cenozoic.

The crystalline basement of the southern part and that of the northern part differ in tectonic structure and metamorphic grade of the Proterozoic rocks. The rocks of the northern (elevated) block exhibit a higher metamorphic grade than those of the southern block separated from the northern one by the milonitization zone, coincident with the Litomerian fault. The northern block is composed of granulite, gneiss,

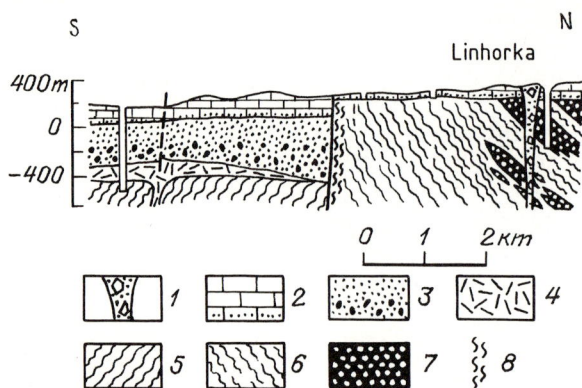

Fig. 41. Sublongitudinal geological section through the pyrope-bearing diatremes of the České Středohoři mountains. (Kopecky and Sattran 1966). *1* Pyrope-bearing vent breccia; *2* Upper Cretaceous deposits with pyrope-bearing sandstone at the base; *3* Upper Carboniferous sandstone with pyrope-bearing basal horizon; *4* quartz porphyry; *5* phyllite, micaceous shale, gneiss; *6* granulite, migmatite, catagneiss; *7* garnetiferous peridotite; *8* mylonite

micaceous shale, pyrope peridotite, and migmatite. Pyrope peridotite is heavily serpentinized. It forms small lenticular bodies among granulites and garnet gneisses.

The rocks of the southern block are dominated by garnetiferous micaceous shale and phyllite which, approaching the northern block, give way to carbonate phyllites and metabasites. In the southern block comparatively weakly metamorphosed rocks are overlain by Upper Carboniferous sedimentary and volcanic rocks, absent from the area north of the Litomerian fault. The fault forms the boundary of the Permo-Carboniferous sedimentary basin of Central Bohemia. In the north, clastogenic sediments are underlain by quartz porphyries and tuff. Terrigenous rocks are relatively rich in pyrope, almandine, and other heavy minerals released by the erosion of pyrope peridotites and garnetiferous crystalline rocks. Pyropes occurring in vent breccia south of the Litomerian fault are believed to have been derived from the sedimentary rocks (Kopecky et al. 1967).

The Mesozoic is represented exclusively by the Upper Cretaceous platform deposits common throughout the area. The rocks are complicated by faults, the persistent Litomerian fault being the major one. The Cretaceous is represented by the Cenomanian calcareous sandstones and conglomerates, Lower Touronian marly sandstones, Middle and Upper Touronian clay limestones, and Coniacian claystones. Terrigenous varieties of the Upper Cretaceous rocks contain pyrope and might well supply this mineral for the vent breccia of some diatremes.

The Paleogene-Neogene is represented by the Lower Miocene sedimentary and volcanic rocks. Sediments of this age are known only from the north-western part of the district; they are sands, clays, diatomites, and tuffites. Some tubular bodies and dykes are overlain by Lower Cretaceous deposits, others intrude them, suggesting two phases of volcanism, viz., Early Miocene and probably Late Pre-Miocene or Pliocene.

The number of Paleogene-Neogene eruptive cone-shaped bodies, lenticular in plan, and dykes over the pyrope-bearing diatremes of the České Středohoři mountains totals a few tens. Most of them occur north of the Litomerian fault. The diatremes of the

Sviňky

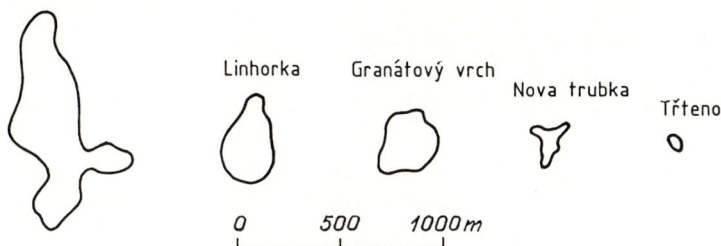

Fig. 42. Shape in plan of some diatremes of the České Středohoři mountains. (Kopecky et al. 1967)

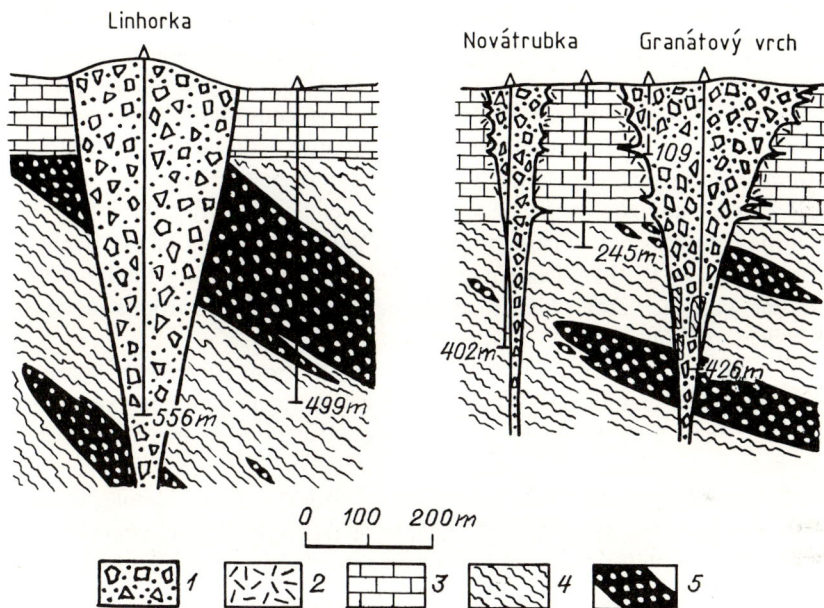

Fig. 43. Geological sections through pyrope-bearing diatremes of the České Středohoři mountains. (Kopecky et al. (1967). *1* Pyrope-bearing volcanic breccia; *2* contact shatter zones; *3* Upper Cretaceous sedimentary rock; *4* granulite-facies metamorphic rock; *5* serpentinized pyrope peridotite

České Středohoři mountains are similar in shape and size to kimberlite pipes (Fig. 42), they also become narrow with depth. It is noteworthy that the decrease in cross-sectional area with depth is different in various diatremes (Fig. 43). Country rocks are shattered at the contacts with diatremes. The width of the shatter zone varies from several meters to 50 m. Fine clastic rocks of eruption breccia are present in country rocks. For example, the Cretaceous shattered claystone contains small fragments of granulites of a vent breccia some 20 km from the southern contact of the Granátový vrch Pipe.

The exocontact zones of some diatremes also contain sill-like extrusions similar in composition to vent breccia. Apophyses extend for many tens of meters from the diatreme contacts. One hundred meters north-east of the Nová trubka deposit, the borehole penetrating fine-grained Cenomanian sandstones encountered (at a depth of 196.2–197.0 m) a fine-grained breccia of the same composition as that in the pipe funnel. The other sill-like extrusion breccia is found at a depth of 14.1–15.8 m, some 60 m south-west of the Granátový vrch body. A breccia sill also occurs along 80 m of the Šibenice Pipe.

Most diatremes are composed of vent breccia, which is a mixture of fragments derived from igneous rocks and rocks intruded by diatremes. Fragments of enclosing rocks account for 90% of the breccia volume. It may also be a combination of fragments from ancient vent breccia and piercing younger igneous rocks forming one or several hypabyssal bodies (commonly it is an elliptic or round stock or dyke). Scoria rocks are considered as intermediate between breccia and massive magmatic rocks. There are a few diatremes filled exclusively with massive igneous rocks. Boreholes often penetrate narrow rims of vent breccia at boundaries of such bodies. We may say that only dykes which, unlike bodies containing vent breccia, give rise to contact metamorphism of enclosing rocks, are filled with massive igneous rocks.

The composition and petrology of rocks filling eruptive bodies within the district discussed form a fairly mixed pattern. The workers who have studied the district classify augitite and limburgite as the glassy facies of compositionally different basalt rocks, with forms transitional to leucite, nephelinite, sodalite tephrite, sodalite nephelinite basaltic rocks, and the like. Diatremes filled with compositionally similar rocks are often in line with each other. Similar regular features are known from structures of kimberlite fields (see Chap. 9).

Chemical analyses were carried out mainly on samples from altered vent breccias (Table 21). Despite the fact that only hand specimens containing no large xenoliths (Kopecky et al. 1967) were analyzed, the presence of small fragments from rocks intruded by diatremes undoubtedly makes results less accurate. From three analyses of massive (?) igneous rocks, two were performed on strongly altered samples whose petrology has not been specified. A single sample of relatively slightly altered rock from the Sviňki Pipe was determined as picritic basalt.

Northern Tanzania

From the large number of volcanoes of Northern Tanzania we shall discuss two, Oldoinyo Lengai and Lashaine; these are the best known and most informative for the topic in question.

Oldoinyo Lengai is the youngest, still active volcano in the Neogene volcanic province of Northern Tanzania. It is situated in the Gregory Rift valley, 16 km south of Lake Natron (Fig. 44). The absolute and relative heights of the volcano are 2900 m and 1950 m, respectively. The summit area of the mountain is occupied by two craters. The older southern crater is inactive and is now a shallow oval depression filled with grey ash and scattered volcanic bombs. The active northern crater is roughly elliptical and measures some 500 by 640 m. The precipitous northern, eastern, and western walls of the northern crater are 130 m in height but the 240-m-high southern wall is broken by two terraces, the lower one being 60 m above the crater floor. The

Table 21. Chemical composition (wt.%) of pyrope-bearing vent breccias of the České Středohoři mountains. (Kopecky et al. 1967)

Oxide	Vent breccias								Picrite basalt
	Linhorka			Nová trubka		Granátový vrch			Sviňky
	58.0– 89.5 m (3 analyses)	397– 425 m (2)	515– 550 m (2)	70– 86 m (2)	116.8– 159.0 m (2)	90.1– 100.0 m (2)	114.0– 173.7 m (4)	245.6– 281.0 m (3)	14.3 m
SiO_2	46.76	45.31	38.68	36.19	35.12	44.58	37.76	39.13	43.39
TiO_2	1.78	1.72	1.40	2.10	1.46	1.58	1.30	1.30	2.53
Al_2O_3	11.48	9.78	9.66	10.38	7.68	9.71	8.71	8.38	7.35
Cr_2O_3	0.01	–	–	–	0.16	0.04	0.04	0.04	–
Fe_2O_3	4.12	2.28	1.63	2.87	4.03	3.10	2.85	3.24	3.26
FeO	4.28	4.70	4.50	4.03	4.05	4.10	4.58	4.84	5.32
MnO	0.13	0.15	0.13	0.06	0.10	0.06	0.07	0.08	0.16
MgO	6.97	10.74	6.03	9.06	14.30	5.46	13.80	9.22	13.79
CaO	11.14	9.36	11.90	12.44	6.40	6.48	7.60	10.32	15.27
Li_2O	0.005	–	–	–	0.01	–	–	–	0.01
Na_2O	3.25	1.70	1.58	0.42	0.49	0.55	0.62	0.63	0.71
K_2O	2.11	1.60	1.59	1.57	1.54	1.58	1.79	1.27	1.08
P_2O_5	0.34	0.28	0.35	0.34	0.28	0.38	0.32	0.39	0.29
CO_2 H_2O^+	6.03	11.47	21.74	19.88	18.39	17.14	15.94	14.90	5.71
H_2O^-	1.22	–	–	–	4.85	4.44	4.12	6.07	0.54
Total	99.625	99.09	99.19	99.34	98.85	99.20	99.50	99.81	99.41

main vent is situated eccentrically in the eastern half of the crater floor, and is surrounded by an area 90 m in diameter consisting of scoria cones, recent ejectamenta, and minor lava flows. In addition to the main volcano, there are smaller cones and explosion craters on almost all the flanks of the main cone.

Detailed study of the structure and stratigraphy of rocks of the volcano has revealed the main trends in the evolution of its volcanism. The first phase, during which the yellow pyroclasts and associated lavas were ejected, was the most active phase. During this period no less than 20 km^3 of lava and pyroclasts were ejected. Lava flows radiated out from the older southern crater, which must have been a vent of considerable size. The major volcanic event is related to this vent when the eruptions of yellow ijolitic tuffs and agglomerates interbedded with lava took place.

The tuffs are composed of crystalline nepheline and pyroxene set in fine-grained groundmass of zeolite, limonite, and carbonate. Agglomerates resemble tuffs, but the former contain large fragments and blocks of rocks varying in lateral and vertical distribution. Fragments are commonly round and no more than 30 cm in diameter. Larger fragments are represented by lava interbedded with pyroclasts. There are phonolites, nephelinites, ijolites, urtites, melteigites, jacupirangites, biotitic pyroxenites, and pyroxene-feldspar rocks (fenite) among the fragments. Lava flows contain nephelinites, feldspar nephelinites, and phonolites.

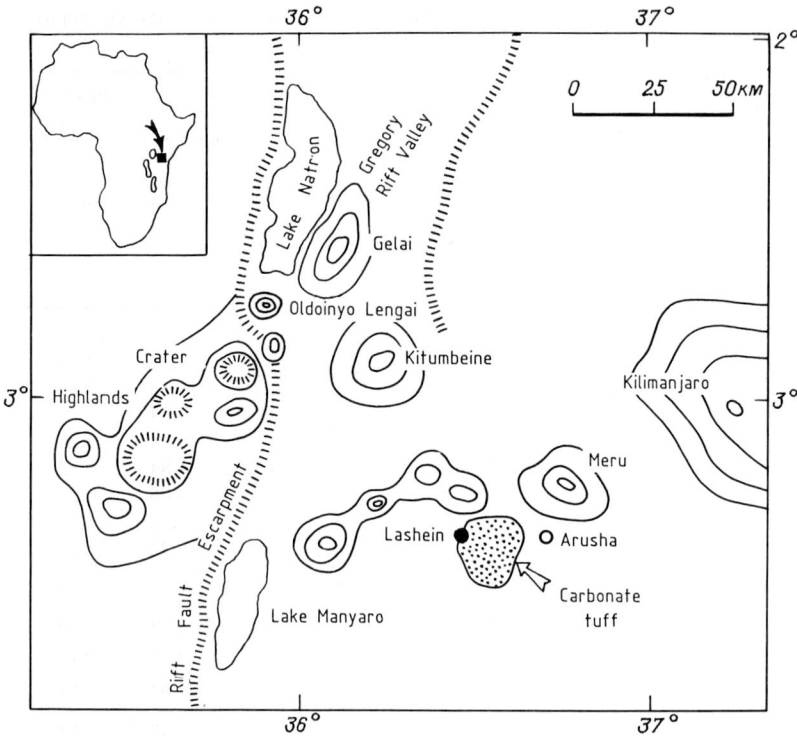

Fig. 44. Locality map of the Oldoinyo Lengai and Lashaine Volcanoes. (Dawson et al. 1970)

After the major phase, strong erosion took place, followed by a minor eruptive phase during which subsidiary cones and craters were formed on the south-western, southern, eastern, and south-eastern flanks of the volcano at an evelation of 1200–1800 m. Grey litho- and crystallo-clastic tuffs and agglomerates were ejected during this phase. Lithoclastic tuffs consist of small nephelinite lava lapilli set in white carbonate matrix with numerous mica laths 2.5 cm long. Crystallo-clastic material fills thin beds of tuff in which crystal fragments of mica, pyroxene, nepheline, and olivine dominate nephelinite lapilli.

The eruptions of the third phase took place from a newly created northern vent, while the old southern crater was gradually filled with material ejected during the later activity of the volcano. Like yellow pyroclasts of the first phase, black tuffs and agglomerates of the rhird phase cover not only the flanks but extensive sites around the volcano, where they rest on a strongly eroded surface of yellow pyroclasts. The black tuffs consist of crystals of nepheline, pyroxene, and mica, intermingled with lapilli of nephelinite, ijolite, and fenite. The material is cemented by carbonate. The agglomerates, whose groundmass is similar to tuff, contain blocks of nephelinite, phonolite, urtite, ijolite, melteigite, jacupirangite, biotite pyroxenite, and fenite, with the addition of wollastonite and carbonatite, which were not found in the older yellow pyroclasts. Some of the tuff horizons are very rich in mica, the latter being locally associated with numerous blocks of fenite. Some tuff and breccia horizons of the third phase, like yellow pyroclasts of the first phase, are coated with encrustations of soda.

The fourth phase is marked by the extrusion of melanephelinite lava. The lava rests on the black pyroclasts of the third phase and is overlain in turn by young loose black ash. Melanephelinite lava is found in several sites. The thickness of lava flows does not exceed 3.6 m.

The fifth phase was witnessed and repeated ejection of grey ash, which fills the southern, inactive crater. Reck (1914), who visited Oldoinyo Lengai in 1913, was the first to observe the pyroclasts. The ashes contain ejected blocks of biotite pyroxenite, biotite ijolite, ijolite, nephelinite, fenite, and blocks of soft crumbly carbonate-rich alkali rock.

The sixth phase is characterized by explosions which threw up semi-cemented grey tuff from the active crater during the 1917 eruption (Hobley 1918). It consists of lapilli of nephelinite and laths of mica set in a carbonate matrix.

The seventh phase resulted in the extrusion of black ash consisting of small lapilli of nephelinite and subordinate large (up to 3.5 cm) plates of biotite. The ash is widespread (it occurs for a distance of many tens of kilometers) and found north, west, and east of the volcano. The activity of the 7th phase lasted from late June 1940 to January 1941. Richard (1942), who studied the eruption, recognized three stages: a preliminary stage characterized by occasional small explosions which threw up old material from the funnel, which lasted from 24 to 30 July, 1940. The second step, the most active, started on about 31 July and continued for 1 or 2 weeks. There were powerful explosions which threw up bolders and bombs, together with sand and ashes, and released great quantities of gas. Ejecta consisted mainly of large quantitites of fine ashes during the third stage, which lasted from mid-August 1940 to January 1941.

The eighth phase of the activity spans the period from 1954–1955, and was on a much smaller scale. Only two small cinder cones were created on the crater floor. In the course of the eruption, alkali tuff was thrown up into the air to a height of about 10–12 m (Guest 1956).

Between January 1955 and the beginning of 1960, no systematic observations were carried out, aerial photographs of the volcano being taken only in February 1958. They convincingly show the continuation of activity. The aerial photographs taken on March 15, 1960, showed a considerable alteration of the cone and the crater floor since 1958. The main vent had shifted further toward the eastern wall of the crater and numerous lava flows had spread over the southern, western, and northern parts of the crater floor. Although the main activity appeared to have been from the main vent, lava had flowed also from a smaller parasitic vent farther onto the southwestern margin of the cone area.

In 3.5 months, one main vent had widened out and was now surrounded by at least nine parasitic spatter-cones, all of which emitted vapor streams and were covered by fresh black ash. The observations of September 18 revealed that there had been little change in the crater since June 26. Small amounts of black ash were being ejected from one of the parasitic cones and gas was discharged from the main vent at intervals of 20–30 s (Dawson 1962b).

However, on 23 September 1960, the activity increased considerably. The vent was filled with black lava which flowed continuously. The lava was constantly bubling

Table 22. Chemical composition (wt.%) of silicate and carbonatite lavas of the Oldoinyo-Lengai Volcano. (Dawson 1962b)

Oxide	Phonolite (3 analyses)	Nephelinite-phonolite (2)	Nephelinite (2)	Melanephelinite (2)	Carbonatite lava pahoehoe	aa
SiO_2	51.97	47.73	41.05	40.53	tr.	tr.
TiO_2	0.99	0.81	2.08	3.34	0.10	0.08
Al_2O_3	19.67	18.96	11.84	12.98	0.08	0.09
Fe_2O_3	4.13	4.18	6.52	4.82	0.26	0.32
FeO	1.48	2.46	3.35	4.44		
MnO	0.26	0.19	0.34	0.32	0.04	0.24
CaO	3.22	4.96	14.14	16.88	12.74	12.82
BaO	–	–	–	–	0.95	1.05
SrO	–	–	–	–	1.24	1.20
MgO	0.36	0.54	1.46	5.87	0.49	0.41
Na_2O	9.58	10.98	11.04	5.87	29.53	29.70
K_2O	5.25	4.40	4.36	2.34	7.58	6.58
P_2O_5	0.21	0.18	0.82	1.66	0.83	1.06
H_2O^+	1.91	1.98	0.25	0.42	8.59	8.27
H_2O^-	0.78	1.45	0.88	0.13		
CO_2	0.22	1.01	1.52	–	31.75	32.40
F	–	–	0.36	–	2.69	1.84
Cl	–	–	–	–	3.86	2.64
SO_3	–	–	–	–	2.00	2.18
S	–	–	0.14	–	–	–
Equiv. O for Fe, Cl, S	–	–	–0.27	–	–2.00	–1.36
Total	100.03	99.83	99.88	99.59	100.73	99.52

and explosions occurred at intervals of some 10–12 s. Masses of lava were hurled to a great height into the air. Between 23 September and 8 October repeated extrusions of ropy (pahoehoe type) and aa lava were observed. The lavas consist of small crystals set in a microcrystalline matrix. Optically, crystals have a small (approximately 15°) negative 2 V, moderate to high birefrigence (third-order greens and reds); a minimal index of refraction is 1.527. The pattern is of no known mineral. The microcrystalline groundmass is composed of highly birefringent sodium hydrocarbonate. The chemical composition of pahoehoe lava and that of aa lava is very similar (Table 22).

The occurrence of soda-rich carbonate lava among recent products of eruption of the nephelinite-ijolite volcano (judging from fragments extruded from depth) under-lain by a carbonatitic complex contributed greatly to the understanding of the genesis of carbonatites and associated alkali rocks. There are two major groups of hypotheses. Those of the first group assume a basic or ultrabasic parent magma which differentiates to carbonatite; it was interpreted as pyroxenite (Davies 1952), peridotite (Strauss and Truter 1951), kimberlite (Saether 1957), and alkali peridotite nephelinite magma (King and Sutherland 1960).

Hypotheses of the second group relate the genesis of carbonatites to primary carbonatite magmas or to cabon dioxide of unknown origin. Some workers postulated a carbonate liquid which was predominantly potassic containing subordinate amounts of Ca, Fe, Mg, Al, Ti, P, F, and H_2O. The reaction between this liquid and granitic country rocks gave rise to the ultrabasic alkaline plutonic suite with residual calcium and magnesium carbonatites (von Eckermann 1948). For the ultrabasic potassic rocks of Uganda the reaction of a calcium-magnesium-iron carbonatite magma with granitic country rock was suggested (Holmes 1950). Many researchers adhere to the concept of magmatic genesis of carbonatites.

The Oldoinyo Lengai lavas are the first recorded occurrence of magmatic alkali carbonate. In all previous cases the residual carbonatite has been calcitic or dolomitic, and this factor has undoubtedly influenced workers who seek to derive the carbonatite from ultrabasic silicate melts (parents). The presence of alkali carbonate on Oldoinyo Lengai supports von Eckermann's contention that alkali carbonate was instrumental in the formation of the alkaline rocks in carbonatite complexes, although the carbonate magma proposed by von Eckermann are dominantly potassic.

Dawson (1962a) ascribes the emplacement of Oldoinyo Lengai to a sequence of carbonatite magma extrusions. In this case each successive batch of magma had contaminated a smaller proportion of sialic material than the previous one. Alkalic ashes and newly formed magma are considered as products which compositionally resemble the primary magma.

Lashaine is a group of small volcanoes represented by tuffaceous cones and tuff rings on the plains to the west and south-west of the town of Arusha (Fig. 44). Their youthful morphology shows that they are the products of some of the most recent activity in the volcanic province. These small cones and tuff rings are unusual in that they are composed to a greater or lesser extent of ejected carbonatite tuff (Dawson 1964). The most interesting of these small volcanoes is Lashaine itself, a small cratered tuff-cone that stands prominently 150 m above the flat plain 32 km west of Arusha. The pyroclastic ring surrounding the crater is strongly asymmetric due to the greatest deposition of ejectementa west of the crater. It is oval in plan, with dimensions 760 m by 1050 m, the maximal dimensions being along the east-west axis.

The tuff cone was built in two stages, the first stage being an eruption of glassy, olivine-phyric scoria, small outcrops of which are seen in the lower parts of the crater wall. Following this was an eruption of carbonatite ash within which is embedded a suite of xenoliths together with mineral grains derived from the fragmentation of the xenoliths and fragments of the earlier scoria.

The lava consists of abundant fresh olivine phenocrysts and rare clinopyroxene phenocrysts in a microcrystalline to glassy vesicular matrix. The chemical analysis of the lava suggests that the rock is an alkali ultrabasic type similar in many respects to slightly carbonatized kimberlites, pircite porphyries, and other porphyritic alkali ultrabasic rocks (Table 23).

The xenoliths consist of angular blocks of country rock basalts, and rounded or discoidal blocks of calc-silicate granulites, rare anorthosite, and a variety of ultramafic rocks: garnet lherzolite, lherzolite, wehrlite, harzburgite, mica dunite, pyroxenite, and pyroxenite containing small amounts of olivine, amphibole, and mica. It is noteworthy that all the granulite and most ultramafic blocks are enclosed directly by carbonatite tuff; some ultramafic blocks are coated with a thin layer of fine-grained

Table 23. Chemical composition (wt.%) of lava from Lashaine Volcano, some similar rocks and peridotite inclusions contained in the lava

Oxide	Lashaine Lava (Dawson et al. 1970)	Dutoitspan Alkali kimberlite (Williams 1932)	Maime-Kotui Picrite porphyry (Egorov 1969)	South Africa Olivine melilite (Taljaard 1936)	Lashaine (inclusions in lava). (Dawson et al. 1970)			
					Garnet lherzolite	Lherzolite	Wehrlite	Spinel harzburgite
SiO_2	39.44	40.56	38.73	37.13	44.37	40.47	40.53	42.41
TiO_2	2.37	0.86	2.08	3.61	0.08	0.22	0.19	0.04
Al_2O_3	5.84	4.67	5.42	10.58	2.44	1.96	2.54	1.85
Cr_2O_3	0.11	–	–	–	0.48	0.32	0.21	0.24
Fe_2O_3	9.21	6.25	6.95	2.00	0.85	1.33	0.79	1.43
FeO	4.59	4.46	6.04	10.43	6.42	7.02	7.16	5.38
MnO	0.17	0.02	0.17	0.26	0.09	0.09	0.10	0.10
NiO	–	–	–	–	0.35	0.35	0.18	
MgO	17.67	21.88	21.99	19.12	42.14	45.12	44.74	46.48
CaO	12.24	6.44	11.00	13.02	1.45	1.58	1.70	0.50
Na_2O	1.97	2.68	0.33	1.32	0.25	0.39	0.42	0.30
K_2O	0.99	1.49	0.96	0.51	0.08	0.10	0.13	0.06
P_2O_5	0.81	0.49	0.24	–	0.05	–	0.05	–
H_2O^+	0.80	5.65		0.83	0.39	0.23	0.31	0.30
H_2O^-	2.26	3.06	6.09	0.37	0.18	0.19	0.22	0.11
CO_2	1.41	1.30		0.47	0.25	0.26	0.27	0.24
Total	99.88	99.81	100.00	99.65	99.52	99.63	99.71	99.62

or glassy olivine-phyric lava that resembles the early scoria. In addition, a large block of this lava has been found that contains rounded xenoliths of peridotite.

Peridotite from the Lashaine lava and tuff does not differ greatly in petrography, mineralogy, and chemistry from those occurring as xenoliths in kimberlites (Dawson et al. 1970). The olivine is highly magnesian, containing 90% Fo; the clinopyroxene contains 89–90% En. The clinopyroxene is classed as diopside, containing a small proportion of hedenbergite (5–6%), jadeite (5–11%), and enstatite (6–14%). Apart from predominant pyrope (68%), grossular (16%), almandine (11%), uvarovite (4%), and andradite (about 1%) components are present in garnets.

CHAPTER 13

Trap Formation Diatremes

Cone-shaped orebodies of trap rock (i.e., flood basalt) formation are known from many regions of the world. Among the best known are the basalt pipes of the Siberian platform and adjacent intermontane troughs which have recently been intensively studied. Primarily involved are pipes to which magmatogenic or hydrothermal mineralization is related. First to be mentioned are volcanoes of the Angara-Ilim iron ore province.

Angara-Ilim Province

The Angara-Ilim iron ore province occupies the south-western Siberian platform. Several tens of volcanic pipes, incorporating most of the orebodies, and therefore the best known, are found here (Fig. 45).

A structural control of the diatreme distribution, as in the case of other volcanoes, remains debatable. Strakhov (1978), analyzing data on the crack-channel orientation determined from the projection of pipes of the Angara-Ilim province, concluded that most pipes in the eastern part of the province are located within ring structures from several tens of kilometers to 120 km in diameter. Emerging, these structures form in turn linear or arcuate clusters. However, there are not distinct ring structures in the western part of the province, therefore, the researchers have to ascribe the structural position of many diatremes to arcuate lines; some of them follow the strike of the Angara folds, while the geotectonic position of others is still vague. Among tens of ring and arcuate structures of the Angara-Ilim province only one ring (Ilim) is reflected in the magnetic field as chains of linear and isometric positive anomalies. The remaining structures are reflected in neither the magnetic nor gravity fields (Strakhov 1978).

Cone-shaped bodies with trap rock formation do not differ in shape and size from kimberlite diatremes. These are isometric and elongate in plan bodies several tens of meters to 2 km in diameter. Like kimberlite pipes, they exhibit a close relationship between cross-sectional shape and depth to the erosion surface (Fig. 46): at a depth of 100–650 m most bodies are isometric or ellipsoid at the present exposure surface, at a depth of 700–1500 m they are strongly elongate and only at great depths do pipes pass into dykes, for example, in the northern Siberian platform. The cross-sectional area of all pipes decreases with depth. Trap diatremes owe their shape not only to the ejection of material from a volcanic vent, but to subsidence and displacement of country rock blocks along fissures dipping inward (Fig. 47).

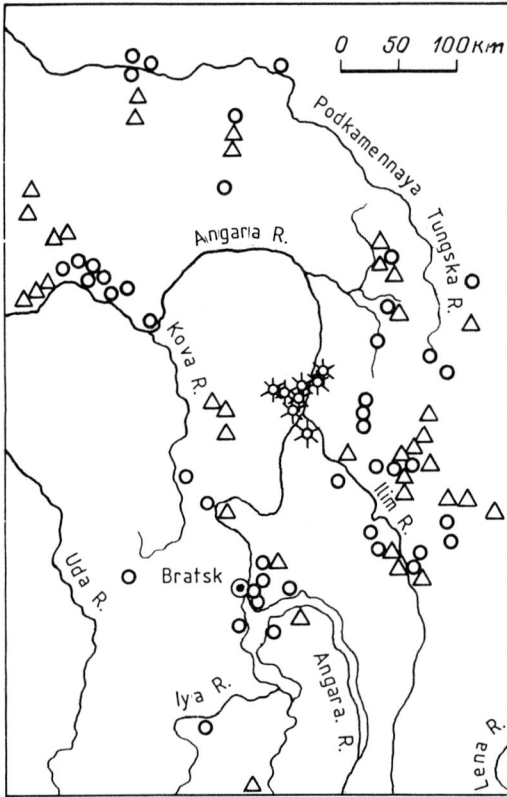

Fig. 45. Generalized map of the Angara-Ilim iron ore province. (Strakhov 1978). Magnetite deposits: *1* known pipe structure; *2* uncertain structure; *3* trap volcanoes

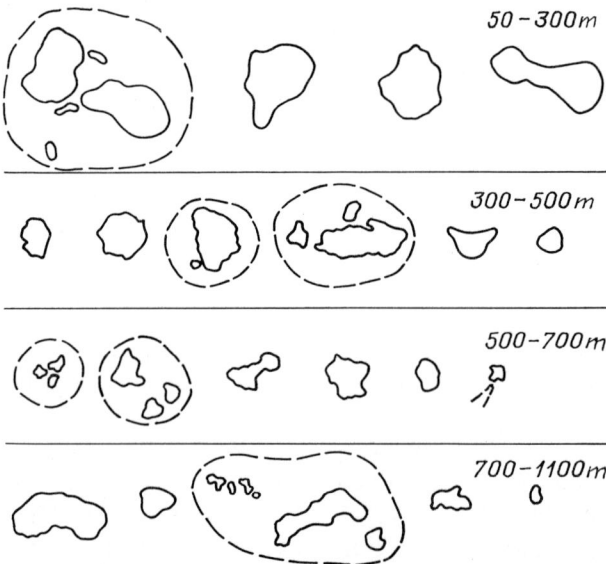

Fig. 46. Shape in plan of volcanic pipes in the Angara-Ilim province. (Strakhov 1978). Groups of bodies whose relative distribution and size are given to the same scale are *circled by dashed line*

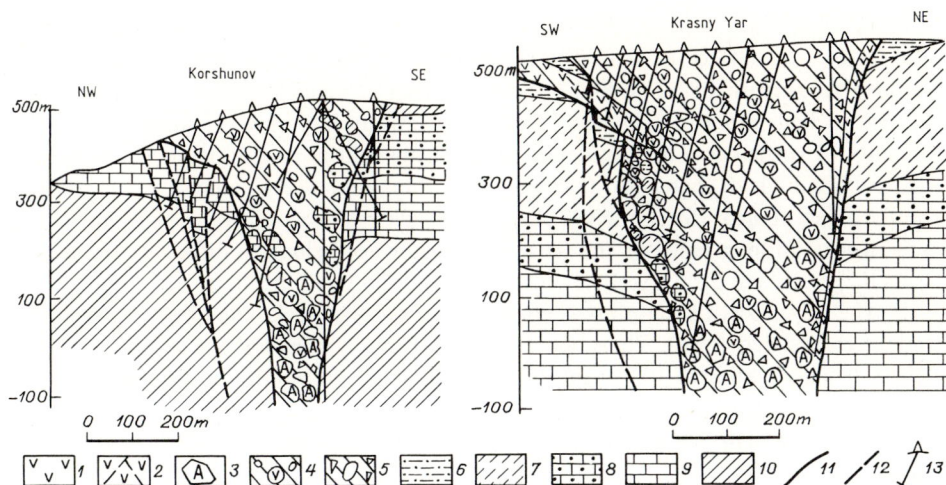

Fig. 47. Geological sections of magnetite deposits showing primary rocks in place of skarn ore bodies. (Strakhov 1978). Trap: *1* holocrystalline; *2* brecciated with calcite veinlets; *3* aphanite; *4* vent-facies tuff; *5* tuffisite breccia; deposits: *6* Silurian; *7* Bratsk Formation (O_{2-3}); *8* Middle Ordovician; *9* Lower Ordovician; *10* Upper Cambrian; *11* limits of volcanic pipe; *12* fractures; *13* boreholes

Some volcanic pipes form closely spaced clusters. It is noteworthy that when diatremes are clustered, the distance between them is neither larger nor smaller than when otherwise measured (see Fig. 46). Despite such a close spacing in plan pipes never join or intersect at depth. On the contrary, the upward widening leads to the convergence of the upper parts of most closely spaced pipes. For example, the Rudnaya Gora deposit has three closely spaced pipes; the largest of them (1800 by 560 m) splits into four separate channels (at a depth of 300 m) encountered by boreholes to a depth of 600 m from the surface (Roslyakov 1960). Two of the channels situated in the western part are lenticular (70 by 260 m and 20 by 100 m). The third and the fourth channels are oval and are situated in the eastern and north-eastern part, respectively (40 by 110 m and 200 by 250 m).

Enclosing sedimentary rocks are fissured in sites adjacent to pipes. Concentrically oblique and radial (relative to a diatreme or diatreme cluster) subvertical cracks are dominant. Subsidence of blocks of enclosing rocks often takes place along the largest concentric fissures; the latter locally contain traps and are marked by hydrothermal mineralization. Radiating joints are often related to trap dykes, magnetite, and hydrothermal mineralization. Radiating cracks open upward and increase in extent, so that ore veins filling the cracks are longer in the near-surface parts than those at deeper layers of deposits (Strakhov and Shiryaev 1975).

In the case of a clustered pattern of diatremes, the fissure zones around them emerge and, as the result, the entire interpipe space becomes dissected. Subsidence along concentric cracks has contributed to the dislocation of country rocks. As the result, depressions are formed around diatremes, the most extensive being next to larger pipes and their clusters. The extent of the depressions determined by the bed dipping toward diatremes and from the subsidence amplitude relative to a level of normal occurrence depends not only on the size of the vocanoes, but on the depth to the erosion surface as well. The comparison of pipe clusters almost equal in area eroded to a depth of 50 to 800 m shows a mean diameter decrease from 10—12 to 4—5 km. Beds dip at 3—10° and 12—13° respectively in peripheral zones and in the vicinity of volcanoes. The subsidence depth in the central part of depressions reaches several hundreds of meters, so that we can find here relatively young deposits eroded in the adjacent areas (Strakhov 1978).

Rocks infilling pipes of the Angara-Ilim province belong to different structural-genetic groups and form a vertical succession common to all the diatremes.

The lower part of the pipes is composed of peculiar rocks known as trap agglomerates (brecciated traps). Those are fragments from 1 m and above in size, most of them being varieties of traps: aphanite trap, dolerite porphyries set in a hyalopilitic and microlitic groundmass, holocrystalline dolerites, dolerite-pegmatites, and gabbro-dolerites. The matrix consists of strongly altered fragments of basalt glass, grains of plagioclase, and other minerals. Locally, fragments are cemented by glassy basic material. In marginal parts of pipes trap agglomerates give way to agglomerates composed mainly of blocks of sedimentary rocks. Fragments of sedimentary rocks incorporated into agglomerates are displaced by 100—200 m downward with respect to the corresponding stratigraphic layer at the margin. The farther a xenolith is located from the margin, the greater the subsidence.

The minimum depth of trap agglomerates from the present exposure surface is 800—900 m. At maximum depths drilled by boreholes (1500—2000 m together with restored eroded layers) only agglomerates were encountered. The agglomerates are believed to be upper brecciated parts of trap necks filling pipes at a depth below 1500—2000 m (Strakhov 1978).

Tuff and tuffisite breccia occur, starting from th deepest levels of diatremes dominated by trap agglomerates. A relative amount of tuffs and tuffisite breccias increases at higher levels, and at the 800—900-m level they commonly replace trap agglomerates completely. A depth of 300—400 m is chosen as the upper boundary of tuff and tuffisite breccia distribution.

Tuff occurs mainly in the inner zones of diatremes at a depth of 800—300 m. They increase in abundance up the section. Tuff consists of rounded and pancake-like agglomerates of dolerite porphyries and altered vesicular and amygdaloidal basalt glass or of a few angular fragments of sedimentary rocks (several millimeters to several centimeters in size) set in a matrix of fine and tiny particles of altered glass, dolerite porphyries, and sedimentary rocks. Unlike tuff, tuffisite breccia contains a noticeable amount of large fragments (1—1.5 m) of trap and sedimentary rocks.

Endocontact zones of diatremes contain agglomerates consisting of angular blocks of sedimentary rocks, from tens or even hundreds of meters in size. The space between the blocks is filled with fine clastic material of the same composition with a different proportion of rounded and angular agglomerates of dolerite porphyries and basalts. Fragments and blocks of sedimentary rocks in agglomerates are found below the original level of beds exposed in rocks filling the pipe walls. However, there are occasional fragments elevated by 100—200 m above their original level of occurrence.

Apart from these tuffs, undoubtedly belonging to the vent facies, the upper part of the diatremes eroded to a depth of less than 300—400 m contains basalt tuff and tuffstone, which on the complex data can be identified as similar to that of the Lower Triassic Korvunchan Formation (Strakhov 1978). The tuff is similar to that of the vent facies in composition of clastic material and matrix. They differ in angular to angular-rounded fragments of traps and in the presence of randomly scattered (locally abundant) grains of quartz and small coal fragments.

The upper part of the least eroded diatremes is composed of bedded and obscurely bedded rocks consisting of fine-grained carbonate-chlorite-serpentine groundmass in which are set grains of quartz, feldspar, decomposed volcanic glass, coal seams yielding gastropods, and pelecepods typical of the Late Jurassic-Early Cretaceous deposits. The deposits are thought to have accumulated in crater lakes. The depressions of crater lakes in known pipes are 200–400 m deep with a flank slope of 30–50°. Strakhov (1978) believes that the higher slope of the lake flanks compared to that of erosion basins suggests that subsidence took place during the period after the accumulation of volcanic and volcano-sedimentary formation.

Small trap intrusions are common among rocks filling diatremes. Most widespread are veins and dykes varying in thickness from several tens of centimeters to 30 m. Sill-like bodies are known from only a few diatremes. Infilling and country rocks, including overlying tuff of the Korvunchan Formation, are pierced by intrusions, except for crater lake deposits.

Minusinsk District

In the northern Minusinsk intermontane trough the basalts not only intrude diatremes of alkali basaltoids (see Chap. 12), but form separate necks and dykes. All the known exposures are located within the Chulym-Yenisei basin, where basalt necks and dykes form a semicircle around the extensive Kopiev anticlinal dome in the center of the basin (Luchitsky 1957). Alkali basalt diatremes are confined to the flanks of the dome.

Basalt pipes and dykes are common in the Upper Devonian redbeds. Locally, dykes and necks cut the Lower Carboniferous deposits; there are also dykes intruding coal-bearing Middle Carboniferous-Lower Permian rocks. No basalt dykes and necks were found in the Jurassic deposits. The age of basalts ranges from Late Permian to Triassic, so that they resemble the traps of the Siberian platform (Luchitsky 1957).

This formation has the appearance of denudation cones up to 150 m high, and can thus be distinguished against the background of the undulating plain. The upper part of the cones is composed of steep-lying rocky basalts with complex outlines. The basalt exposures are surrounded by a train of volcanic tuffs building up gentle (less than 20°) soddy and talus slopes. Tuff terrain is oval or almost round in outline up to 160 m in diameter. The lower part of the cones consists of flat-lying Upper Devonian terrigenous redbeds. Errosion-resistant basalt dykes occur on the slopes of mud cones. The dykes are separated by tuffs from basalts composing cone summits. Some cones have up to three radiating dykes.

Special studies show that basalts and tuffs have a subvertical uneven contact. The contact between the tuffs and the underlying rocks is even, gently (5–15°) dipping to the center of the pipe, and forms an almost regular funnel. Lower tuff beds com-

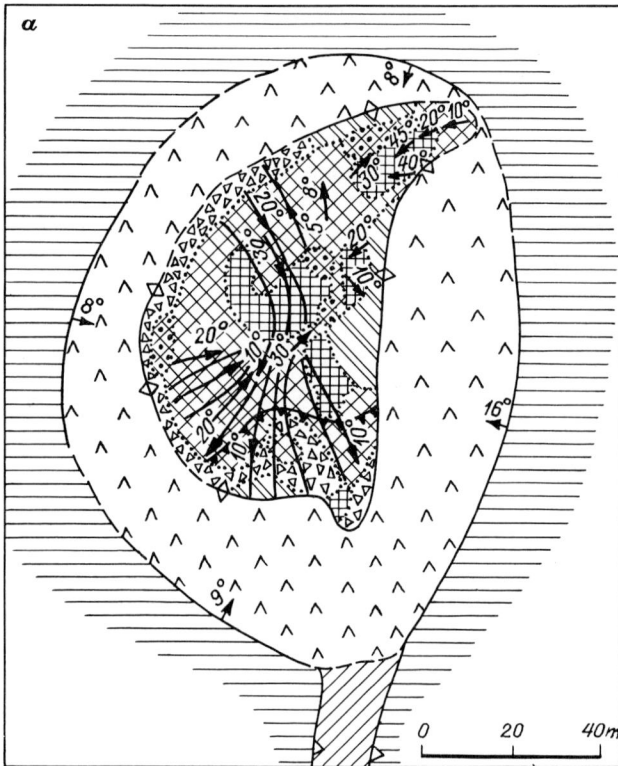

Fig. 48a,b. Structural-geological sketch maps of the Baradzhulskaya and Krasnoozerskaya Pipes. (Kalmykov 1963). Basalt: *1* intersertal fine-grained porphyritic (**a**), with glassy groundmass (**b**); *2* fine-grained porphyritic melanocratic; *3* fine-grained prophyritic (**a**), fine-grained basalts (**b**); *4* medium grained porphyritic with microdoleritic groundmass; *5* fine-grained brecciated oligophyric; *6* fine-grained brecciated aphyric amygdaloidal (**a**), medium grained porphyritic with pilotaxitic groundmass (**b**); *7* basalt tuff; *8* Upper Devonian redbeds; *9* high concentrations of xenoliths (with indentations directed into the area); *10* boundaries separating basalt varieties; *11* orientation of basalt columnar jointing; geological boundaries: *12* observed; *13* inferred; contact surface: *14* vertical and subvertical (85–90°); *15* steep (75–85°); *16* gentle (up to 15–20°)

monly contain abundant large (up to 0.5 m) angular fragments of sedimentary rocks. Such fragments, although less abundant, also occur higher in the section.

Pyroclasts are lithoclastic basalt tuff. Clasts composing tuff vary in size from a fraction of a millimeter to several centimeters. Clasts of enclosing and underlying rocks are also common. The cinder matrix of tuff is locally replaced by carbonate, iron hydroxides, and analcite.

Basalts both within a single diatreme and in different pipes are almost similar in composition. The distribution of structural varieties of basalts within diatremes exhibits more or less regular trends (Fig. 48).

Basalts are commonly quite fresh, rich in glassy material containing scattered fine (less than 3 mm) idiomorphic crystals of olivine and sporadic labradorite and microlites of plagioclase and clinopyroxene. Olivine contains about 65% of fayalite. Sub-

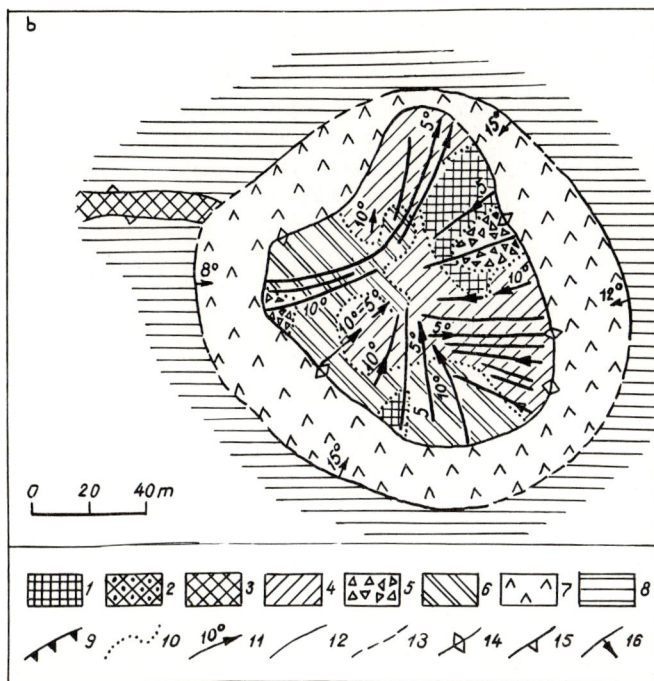

Fig. 48b

ordinate apatite, magnetite, and secondary minerals (serpentine, chlorite, carbonate) are commonly present. The chemistry of the rocks (Table 24) is identical to that of basalts intruding alkali basaltoid pipes within the Minusinsk depression (see Table 19) and the rocks resemble typical plateau-basalts.

Contraction fissures in the form of columnar jointing are very common in basalts. They dissect the rock into penta-hexahedral prisms 5–25 cm in diameter (with a cross-section 5–25 cm wide). The long axes of prisms are oriented toward the middle part of a diatreme and always subnormal to the contact surface separating basalts and tuffs. It is interesting that boundaries between individual basalt varieties do not considerably affect the columnar jointing orientation. This means that successive batches of basalt magma have intruded during short time intervals when the earlier rocks could solidify, but their temperature was still too high to inhibit cracking.

Basalts composing necks or dykes contain xenoliths derived not only from enclosing but from deep-seated and presently eroded rocks. Besides, basalts contain (locally abundant) peridotite xenoliths up to 25 cm in diameter. Peridotite is bottle-green olivine and clinopyroxene with subordinate spinel and chromite.

Despite the similarity in rock composition and common structural features, the volcanoes of the Minusinsk district and those of the Angara-Ilim province are different. For example, pipes of the Minusinsk district do not exceed 200 m in cross-section, while those of the Angara-Ilim province are giants, reaching 1500–2000 m in size. All the Angara-Ilim pipes, including the least eroded, have steep vent walls whose slope

Table 24. Chemical composition of rocks from basalt pipes of the Minusinsk trough. (Luchitsky 1957)

Oxide	Basalt				Analcime diabase	Trachy-dolerite
	Uchum Lake	Chernoe Lake	Kopievo	Western Chulym-Yenisei basin	Kopievo	Baradzhul Ulus[a]
SiO_2	42.58	43.92	45.02	43.29	41.81	44.15
TiO_2	3.15	3.22	2.91	1.74	2.52	2.18
Al_2O_3	14.03	14.18	15.05	18.61	12.01	12.00
Fe_2O_3	6.66	7.21	7.65	8.03	8.30	4.29
FeO	5.23	6.84	6.65	3.31	7.10	9.82
MnO	0.08	0.25	0.21	–	0.07	0.10
MgO	12.54	6.24	6.25	5.46	6.46	7.81
CaO	5.91	11.04	9.41	12.40	8.65	8.96
Na_2O	2.39	2.56	2.42	2.90	1.97	4.44
K_2O	0.52	0.76	0.92	1.00	1.53	9.56
P_2O_5	0.34	0.36	0.31	–	0.44	0.76
H_2O	6.70	2.63	3.39	2.51	7.90	4.53
CO_2	–	1.63	–	1.61	–	–
Total	100.13	100.84	100.19	100.86	98.76	99.60

[a] Ulus (Mongolian) – camp of nomad tents.

increases with depth. The Minusinsk volcanic pipes are crowned by a funnel plunging gently centerward. Basalt tuffs in the Minusinsk pipes compose only peripheral parts of the funnels, while the Angara-Ilim diatremes are filled with tuffs to a depth of many hundred of meters. In contrast, basalt and other basic igneous rocks comprise most of the volume even of weakly eroded Minusinsk pipes, while massive and brecciated varieties of traps are found only at great depth in the Angara-Ilim diatremes. Finally, the Minusinsk pipes have not suffered the extensive superposed processes (e.g., skarn formation, magnetite mineralization, and the like) typical of the Angara-Ilim volcanoes.

Such important differences in structure and hence in environment during the formation of the volcanoes, despite the similarity in rock composition, may be ascribed to differences in geological setting: the Minusinsk pipes were formed within the intermontane trough, while the Angara-Ilim pipes were emplaced under more conservative platform conditions.

CHAPTER 14

Trachyte Diatremes

Although alkali magmatism is widespread over the earth, alkali volcanic rocks occur fairly rarely in platform areas. Two trachyte terranes on the Siberian platform provide examples.

Kuonamka District

Alkali volcanic rocks were found in 1960 by Dukhanin and Lopatin (1961) in the lower Bolshaya Kuonamka River. Pyroclasts resting on the irregular surface of dolomites of the upper member of the Billyakha Formation (PR_3) crop out on the left precipitous side of the river. They are confined to a fault zone where dolomite beds dip at $10°$ to the axial part. Dolomite in the northern part of the zone is shattered and forms differently oriented blocks up to several meters in diameter. The space between the blocks is filled with pyroclasts.

Pyroclasts form a deposit vague in shape. The maximal visible thickness is about 20 m. Talus covering tuff and tuffisite breccia contains rock fragments of tuff; feldspar quartz sandstones occur in the watersped area. This suggests that pyroclasts give way to sedimentary rocks up the section. Sandstones in talus covering the pyroclasts are similar in composition to those of the basal beds of the Staraya Rechka Formation (Vendian) exposed farther north (200 m) at the same elevation; this implies a probable Sinian age of the volcanic rocks.

The structural zonation of the territory revealed that the volcanic rocks are related to a discontinuity between the crustal blocks differing in the nature of mega- and macrofissures (Fig. 49). More comprehensive studies provided data allowing the compilation of a schematic geological map for the Vulkan site (Fig. 50), interpretation of the genesis of volcanic rocks from various outcrops, and determination of the absolute ages of the rocks.

The site is underlain by medium- to thick platy dolomites of the upper member of the Billyakha Formation (PR_3), overlain by terrigenous rocks of the Staraya Rechka Formation (Vendian) in the watershed area. Volcanic rocks crop out, starting from the edge of the water of the Bolshaya Kuonamka River and can be traced intermittently along the valley slope to a height of about 20 m. Dolomites underlie the river bed and most of the coastal cliff. There are only three localities where volcanic rocks descend to the water level; mainly they rest on dolomites 6–12 m above the river level and reach the upper edge of the valley slope throughout the entire extent of the site discussed. In the middle of the cliff, tuffs onlap a fairly uneven, hummocky surface of dolomites.

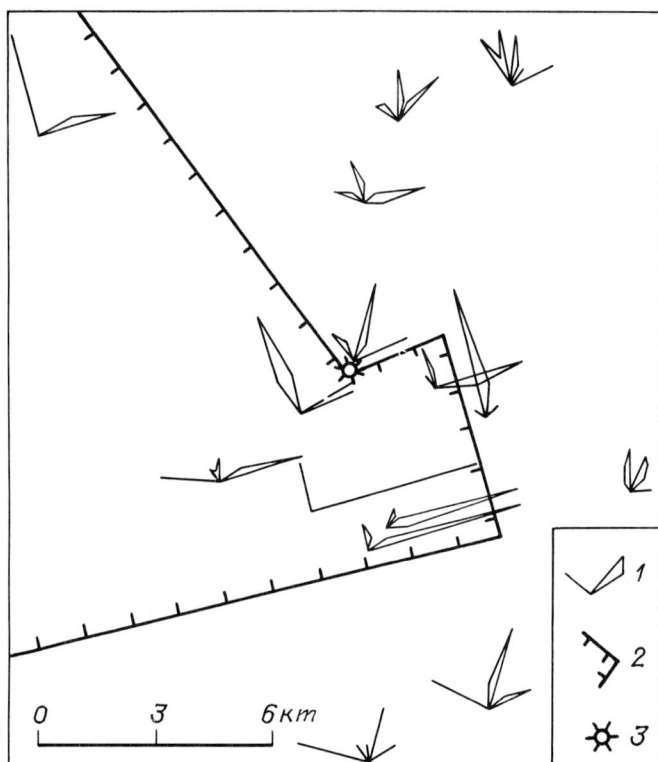

Fig. 49. Macrofracture pattern and structural setting of a Late Proterozoic volcano in the southern Nizhnyaya Kuonamka basin – Vulkan site. (Milashev 1979). *1* Rose diagrams of macrofracture system in the Upper Proterozoic dolomites; *2* boundaries separating crustal blocks considerably varying in mega- and macrofracture systems; *3* Late Proterozoic volcano

Here we find two major groups of volcanites: tuff and volcanic breccia. Tuff beds lie on the eroded surface of dolomites of the Bilyakha Formation. These ataxitic rocks change their color from greenish-grey to buff along the strike and up the section. Tuffs are not stratified. Macroscopically, taxitic structure is seen owing to a large proportion (averaging 50%) of pale-colored large fragments from enclosing and intruded rocks. The fragments measure 0.1–0.2 m, some reaching 0.5 m. Tuffs exposed in the center near the water edge may be considered as part of covering pyroclasts on the basis of their structural and textural features and owing to the presence of an erosion "window" where underlying dolomites are exposed.

Coarse clastic volcanic breccia occur is occursin in the exposures at low elevation. Here dolomite blocks reach 3 m by 6 m in size. The orientation of blocks does not show any regular trend: some lie subhorizontally, others dip randomly, and some others stand on their heads. The outline and enclosing rocks of the outcrops resemble those of volcanic vents. The southern, smaller, vent (22 m by 27 m) is situated on the shore, while the larger vent, 100 m farther north, is half hidden under the water.

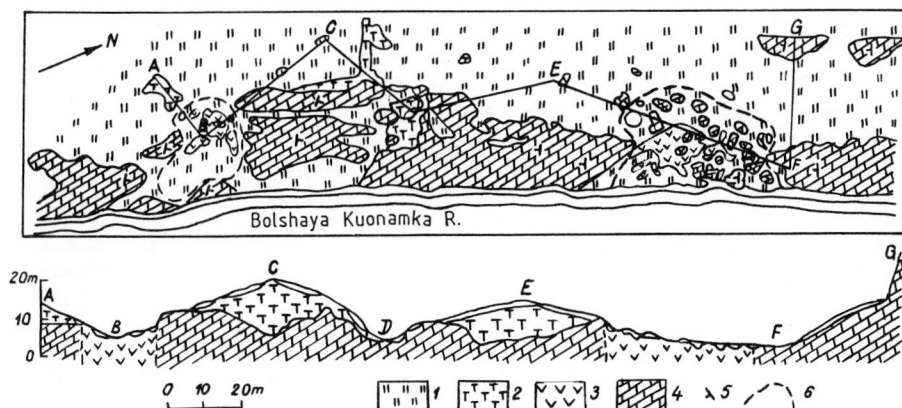

Fig. 50. Geological map and section of the Vulkan site at the lower Bolshaya Kuanomka River. *1* Loose Quaternary deposits; *2* trachyte tuff; *3* trachyte volcanic breccia; *4* Late Proterozoic dolomite; *5* attitude; *6* inferred boundaries of volcanic vents

The tuffs and volcanic breccia have a fairly motley composition. Clastic material is mainly carbonate and terrigenous-carbonate rocks, sandstone, quartzite, basalt, dolerite, volcanic glass, and trachyte. Glass is amygdaloid, commonly cryptocrystalline with discrete pseudomorphs after acicular microlites. Glass is replaced by secondary products varying in color from brownish-green to light brown. Amygdules, as a rule, are oval-elongate, infilled with zeolite, chalcedony, quartz, or mica-chlorite material.

Trachytes (trachytic porphyries) consist of phenocrysts of potash feldspar (orthoclase) and xenogenic (?) grains set in a completely altered relic microlitic groundmass. Decomposed volcanic glass of the groundmass contains randomly oriented pseudomorphs of pale green micaceous mineral after rectangular elongate and square microlite (potash feldspar).

Fragments of different composition exhibit considerable change in quantity both in lateral and vertical direction, so that volcanic rocks even from the nearest sites differ in abundance of fragments of enclosing rocks and pyroclasts. Additionally, the most coarse clastic components of tuffs and volcanic breccia are dominated by dolomite and other carbonate rocks composing most of the sedimentary cover, including its upper part on this platform area.

On the basis of chemical analyses, massive varieties of igneous rocks filling the volcanoes in the lower Bolshaya Kuonamka basin may be classified as ultrapotassic trachytes (Table 25). Compositionally, these rocks are similar to trachytes dominating the volcanic rocks from the eruption center at Saltpetre Kop, South Africa (McIver and Ferguson 1979); they are spatially and genetically related to olivine melilitites and carbonatites.

A K-Ar age was determined on two trachyte breccia samples and one tuff sample from the author's collection. The ages at 1210 ± 20 and 1280 ± 70 m.y., and 1440 ± 80 m.y. were obtained for the first, second, and the third sample, respectively. The oldest age of tuff is ascribed to the presence of xenogenic material from shallow

Table 25. Chemical composition of ultra-potassic trachyte from the Kuonamka district, Siberian platform, and trachyte from Saltpetre Kop, Southern Africa

Oxide	Ultrapotassic trachyte Breccia		Tuff	Trachyte
	Author's collection	(Vaganov and Konstantinovsky 1978)	Author's collection	(Rust 1937)
	(1 analysis)	(2)	(4)	(2)
SiO_2	57.18	61.50	45.00	51.88
TiO_2	1.08	0.97	1.56	0.55
Al_2O_3	13.94	14.35	12.23	16.02
Cr_2O_3	0.01	–	0.02	–
Fe_2O_3	4.36	2.27	5.31	0.56
FeO	2.51	2.41	4.50	4.39
CaO	0.47	0.32	3.35	8.17
MgO	6.14	5.00	11.16	0.38
MnO	tr.	0.03	0.08	0.23
K_2O	9.20	9.34	5.61	10.91
Na_2O	0.13	0.10	0.10	0.47
P_2O_5	0.20	0.20	0.20	1.14
Ignition losses	4.58	3.26	10.95	5.68
Total	99.80	99.75	100.07	100.38
CO_2	0.16	0.07	4.77	–

levels of the Archean crystalline basement. Thus, the data confirm a Late Proterozoic age of the volcanic rocks described.

Central Aldan District

The Central Aldan district occupies the central Aldan shield in the area where the alkali ultrabasic Mesozoic magmatism seems to have been most intense, i.e., on the northern slope of the Median Arch. A peculiar feature of the Mesozoic magmatism is that volcanic, intrusive, and metasomatic masses of different age groups are related to single eruption centers at the intersection of fracture zones. The occurrence of bodies within such clusters is attributed to the vertical ring and radial fissures contributing to the formation of the central type massifs from a thousand square meters to several tens of square kilometers in size. All the bodies are typical hypabyssal structures: sills and sheets, laccoliths, stocks, intrusions and extrusions of central type; they occur within the sedimentary cover or at the cover/basement boundary, i.e., under near-surface conditions (Bilibina et al. 1967). A succession of volcanic extrusions and multiphase intrusions with pronounced young alkali metasomatism when alkali metasomatites occur at margins of bodies is found both within the entire Mesozoic magmatic terrane and within its age groups.

Bilibina et al. (1967) recognized five age groups within the Mesozoic complex of the Aldan shield. The first group (T_2-J_1) comprises low alkali trachytic and quartz porphyries and their pyroclasts. The second group (J_{2-3}) includes potassic alkali syenites forming laccolith-like, central, often intraformational bodies, dykes, and stocks. The oldest of them is alkali lava breccia. The third group ($J_3 - K_1$) consists of volcanic and subvolcanic equivalents of potassic syenites, pseudoleucitic rocks, alkali trachytes, and basalts. Clastic lava and tuffisite breccia of the rocks are very common. The fourth group (K_1) is represented by low alkali syenites and syenite porphyries. The fifth group (K_2) incorporates the youngest intrusions and metasomatic rocks of aegirine granite and grorudite type; the outline of the bodies repeats that of the fracture zones in the crystalline basement, eruption and metasomatic breccia are very common here as well.

The rocks of the first to fourth groups were derived from magmatic centers within the long-living and high amplitude fault zones commonly bordering block structures, while the dyke complex of the fourth and fifth groups and the remaining rocks of the fifth group occur within the low amplitude fault zones of the axial part of the Yakutia graben and horst structures. The Mesozoic complex also contains hyperbasites of the Inagli Massif and kimberlite-like rocks of the Central Aldan district.

Three groups of cone-shaped bodies (Fig. 51) filled with breccia of trachyte-mica porphyritic rocks, orthoclase tuffisite breccia, and trachytic porphyries are known as the Tobuk-Khatystyr field from the north-western part of the district. All groups strike sublongitudinally and form an en-echelon north-east-trending belt. Such an arrangement of diatremes is attributed to their occurrence within the clusters where the Inagli fault zone is intersected by longitudinal fractures of the thrust fault type (Zuev 1973a).

All pipes occur within a platform sedimentary terrane of Lower Cambrian to Lower Jurassic rocks totalling 50–100 m in thickness. Diatremes are 50–150 m in diameter and are spaced at 100–300 m. The enclosing carbonate rocks of a 2–3 m zone are faulted and injected by material which had intruded along multiple fissures. There is no evidence of high temperature processes, rocks are silicified to a different degree. The fault zones incorporating explosion pipes contain quartz-orthoclase, feldspar-carbonate, and quartz-carbonate metasomatites of breccia and lava breccia type, also present east of the terrane in the faulted and intraformational fracture zones.

The Tobuk group contains three multi-step orebodies of volcano-metasomatic eruption breccia (Zuev 1973a). The central parts of the diatremes are filled with porphyritic melanocratic breccia containing fragments (10–30%) derived from the altered Archean crystalline rocks and enclosing sedimentary rocks cemented by altered magmatic rock. The rock consists of numerous quartz-hydromica-carbonate pseudomorphs and porphyritic phenocrysts set in a mica-carbonate fluidal base. Marginal zones of diatremes are composed of light-colored breccia with abundant (40–50%) xenoliths. The matrix contains numerous (up to 30%) quartz-carbonate pseudomorphs set in the quartz-feldspar-mica-carbonate relic fluidal groundmass.

Pyrope (violet and orange), picroilmenite, chrome diopside, alumochromite, and chrome picotite are found in diatremes of the Tobuk group. The minerals are derived from xenoliths of garnetized basement rocks and peculiar symplectite nodules of quartz-, pyroxene- and pyrope-spinel composition and occasional inclusions of pyrope serpentinite type.

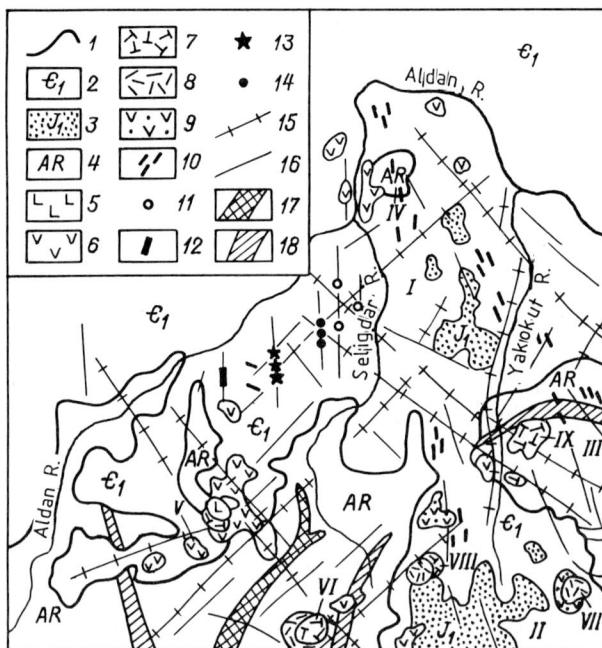

Fig. 51. Geological-structural map of the Central Aldan district. (Zuev 1973a). *1* Platform cover boundary; *2* Lower Cambrian carbonate deposits; *3* Lower Jurassic terrigenous rocks; *4* Archean crystalline rocks; Mesozoic magmatic complex: *5* dunite and peridotite; *6* quartz and trachyte porphyries and other rocks of the first age group; *7* alkali syenite and gabbroids of the second group; *8* alkali volcanic rocks of the third group; *9* syenite, monzonite and other rocks of the fourth group; *10* dyke complex of the fourth and fifth groups; *11* orthoclase tuffisite breccia pipes; kimberlite-like rocks of the Tobuk-Khatystyr field: *12* pyroxene minette veins, western group; *13* kimberlite-like breccia pipes, Tobuk group; *14* pipes infilled with trachyte porphyries, pseudoleucitic porphyritic breccia, orthoclase tuffisite breccia, Khatystyr group; fault tones: *15* Proterozoic; *16* Mesozoic; *17* anticlinorium axes; *18* synclinorium axes. Grabens: *I* Kuranakh, *II* Yakokut; horsts: *III* Elkon; *IV* Bainai; massifs of the central type: *V* Inagli; *VI* Tommot; *VII* Yakokut; *VIII* Ostry; *IX* Ryabinovy

Cone-shaped bodies of the Khatystyr group display a more complex structure and composition. The central parts of diatremes are infilled, as a rule, with melanocratic eruption breccia. Clastic material (10%) is set in a matrix of heavily corroded porphyritic phenocrysts of mica, pyroxene, and arthoclase set in turn in the trachyte-mica-carbonate groundmass, containing microlites of mica and pyroxene. Orange pyrope, chromspinelids, and quartz-spinel symplectites also occur. Blocks of pseudoleucitic porphyritic breccia with a few grains of orange pyrope and chromspinelids are present in the inner parts of the cone-shaped bodies. Pseudoleucitic porphyry of the second age group are considered to be the equivalents in the Aldan district (Zuev 1973b). Marginal zones of the pipes are composed of orthoclase tuffisite breccia predominantly containing xenoliths of garnet-gneiss and minor carbonate rocks. The metasomatism is expressed in silicification and feldspatization of the rocks. The rocks may be equivalent to agglomerate tuff of the third age group.

Table 26. Chemical composition (wt.%) of rocks from the Tobuk-Khatystyr field (Zuev 1973a)

Oxide	Tobuk group		Khatystyr group		
	Eruption breccia of trachymicaceous rock (8 analyses)	Tuffisite breccia of trachymicaceous rock (6)	Trachyte porphyry (6)	Pseudoleucite porphyritic breccia (1)	Orthoclase tuffisite breccia (1)
SiO_2	39.64	35.37	45.45	53.20	58.64
TiO_2	0.67	0.47	0.70	0.60	0.44
Al_2O_3	6.22	4.33	8.10	12.47	12.48
Fe_2O_3	6.19	6.89	6.97	5.95	3.83
FeO	1.36	0.65	1.45	1.43	0.20
MnO	0.08	0.12	0.14	0.18	0.06
MgO	13.23	5.48	11.03	6.88	1.29
CaO	13.53	21.41	9.99	3.98	5.90
Na_2O	0.21	0.22	1.22	2.07	0.72
K_2O	0.74	2.20	4.40	6.02	9.79
Cr_2O_3	0.09	0.09	0.10	0.08	0.03
P_2O_5	0.38	0.29	0.84	0.57	0.18
CO_2	11.90	18.34	3.21	–	–
Ignition losses	5.64	4.27	7.03	6.28	6.09
Total	99.88	99.93	100.63	99.71	99.65

The rocks of the Tobuk group have been subjected to intense carbonatization with attendant replacement of most silicate minerals and release of silica, as is evidenced by chemical analyses (Table 26). The rocks of the Khatystyr group have suffered from secondary processes to a lesser degree (they contain abundant porphyritic phenocrysts set in the microlitic groundmass), so that chemical analyses, to a first approximation, reflect their primary composition. Zuev (1973b), using L.L. Perchuk's pyroxene-garnet geothermobarometer, determined that inclusions in the trachymicaceous and trachytic rocks of the Tobuk-Khatystyr field had formed at 620–800°C and 200–800 MPa, i.e., under conditions typical of the basement crystalline rocks of the district metamorphosed at 690°C and 850 MPa. Zuev (1973b) emphasizes a genetic relationship between the volcanic rocks of the Tobuk-Khatystyr fields and Mesozoic magmatism of the Aldan shield, and hence relates them to mean pressure explosions within the crustal substratum. Their peculiar features may be attributed primarily to the considerable effect of a chemically bound water fluid and its partial pressure, as well as to various processes responsible for the metamorphism and magmatic replacement of the crystalline basement rocks, and the differentiation of a magmatic system.

In conclusion it may be said that the distribution pattern, geotectonic setting, relations to enclosing rocks, outlines in plan and in vertical sections, internal structure, and some other peculiar features of diatremes filled with a wide range of rocks (from kimberlite to trachyte) not only show similar trends, but to a first approximation may be considered identical. However, despite an almost complete identity, compositionally different volcanic rocks of a diatreme exhibit their peculiar features when averaged values are compared. For example, the sizes of individual pipes, including the area in plan and depth of the emplacement, do not directly depend on the composition of enclosing rocks. However, the average size of trap diatremes is several times larger than that of pipes filled with other rocks, and the latter show a clear tendency to a successive decrease within the series kimberlite – alkali basaltoid – carbonatite – trachyte.

We cannot say that this is a mere accident. Most probably, these phenomena may be ascribed to different volumes of magmatic melts which had been generated in subcrustal layers and brought up to the Earth's surface, and not to the differences in specific energy potentials of the kimberlite, alkali ultramafic, trap, and trachytic magmas. This conclusion complies with the distribution of appropriate magmatites within the Earth's crust. As a consequence the fact may be also mentioned that the trap formation rocks filling the pipes account for a small proportion of the entire volume of basic magmatites found in the near-surface beds of the Earth's crust, while in the platform areas ultrabasic, alkali ultrabasic, and intermediate magmatites are mainly vent facies volcanites.

The differences observed in mean sizes of diatremes may well be attributed to the variability of thermodynamic parameters (pressure, temperature, and the like) of magmatic melts; for example, the mean area of diamondiferous diatremes is much larger than that of the pyrope subfacies kimberlite pipes, and the latter are larger than picrite porphyritic diatremes within a single field (see Chap. 6).

The initial variability in composition and volume of magmatic melts responsible for these differences in mean sizes of diatremes and the abundance of rocks filling the pipes (with respect to a total volume of igneous rocks of a certain type) might be ascribed to geodynamic processes operative at depth, which had not considerably affected diatreme formation. In contrast, the similarity of geological setting, internal structure, and pipe geometry, as well as important structural and textural features of rocks filling the bodies provide strong evidence for a close relationship between major mechanisms and physicochemical conditions under which pipes were formed, from which we can analyze the conditions of their emplacement, irrespective of the petrography of volcanic rocks.

Part III Mechanism and Conditions of Pipe Formation

CHAPTER 15

Hypotheses on the Genesis of Pipes

Kimberlite explosion pipes and basaltic explosion pipes are set expressions known in geological literature since the late 19th and mid 20th centuries respectively; they reflect a genetic nature and, in fact, predetermine the ideas of geologists as to the origin of these cone-shaped bodies. A. Daubrée (Rust 1937) contributed to the idea, by running an experiment in which tubular openings in granites were formed by an explosion, and using the results obtained to explain the origin of volcanic necks. He proposed the term diatreme for such vertical tubular bodies (Gk. $\delta\iota\alpha$ – through, via; $\tau\rho\eta\mu\alpha$ – opening) (Rust 1937). For these reasons and from the knowledge of the shattering of rocks within a comparatively small cross-sectional area by a considerable vertical extent of tubular-shaped zones, almost all the researchers have accepted a priori the explosive origin of diatremes. The causes for giving way to extremely high pressures in the near-surface layers of the Earth's crust, resulting in the outbreak of deep-seated material to the surface, were debated, rather than mechanisms of diatreme formation.

The explosive character of the formation of pipes seemed quite obvious, and therefore most investigators did not seek to prove it, and even used it as a basis for other original hypotheses. For example, Lodochnikov (1936), sharing the opinion about excessive pressure of kimberlite magmas near the Earth's surface, and comparing the vent of a diatreme with a muzzle, put forward a hypothesis which relates the origin of some stony meteorites to the extrusion of deep-seated matter of planets in the course of volcanic eruptions from edifices of kimberlite pipe type.

Not all the authors discuss in detail the conditions under which explosion pipes are formed. Some only mention a possible cause which led to the eruption of the subcrustal matter, others concentrate on the regime and successive stages in the diatreme formation, still others confirm their statements by quantitative assessments of physico-chemical parameters and energetics of the processes.

The genesis of kimberlite and some trap diatremes are the most debatable problems. Based on the probable mechanisms responsible for diatreme formation and, primarily, the sources producing adequate pressure, we recognize several major groups of hypotheses. The most numerous are those which relate the formation of cone-shaped bodies in the upper layers of the crust to the active influence of ascending subcrustal melts, their emanations, or both.

Wagner (1909), the first to state clearly the volcanic origin of kimberlite pipes, wrote that the present kimberlite pipes are eroded volcanic vents resulting, as was proved by Daubrée (Rust 1937), from explosive emanation of gases and vapors under high pressure. The vents are filled partly with fragments and comminuted products of intruded rocks fallen or washed back into the vent and partly with material, derived from a kimberlite magma, and with solidified magma.

Sobolev (1960) believes that kimberlite pipes are formed not as a result of a real explosion due to a drastic increase of pressure in a magmatic chamber, but due to the crust outburst when magma moves under very high pressure. First, it moves along vertical fissures at a great depth, where a peculiar breccia pattern already exists. When magma reaches a relatively small depth, some plumes break through the crust like a fountain and come to the surface; later some are sucked into the funnel created. In this case the pipe is either completely filled with clastic material or the material is cemented by magma coming from below. Sobolev, comparing the data on a depth of pipes and character of fragments in some pipes of Yakutia, suggests that the explosion, i.e., the transition from a vein to a pipe takes place at a depth of 1.5—2.0 km.

Many Soviet geologists have shared this viewpoint. However, no special studies were carried out for many years. Some workers claim to have recognized a separate explosive stage in a complex process of the formation of diamondiferous deposits, but in fact this was a general statement.

In 1976, Kostrovitsky published a monograph devoted to physical conditions, hydraulics, and kinematics of kimberlite pipe filling. The author, holding to the explosion hypothesis, made an attempt to explain the mechanism of kimberlite diatreme formation by means of energetics estimates and analyzing the shape and size of diatremes. Because from the explosion theory we know that ejection vents have an explosive index equal to or above 0.7 and that of kimberlite pipes is 0.1, it is concluded that kimberlite diatremes must be classified as shatter channels. The gradual narrowing of pipes with depth is inconsistent with the presence of an explosive chamber of adequate energy capacity, and hence rules out the interpretation of pipe formation as a result of explosions from satellite chambers. It is concluded that the pipe vent formation is extended in time and results from repeated explosions whose front moves toward the earth surface.

Kostrovitsky provides the following schematic description of explosions. Explosions take place only when gases reach a critical state (in the case of the chemical nature of an explosion, the active gases should reach a threshold concentration) and at a certain moment of time are confined in nature. Their destructive effect is normal to the earth surface. Above the explosion center and level with it, there formed a shatter zone, which later often collapses, and a vertically emplaced system of radial and concentric cracks (effect of confined explosion). Thus, in the vertical direction a new chamber for a subsequent explosion is created. The explosive activity extends mainly along the initial deep fissure. Pipes remain elongate until the explosions take place within the fissure. If explosions occur far away from the primary fissure, moving on their own, the pipe becomes isometric in shape. When explosions approach the Earth's surface, at a certain depth they acquire energy sufficient not only to shatter rocks but also to eject them. As a result, in the upper parts of diatremes there are created "blows", which are typical ejection funnels with an explosive index of no less than 0.7 (Kostrovitsky 1976). Later, Kostrovitsky renounced this point of view, and now believes that explosions are not very important in kimberlite pipe formation, and that shattering follows the mechanism of fluid brecciation (Kostrovitsky 1980).

The idea of the explosion outburst of volcanic gases from a trap magmatic chamber to the Earth's surface was used as a basis for the most popular and substantiated hypothesis on the genesis of ore-bearing cone-shaped bodies on the southern Siberian platform (Antipov et al. 1960; Strakhov 1978). These pipes, like kimberlite basalt, dolerite, and other pipes, have no widenings which can be identified with explosion chambers either in the lower parts or in fissures in which they pass at depth. Moreover, fissure walls have no traces of intense shattering, an inherent feature in the case of explosion. Hence, the basement fissures which are projections of pipes at depth were only conduits for volcanic gases and pyromagma.

The above allows us to conclude that volcanic pipes of the Angara-Ilim type have resulted from outbreaks of volcanic gases from a magmatic chamber along the basement fissures and due to the shattering effect of gases and pyromagma only within a pipe vent; however, gases do not continue to expand spasmodically below the pipes in fissures projecting at depth. It is assumed that the "explosion" is preceded by the filling of fissures with fluids and pyromagma under high pressure, and as a result fissure walls move slightly apart and then return to the point of departure. This mechanism was proposed to explain the presence of deep-seated rock fragments larger in diameter than the thickness of kimberlite veins in which they are found (Strakhov 1978).

Ivankin and Musatov (1973) have proposed a different mechanism for the formation of the Angara-Ilim pipes. In their opinion, the tubular bodies are formed due to buried explosions of a trap magma oversaturated with ore fluids. Their magmatic chambers are thick trap sills at the boundary of the basement and platform sedimentary cover or in the lower parts of the cover at a depth of 3—4 km from the present exposure surface. The pipes are supposed to be filled with massive and breccia traps from below, some 0.5—2.5 km above the magmatic chamber, and higher they are composed of explosion skarned trap agglomerates.

The geologists who studied the Angara-Ilim deposit in detail had paid attention to some important errors in the interpretation proposed by Ivankin (Strakhov 1978). In particular, a heterogeneous material containing vent tuff, large blocky breccia of sedimentary rocks, metasomatic ores, and skarn is classified as "eruptive explosion trap agglomerates subjected to skarnization". However, even the most mineralized and skarned pipes have extensive parts filled with agglomerates and tuffs almost not skarned. In many bodies such tuffs and agglomerates prevail in quantity over ores and skarns. Additionally, in this district there are pipes filled with hardly mineralized and not skarned breccia. Hence, the formation of diatremes cannot be attributed to buried explosion of ore solutions or to a mixture of ore-bearing hydrotherms and magma.

The statements of Nekrasov and Gorbachev (1978) and Lorenz (1975, 1979) resemble these hypotheses. These authors ascribe the diatreme formation to the outburst of highly concentrated fluid generated in the subcrustal layers without the contribution of alkali-ultrabasic or other melts. According to the hypothesis of Lorenz, cone-shaped bodies are formed due to the emission of water vapors generated when magma reaches the near-surface water-saturated layers of the crust.

In the sense of Williams (1932) and Menyailov (1962), diatremes are enlargements at the intersection of two or more fissures where clastic material from fault breccia is partly expelled and partly caught by magmatic melts ascending along fissures. These ideas, which seem reasonable and were confirmed by several examples (when explosion pipes occur at fault intersection), are in fact inconsistent with original data because many hundreds of kimberlite, trap, and other pipes are not related to any faults inferred from geological or geophysical data.

Mikheenko (1972) (Mikheenko and Nenashev 1961) also related diatreme formation to tectonic stresses and classified kimberlite pipes as diapiric structures. He emphasized the low temperature of a squeezed rock "where the role of the cement composed of serpentine and carbonate was given to a hydrodynamically active fluid of a media"; therefore he proposed the name kimberlite diapirs for kimberlite pipes. It is noteworthy that petrographical and structural studies of kimberlitic rocks suggest the infilling of pipes (and dykes) with a material of high plasticity which might be typical of fluidized liquid magma, but not of a cold serpentine-carbonate mass (see Chap. 2).

Vasiliev et al. (1961) believe that the very high pressure required for diatreme formation and diamond crystallization is produced at the boundary of the platform sedimentary cover and crystalline basement where magma withdrawn from the Earth's interior creates a lentiform chamber. Separate batches of magma fill the cracks in the chamber vault, cementing it and creating dykes. Then the incandescent kimberlite magma comes into reaction with water and hydrocarbon compounds of enclosing rocks, producing self-detonating gas mixtures whose explosion leads to the accumulation of free carbon. As a result of repeated explosions, the media becomes oversaturated in carbon and conditions formed for the alteration of carbon to diamond. The explosions also resulted in the emplacement of diatremes at the apical part of the chamber.

We shall not give a detailed account of the numerous contradictions to the hypothesis to explain the observed differences in the diamond content of kimberlites which have been discussed elsewhere (Milashev 1974a) and the aspects pertinent to the genesis of diatreme will be considered only in passing. One of the main conditions for the formation of explosion pipes, according to Vasiliev et al. (1961), is the presence of intermediate chambers at the basement-sedimentary cover boundary where hydrocarbons are "ingressed" from sedimentary rocks. The authors do not specify the reasons and the mechanism of hydrocarbon "ingress" in the magmatic chamber, where pressure is ten times that of the surrounding rocks. The principal postulate of the hypothesis is inconsistent with the available geological data, viz., the presence of intermediate chambers: no evidence of such chambers was found either in pipes mined to the roots (Kimberley, De Beers, and other mines) or in diatremes mined to a depth of several hundreds of meters below the basement-cover boundary (the Kharakhtakh pipe in Siberia and many African diatremes).

Different again are hypotheses relating the formation of cone-shaped bodies to the effect of ascending (Locke 1926; Sawkins 1969) or descending (Bessolitsyn 1970) solutions on country rocks. According to Locke (1926) the ascending solutions partly dissolve country rocks along subvertical channels, leading to the collapse of unsupported blocks, later cemented by hydrothermal minerals. The descending transport of clastic material and the absence of newly formed magmatogenic products are peculiar features of such pipes. Sawkins (1969) suggests that ascending solutions can give rise to chemical brecciation of surrounding rocks similar in mechanism of fissure formation to the injection of chemical reagents into petroliferous dolomites. The resultant fine-clastic breccia is marked by closely spaced fragments as if displaced in the direction normal to the front of intruding vein material and also by the presence in interstices of newly formed minerals.

Bessolitsyn (1970) suggests that the Angara-Ilim pipes were formed due to the effect of descending solutions of country rocks, and classes the pipes as ancient karst formation. He refers to a high slope (larger than the natural slope) of rims of the crater lake in a pipe, which implies the collapse and presense of large blocks derived from the terrigenous and sedimentary rocks lying above which, in his opinion, had fallen from the roof of the karst cave. However, there is no evidence for the karst origin of the pipes. If we take into account that deposits of the Korvunchan Formation are underlain by the vent-facies tuffisite and trap agglomerates (and not exclusively sedimentary breccia) in all the diatremes, the Angara-Ilim pipes are of volcanic rather than of karst origin.

The hypotheses ascribing the formation of cone-shaped bodies to a partial dissolution of country rocks may be used only for elucidation of the genesis of certain types of hydrothermal ore deposits, and are in fact invalid for the recognition of even the initial phase in the formation of diatremes having some traces of volcanic edifices and, in particular, mainly or completely filled with magmatites.

The account of the hypotheses on the genesis of tubular bodies shows that most researchers, in using the term explosion pipes, exclude the explosion itself (i.e., the instant rise due to a drastic drop of pressure in a chamber) from the processes responsible for the formation of the bodies. Almost all the authors (except Vasiliev et al. 1961; Kostrovitsky 1976) mention that the matter withdrawn from depth to the surface reaches a certain level, at which the overlying strata cannot withstand the pressure from below. Some workers consider the outburst as a catastrophic eruption, while others visualize it as the development of a channel over a period of time.

CHAPTER 16

The Connection of Diatremes with the Earth's Surface and Major Stages in Diatreme Formation

The discussion presented in Chapters 1 to 14 implies a principal similarity in the geotectonic setting, internal structure of cone-shaped bodies, and structural features of compositionally different rocks filling the diatremes. Therefore, the mechanisms and main conditions under which pipe bodies of all magmatites are formed may be considered as similar and hence some general rules of diatreme formation may be outlined. To analyze the conditions of cone-shaped body emplacement including the dynamics, energetics, and geological characteristics of processes responsible for their formation, problems such as the relationships between vents and the Earth's surface and major stages in diatreme formation must be considered. Detailed studies of kimberlite, trap, and other diatremes provide reliable answers to these questions.

It is a very important question whether diatremes are exposed or are blind bodies, because the solution of this problem to a certain degree provides a possibility of comparing diatremes and volcanic edifices, and in this context interpreting major genetic aspects. However, opinions differ. Despite the widespread belief about the exposure of pipes, the antitheses are also known. In this case the existence of blind magmatite pipes is either ascribed to a mechanism applied by the author for the explanation of diatreme formation or it reflects the general ideas of the worker. For example, the diapiric mechanism of the kimberlite intrusion into rocks of the platform cover (Mikheenko and Nenashev 1961; Mikheenko 1972) necessitates the existence of a large number of blind pipes. However, this statement is poorly substantiated (as was mentioned in Chap. 15), and cannot be considered as a scientific hypothesis.

The existence of blind pipes may be inferred from the assumption of Kostrovitsky about diatreme formation due to confined explosions. He says: "The imperative effect of such explosions is the formation of a collapse column of shattered rocks which depends on an explosion strength and does not always reach the surface" (Kostrovitsky 1976, p. 26). In this connection Kostrovitsky gives some geologo-petrographical indications which he classes as characteristic features of blind cone-shaped bodies. One of them is the clastic to magmatic material ratio (the amount of the latter is negligible, if present at all) and the gradual increase in the proportion of clastic material in the upper parts of the pipes. Another characteristic feature of blind tubular bodies is a gradual transition from the brecciated material mixed with magmatic material through coarse-clastic breccia composed exclusively of enclosing rocks to undeformed wall and roof rocks. The absence of distinct contacts and polished contact surfaces suggests slight and slow vertical transport of magmatic and clastic material. If these two main features occur, even by the presence of tuff in a pipe, we may classify pipes as blind cone-shaped bodies from which gas, according to Kostrovitsky, has flown to the surface along a fissure.

It is noteworthy that these theoretically probable features of blind cone-shaped bodies have been never recorded from the many hundreds of well-known kimberlite and trap diatremes. One single blind cone-shaped body reported by Kostrovitsky can hardly be assigned to explosion pipes proper. The body is composed of copper-bearing tourmaline breccia. The brecciation progressively decreases upward, and in the apical part there are only vertical fissures which bound the core consisting of hydrothermally altered rocks (Sillitoe and Sawkins 1971). The geologists who studied the deposits suggest that the pipe formed as a result of mineralization stoping, which is quite likely. In the case of kimberlite, trap, and other explosion pipes, the absence of the evidence mentioned by Kostrovitsky implies a free connection of these diatremes with the surface and questions the importance of confined explosion in their formation.

The majority of workers have arrived at their conclusions about the exposures of kimberlite and other diatremes not only on the basis of the absence of the indications of blind pipes, but owing to the geological records implying a free connection of the pipe vents with the Earth's surface. Moreover, there is no doubt, with the exception of some statements (Luchitsky 1971), about the exposures of the basalt, alkali-ultra-basic, carbonatite, and other pipes, which may be compared with feeders of modern volcanoes extruding products of similar composition. It is not that simple in the case of the kimberlite pipes because there are no rocks of such composition among the products ejected from recent volcanoes, so that we cannot draw a direct analogy. Therefore, we use indirect indications to solve the problem of the exposure (outburst) of kimberlite diatremes. The most important of them are the following.

All weakly eroded kimberlite pipes are crowned with funnels (see Chap. 1) which are classed as explosion funnels or as resulting from the collapse of the diatreme walls because their angles tend to become equal with those of the natural slope. During submarine eruptions, the slope of the funnel walls becomes steeper and the diatreme aquires a mushroom shape (see Fig. 2).

The composition of products contained in diatremes and a detailed study of the deformed enclosing rocks argue for the exposure of diatremes. Among products contained in pipes are tuff, eruption breccia, and massive varieties of magmatites. For example, in the kimberlite breccia, xenoliths derived from enclosing rocks average 20%. Because prior to pipe formation the remaining 80% of the volume had also contained enclosing rocks, the question arises how and when were they removed? In fact, the kimberlites, like other magmatic rocks occurring in the form of cone-shaped bodies, might well fill the space left after the ejection, compaction, or assimilation of country rocks. However, the assimilation should be ruled out as a possible mechanism because numerous fragments of sedimentary and metamorphic rocks contained in eruption breccia either have no visible traces of the solution or important reaction rims (see Chap. 4). The assumption of the formation of the diatreme due to the compaction of strata (by confined explosions or due to some other mechanism) is also inconsistent with the data available: not only are the country rocks being compacted, but they are even highly fissured over a radial distance of several tens of meters from the pipe contact.

The contention about the vent formation due to shattering and ejection of most of the country rocks once filling the pipe is at variance neither with the observed proportions and contact relationship between xenogenic material and magmatic rocks nor with the character of diatreme exocontact zones. Nevertheless, it is evident that the consistency of the statement even with very important records does not by itself prove its validity. If the diatremes were actually formed when shattered rocks were brought to the surface, then the extruded material should have accumulated around the pipe funnel. The relics of tuff ejectamenta were found from only one of the diamondiferous pipes, the Mwadui pipe (Edwards and Howkins 1966) which is the largest known, youngest and the least eroded kimberlite diatreme. There are no ejectamenta around the remaining kimberlite pipes even of Meso-Cenozoic age and they could not be preserved because a depth of erosion reaches several tens to several hundreds of meters.

Because the depth from the primary to the present-day surface in areas of kimberlite pipe occurrences is determined by body morphology independent of geological data (xenoliths found in young rocks now absent from the stratigraphic section; paleotectonic reconstructions, and the like), even a single example of almost uneroded diatreme containing relics of erupted material around the pipe mouth may be considered as a strong argument, proving that all or almost all kimberlite diatremes had once been exposed and were formed due to shattering and ejection of country rocks.

As to kimberlite pipes, a probable morphology of surface deposits — the accumulative forms of extruded material — may be inferred only from indirect evidence. The structural similarity of the mouth part of the Mwadui pipe and that of maars suggests that the discontinuous mantle of kimberlite tuff around it is a relic of a tuff ring which is a characteristic feature of maar craters. Referring to approximate estimates of Kostrovitsky (1976), a medium-sized kimberlite pipe could have a tuff ring 65 m high and 500 m wide. Additionally, the fact that some weakly eroded diatremes contain eruption breccia and massive varieties of kimberlites implies that apart from very common and even prevailing craters and surface edifices of maar type, the kimberlite pipes are locally terminated with conical superstructures typical of stratovolcanoes. The recent Ol Doinyo Lengai and Lashaine volcanoes (see Chap. 12) are crowned with tuffisite lava cones composed of pyroclastics derived from alkali-ultrabasic and carbonatite magmas.

The presence of kimberlite tuff, which cannot form in closed vents, in the upper parts of many diatremes, also lends support to a possible pipe exposure. The fact that tuff occurs mainly in the upper parts even of diatremes unexposed at depth is confirmed, on the one hand, by the inversely proportional relationship between the amount and frequency of occurrence of these rocks, and, on the other hand, the depth from the primary to the present-day surface of the pipe inferred from independent geological data.

Thus, the data available on the geology, structure, texture and mineral composition of rocks contained in pipes indicate reliably that actually all diatremes, even kimberlite, were once exposed, and that their vents were formed as a result of country rock ejection.

The number and the peculiar features of the major stages in pipe formation may be determined by using the chronology of the formation and infilling of diatremes. The distribution of reliably located xenoliths across the vertical section of diatremes also provides important information. It will be recalled that not only fragments of rocks, from great, even subcrustal depths occur, but other fragments also occur hundreds of meters below their original position, and even of younger rocks omitted from the section are now present in all diatremes (see Chap. 3).

Small light fragments (belemnite rostra, pieces of charred wood, and the like) and large xenoliths of carbonate and terrigenous rocks ("floating reefs") occurring hundreds of meters below their original position could not subside in basic and ultra-basic magmas of much higher density; but if the fragments had not subsided in the melt, this means that they sank due to collapse, so that diatremes were for some time channels not yet filled with the melt, but by the collapse of the channel walls clastic material could well have sunk for a considerable depth.

In this connection it should be mentioned that all large (with a mass of several tons) and gigantic "floating reefs" of rocks contained in diatremes could move only downward. Had the pipe vents resulted from a magmatic melt, then both small and large fragments of sedimentary rocks having low density compared to basic and ultra-basic melts would move upward. In contrast, had the pipe vent formed due to the out-burst and subsequent emanation of strongly compressed gases, then the largest frag-ments of rocks falling off the walls of the vent should withstand the extending gas stream and sink down.

This means that the complex process of diatreme formation consists of two main stages: the development of the vent, and its infilling. Despite a close interrelation of the stages and partial overlapping in time, the volcanic processes at each stage differed in temperature regime, pressure, composition, properties of a "working body", dynamics, and many other parameters. Therefore, the physical state of substance contributing to diatreme formation, if representative data are available, should be analyzed taking each stage into due account.

CHAPTER 17

Physical State of the Material Involved in Diatreme Formation

As already stated, two main stages are recognized in diatreme formation: the development and infilling of vents; therefore, it would be better to assess the physical parameters of the material separately for each stage. However, the available records are, unfortunately, very scarce, and the techniques used for the determination of temperature and pressure at the time when volcanites were formed are not very sophisticated. Hence, the estimates of the parameters not only for each stage but even for the whole explosive phase of pipe formation are far from being reliable and accurate.

Temperature

The temperature regime of diatreme formation not only displays a wide range, but exhibits considerable variations in the mean and sign of the temperature gradient in time. We do not need especially to confirm the fact that the initial and final temperatures of the crustal matter within a diatreme (country rocks extruded by vent formation and rocks now filling the pipe) depend, in fact, exclusively on the depth of the level studied and the past and present geothermal gradients in the area discussed. Additionally, the volcanic processes resulting in the formation and infilling of the pipes were operating at fairly high but not constant temperatures.

The problems stated in the book do not necessitate comprehensive analysis of the temperature regime of diatreme formation, so we shall limit ourselves to estimates of maximal temperatures characteristic of volcanic processes, which makes the solution much easier. The assessments may be based on the degree of pyrometamorphism of the country rocks and xenoliths and on the probable temperature of the formation of mineral parageneses crystallized by the infilling of a pipe.

A not very intense contact metamorphism of the country rocks and xenoliths is a common feature of diatremes otherwise different in composition. It is thus a widespread opinion that formation and infilling of pipes take place at low temperature, and some workers have even suggested that pipes are formed as a result of diapiric intrusion of the deep-seated cold matter (see Chap. 15). However, eruption and tuffisite breccias in all well-known pipes apart from unmetamorphosed and weakly metamorphosed xenoliths contain some xenoliths (commonly not abundant) with traces of intense pyrometamorphism (see Chaps. 4, 12, 13). Pyrometamorphism (thermal metamorphism), which gave rise to garnet-cuspidine, garnet-monticellite, and monticellite-phlogopite mineral assemblages, is found not only in xenoliths of explosion pipes but among metamorphic and metasomatic rocks at contacts with basic and rarely acid intrusions. Korzhinsky placed these rocks in a special melilite-monticellite facies marked by high temperature and low pressure.

Harker and Tuttle (1956) showed experimentally that monticellite associated with calcite is formed (at CO_2 pressure of 10–70 MPa) within a temperature range of 700–900°C. Cuspidine was synthesized at H_2O pressure of 130 MPa at 500-700°C (Valkenburg and Rynders 1958). The intense contact metamorphism of the country rocks is typical of dykes and not common in pipes; it was found only in lenticular in plan flattened tubular bodies which are root parts of the diatremes exposed by erosion, that is a transitional zone from a pipe to a dyke.

When these facts are interpreted genetically, some peculiar features of the pipes should be taken into account to explain a number of apparent contradictions. Among the most important features of diatremes is a much larger, as compared to dykes, cross-section area which rapidly increases toward the surface, and a high proportion of xenogenic material, locally above 50% of the volume of rock contained in a diatreme.

Obviously, the magma ascent along fractures in the Earth's crust is accompanied by a fairly piecemeal release of pressure. Steady powerful pressure from below allows the magma to reach the surface under conditions when the pressure drop in the apical part of the column follows, in general, the laws of hydrostatics. High pressure impedes the separation of volatiles and magma is apparently a fiery liquid, relatively homogeneous, mass containing a variable amount of phenocrysts separated at the intratelluric, plutonic, and hypabyssal phases of the magmatic stage of evolution. A specific composition of minerals crystallized under near-surface conditions, the character of contact metamorphism, and observation of active volcanoes suggest that the temperature of the trap, trachyte, and probably kimberlite and alkali-basalt magmas is 1100–1300°C at the hypabyssal phase (Bazilevsky 1966; Reverdatto 1968; Feoktistov 1972).

These conclusions are supported by the experimental study of natural silicate systems of various composition (containing H_2O, 0.5–4.0%, often 2.0–2.5%) within a wide range of pressure and temperatures. The range for the co-existance of the melt and crystals at a pressure of 500 MPa is 950–1150°, 1000–1200°, 1150–1230°, and 1180–1400°C respectively for alkali basalt, tholeiite, olivine basalt, and pycrite, respectively (Nasedkin et al. 1969).

The outburst of magma into a diatreme connected with the surface (see Chap. 16) results in a drastic release of pressure. This and some other reasons (discussed in Chaps. 18 and 19) lead to a spasmodic decrease in magma temperature, which is promoted also by the entry of numerous fragments of relatively cold country rocks into the melt. Therefore, the country rocks surrounding the channel and fallen from above could be subjected to intense pyrometamorphism only in the apical part of the feeding dyke, in the transitional zone, and in the lowermost part of the pipe where the temperature of the melt is still fairly high.

Continuing to fill the diatreme and ascending to the surface, the cooling melt entraps new batches of clastic material. Owing to low temperature and continuous cooling, the magma, continuing its way now produces a slight contact effect on both high horizons of the country and newly engulfed xenoliths. Under such conditions country rock xenoliths in eruption and in tuffisite breccias filling the explosion pipes should be composed mainly of unmetamorphosed varieties and subordinate strongly altered fragments withdrawn from the root parts of diatremes. Because the observed proportions of differently metamorphosed country rock xenoliths coincide with the expected ones, these factors may be considered as lending support to the concept of their origin, discussed above.

The initial temperature of the compositionally different melts intruding the pipes was 1100–1300°C, while that of the mass filling a diatreme is lower by hundreds of degrees. The effusive habit with an inconsiderable degree of crystallization of the groundmass of rock filling both small discrete dykes and large diatremes, suggests a similarity in condition of melt crystallization under quite different geological environments. This unexpected result is fully consistent with conclusions drawn from the analysis of thermal conditions imposed by the emplacement of these bodies. The exhaustion of thermal resources and hence the solidification of melts take place during a short time interval owing to a small (averaging 0.5 m) thickness of small basic and ultra-basic dykes. These features of diatremes, later filled with supercooled or almost overcooled melts, allow us to arrive at similar conclusions.

Crystallization of minerals composing the groundmass takes place by the intrusion and solidification of melts filling diatremes. In most kimberlites and comagmatic picrites such a main mineral is in all probability melilite, commonly replaced by serpentine, carbonate, and sometimes by monticellite in the postmagmatic phase. Clinopyroxene rather than melilite should crystallize from the melts with low calcium content. Although kimberlites with pyroxene groundmass have not been found as yet, adequate derivatives of the ultrabasic magma whose solidification gives rise to meime-chites are well known. The subparallel orientation of microlites of melilite and clino-pyroxene being the major elements of the fluidal structure of rocks is strong evidence for their crystallization prior to melting, most probably at the very beginning of its intrusion into a diatreme. On solidification of the melt, perovskite and some other minerals appear to be the first to crystallize.

Monticellite is formed at the first very high temperature (700–900°C) phase of the postmagmatic stage. Then follows the phlogopite phase, not ubiquitous but more extensive than the previous one. The temperature interval of kimberlite phlogopitiza-tion, based on a set of features and thermometrical measurements, is estimated at 550–700°C (Milashev 1965, 1972a) and may be used as the lower temperature limit for the formation of kimberlitic rocks in diatremes. The paleomagnetic study of Savrasov (1963) and, in particular, a distinct concordant orientation of the remanent magnetization vectors found by him in rocks of each generation imply that rocks in diatremes are consolidated above the Curie temperature of magnesian magnetites, that is above 500°C. Bailley (1963–1964) has obtained similar results (600–300°C) for the final emplacement of kimberlite breccia in the course of thermodynamic study.

Pressure

The hypotheses on the origin of pipes reviewed in Chapter 15 show that almost all workers, with rare exceptions, agree on the presence of high pressure as a prerequisite condition for kimberlite breccia formation. Opinions differ as to the reasons and

sources responsible for the production of high pressures at a depth of several kilometers from the surface: some ascribe it to the pressure of the ascending subcrustal
material, others try to produce evidence in support of the explosion of chemical
compounds, including hydrocarbons contained in rocks of the platform sedimentary
cover.

It is difficult to estimate the pressures just preceding the intrinsic volcanic processes
at the time of diatreme formation, that is at the hypabyssal phase of the magmatic
stage and at the explosion phase. At least two different techniques may be used:
(1) estimates based on the calculations of critical pressure by the extrusion of overlying strata at given mechanical properties and thickness of rocks composing the strata;
(2) experimental data obtained for stability parameters of minerals crystallized in
magmas of adequate composition at the hypabyssal phase and by the formation of
rocks within a diatreme.

Let us estimate critical pressures adequate for the formation of tubular bodies at a
depth of 2 km within terrigenous and carbonate rocks typical of most platform areas
with the developed sedimentary cover. Given the atmospheric pressure, the crushing
strength of sandstone and limestone is 74 and 96 MPa, respectively (Birch et al. 1942).
At depth, under conditions of uniform compression, the mechanical strength of rocks
increases, reaching, at a pressure of 138 MPa, in sandstone, limestone, and dolomite
280, 447, and 506 MPa, respectively. At a depth of 2 km, lithostatic pressure is about
55 MPa. Birch and coworkers have reported no experimental data for the mechanical
strength of rocks under such conditions. By extrapolation, we find the crushing
strength for sandstone, limestone, and dolomite of 220, 380, and 470 MPa, respectively, at a pressure of 55 MPa. In a depth interval of 0–2 km and in the case of the
subordinate occurrence of sandstone (5%), the proportion of limestone (45%) and
dolomite (50%) being approximtely equal, the parameter is taken to be 420 MPa.
The rupture strength in most rocks amounts to 0.1–0.15 of the crushing strength, in
our case it is 55 MPa.

With allowances made for the critical rupture strength obtained for the country
rocks surrounding the diatreme, we assess the magma pressure under which rupture
stresses close to breaking are produced in essentially carbonate rocks of a platform
cover. Because with depth pipes pass into dykes several meters thick and hundreds of
meters long, the upper halfspace above dykes where rupture stresses are applied may
be tentatively identified as a laterally lying half-cylinder. Then the highest stress σ
in the vault of rocks overlying the dyke may be derived from the formulae used for
the calculation of thick-walled vessels, and in particular from an approximate equation

$$\sigma = \frac{p_2 r_2^3 - p_1 r_1^3}{r_1^3 - r_2^3} + (p_2 - p_1) \frac{r_1^3 r_2^3}{2(r_1^3 - r_2^3)} , \qquad (17)$$

where p_1 and p_2 are external and internal pressures; r_1 and r_2, outer and inner radii
of the cylinder.

We make the dyke thickness equal to the inside diameter of a half-cylindrical zone
and a thickness of overlying strata equal to the outer radius of the zone. Substituting
corresponding values of Eq. (17) shows the pressure of magma and its emanations
of 206 MPa for the disruption of rocks above the dyke at the time of the formation

of diatremes of given parameters. At a lithostatic pressure of overlying strata of 55 MPa, the total pressure adequate for the diatreme formation is 260 MPa. Strakhov (1978), using only the disruption strength of rocks, arrived at similar conclusions: the total pressure of about 540 MPa is necessary for pipes to be formed at a depth of 5 km (lithostatic pressure of 130 MPa). The values obtained for the melt pressure at the close of the hypabyssal phase and just before the beginning of the explosion stage are consistent with the available petrological data. For example, kimberlites apart from secondary contain also late magmatic monticellite, which, according to the experimental study of Kushiro and Yoder (1964), can crystallize from melts only at fairly low pressure, no greater than 700–800 MPa.

The explosion stage consists of several repeated phases, each in turn being responsible for the roof outburst, formation of the channel owing to volatiles, and infilling of the vent formed by the melt cementing the eruption breccia. The resultant sealing of the channel leads again to increase of pressure and to a new extrusion, and so on. The repeated eruptions often incompletely "clear up" the channel, therefore blocks of earlier generation rocks remain in marginal parts of many diatremes, probably until the gradual extinction of explosive activity. Observations on the distribution of xenoliths of various rocks in successive generations of volcanites filling the complexly built diatremes imply a decrease in depth of the latest eruptions. This fact, along with the presence of volcanic relics of earlier generations, agrees well with the statements about a continuous drop of pressure in the course of volcanic activity.

When pressure reaches values at which the disruption of the overlying strata can take place, the enormous energy concentrated in a narrow zone gives rise to the formation of a tubular-shaped channel. It might be possible to assess the pressure produced by the processes responsible for the diatreme channel formation, i.e., the pressure produced at the first phase of the explosive stage and inferred from data about the extrusion of plutonic and sinking of overlying rocks with due regard for density, size, and shape of their fragments. However, to obtain the most rough estimates based on the calculation of transporting ability of volcanic emanation, we should know, apart from the easily measured parameters of clastic material, also the density and viscosity of a "working body" and at which rate it flows.

Viscosity, Density, and Phase State of the Matter

Viscosity, density, and phase state of the material incorporated in diatreme formation vary over a wide range. Study of the explosion pipes made most workers conclude that diatreme formation takes place by the action of magmatic melts, fluids, and gases. However, opinions differ mainly as to the importance of individual phases in the formation of the vent and the pipe. For example, some geologists relate pipe formation to the outburst of the sedimentary cover rocks mainly by magmatic melts, others ascribe it to fluidization processes, still others adhere to the hypothesis of gaseous explosion by diatreme formation (see Chap. 15). We know that those who advocate the leading role of magmatic melts by diatreme formation do not exclude the fluidization and gaseous phase, just as those who advocate the fluidization and gaseous explosion genesis of pipes also do not rule out the intrusion of magmatic melts in the already formed pipe vents, therefore, without giving preference to anyone phase, we shall assess viscosity and density of each step.

The viscosity of magmatic melts depends on many factors, the most important being proportions of petrogenic elements, volatile content, temperature, and the amount of phenocrysts and xenoliths. The study of lavas extruded from recent volcanoes and metallurgical slag shows a direct relation between the viscosity of silicate melts and their contents of SiO_2, Al_2O_3, Fe_2O_3, P_2O_5 and the inverse relation between viscosity and contents of Na_2O, K_2O, CaO, MgO, MnO, FeO, and TiO_2. This relation is often not strictly proportional but fairly complex, because when some or other components reach "critical" contents, the viscosity of melts begins to vary spasmodically. For example, slag in the system $SiO_2 - FeO - CaO$ with a content of 25% SiO_2 and 30% CaO is very mobile, but with SiO_2 in excess of 25% of its viscosity increases drastically (Selivanov and Shpeizman 1937). Similar relationships were found by observations on natural objects. According to MacDonald (1972) dynamic viscosity of basaltic lavas varies between 10^2 to 10^5 Pa s^{-1}. Of interest are data obtained by Fedotov (1976) on the composition, viscosity, and succession of lava extrusions in the 1975–1976 Bolshoi Tolbachik eruption. The eruptions began at the Severny volcano and lasted for 2 months. Gas emissions fountained from newly formed craters to a height of several kilometers and large scoria cones were created. The ejecta consisted of viscous, highly magnesian basalt lavas of medium alkalinity: SiO_2 49–50; MgO 9.5–10.5; CaO 11–12; Al_2O_3 12.5–13.5; $Na_2O - K_2O$ 3.2–3.5%. The first lava flow had a viscosity of 2×10^6 Pa s^{-1}. At the final stage of the Severny eruption there appeared varieties transitional in chemical composition to sub-alkali-alumina basalts extruded from the Yuzhny volcano, the viscosity of the lavas was 6×10^3 Pa s^{-1}. The eruptions from the Severny volcano were followed by those at the Yuzhny volcano, where activity was mild and effusive; there issued liquid lavas (viscosity $10^3 - 10^4$ Pa s^{-1}) of the following composition: SiO_2 50–51.5; MgO 4.5–6.5; CaO 8.5–9.5; Al_2O_3 16–17; $Na_2O - K_2O$ 5.5–6.0%.

Viscosity of magmatic melts is strongly temperature-dependent. In fact, the temperature effect is unidirectional, but multifaced in mechanism. One of the reasons is melt structure proportions of phyric phenocrysts and volatile constituents whose variability, all things being equal, is directly proportional to magma temperature.

When crystalline components are heated to melting temperature, the accelerated motion of atoms results, as is known, in a disturbance of the crystal lattice. However, it has long been known that if it is slightly higher than the liquidus temperature not all the atomic bonds break. In particular, there is evidence that a latent melting heat amounts to a small proportion from the entire energy required for the disturbance of a crystal lattice (sublimation heat). The values of density, refraction indices and some other physical parameters of silicate mineral glasses are almost the same as those of crystals. Hence, glass, despite the absence of crystal lattices intrinsic in minerals, also has some structural elements characteristic of silicates.

The crystal structure of silicates depends on the presence of silica-oxygen tetrahedrons, which together with the atomic bonds in them, govern the major physical properties of silicates. Therefore, we may assume that analogous silica-oxygen tetrahedrons are present in glass, and at a temperature close to that of the liquidus seem to be present in silicate melts as well. However, if in glasses, tetrahedrons are "tightly" bound, forming a random space lattice, then in melts they can move due to a slight effect resulting in the yield of the melt. The electric conductivity inherent in silicate melts implies an electrolytic dissociation, i.e., the presence of ions in these melts. Petrological, mineralogical, and geochemical observations give grounds to assume that a high proportion of ions in magmas are complex in composition, that is they are complex anions (Milashev 1972a; Shcherbina 1972). The motion of heat atoms is retarded when temperature decreases, and a large number of complex anions form; as a result the structure of the melt is also complicated and its viscosity increases.

Even a slight decrease in temperature of silicate melts considerably increases its viscosity. Apart from experiments (Table 27), it is confirmed by field observations, for instance, the measurements of viscosity of basaltic lavas from Makaopuhi Lake (Hawaii). The viscosity of the lava containing 25% crystals and 2–5% gas vesicles (bubbles) is 60 Pa s^{-1} at 1200°C. However, even at a slight drop of temperature (by 70°C, to 1130°C) the viscosity of the lava doubles and reaches 80 Pa s^{-1} (Shaw 1968).

The decrease in magma temperature not only retards the motion of atoms and makes the melt structure more complex, but results in crystallization of a large number of phyric phenocrysts, which, being suspended, increase in turn the viscosity of magma. The effect of phenocrysts on the total viscosity of magma may be obtained from the Ellers equation for the calculation of suspension viscosity with a volume fraction Φ of the solid phase no greater than 0.5 of the system volume

$$\mu_m = \mu_l \left\{ 1 + 2.5 \, \Phi \left[2(1 - 1.35 \, \Phi) \right] \right\}, \tag{18}$$

where μ_m and μ_l are viscosities of an unhomogenous mixture and pure liquid.

Applying the formula to measure the viscosity of basaltic lava of Makaopuhi Lake, we see that more than half of the values were affected by impurities of 25% of the phenocrysts (27 against 50 Pa s^{-1} at 1200°C and 43 against 80 Pa s^{-1} at 1130°C). It is clear that when the crystalline phase in such lavas is estimated at 50%, their total viscosity at the same temperature will rise to 200 and 320 Pa s^{-1}, respectively. Galimov (1973) has obtained the viscosity of the kimberlite magma at 1000° of 10^4 Pa s^{-1}.

The viscosity of magmatic melts is governed also by the amount of volatiles contained in them and primarily by the H_2O and CO_2 content. The increase of the H_2O content in a melt by only 1–2% leads to a two-threefold decrease in its viscosity. When the proportion of volatiles is above 10%, the melt acquires properties of a fluid whose viscosity drastically decreases, and at a gas content over 75% approaches the viscosity of gases. The viscosity of H_2O and CO_2 at a pressure of 50 MPa and a temperature of 100–700°C is $(10–45) \times 10^{-6}$ Pa s^{-1} (Vulkalovich and Altunin 1965; Vulkalovich et al. 1969).

Among magmatic rocks filling diatremes phyric ultrabasic and alkali-ultrabasic varieties and carbonatites, based on the above features, are to be assigned to a family whose members are derived from the least viscous magmatic melts. The mean viscosity seems to be characteristic of trap magmas with higher SiO_2 content at a lower amount of FeO, CaO, and volatiles as compared to the alkali-ultrabasic melts. The trachytic magmas are to be regarded as the most viscous.

Field observations support this successive increase in viscosity of melts differing in composition. As a rule, there are no trachyte dykes, while trap dykes are not only

Table 27. Viscosity of silicate melts (Pa s^{-1}). (Birch et al. 1942)

Basalt	Temperature, °C			
	1400	1300	1200	1150
Andesite	14.0	26.0	3120	8000
Olivine				
Hembudo	13.7	29.6	318	3790
Konura	12.0	17.3	73	–
Nepheline	8.0	9.7	19	–

extensive but fairly thick. There are many kimberlite and picrite-phyric dykes only 0.1–0.2 m thick and extending for many hundreds of meters. Very thin (up to 1 cm) stringers of kimberlitic and pycritic porphyry are found in rocks surrounding explosion pipes some tens of meters away from the diatreme contacts. The fact that these veinlets have resulted from the solidification of magmatic melts rather than fluids is supported by the presence of the groundmass flow structure caused by the subparallel arrangement of acicular microlites of melilite or serpentine-carbonate pseudomorphs after melilite.

The density of silicate melts is mainly a function of chemical composition. The density of a melt resembles that of glasses of the same composition and is much lower (by 1.0–0.3 g cm^{-3}) than that of rocks formed by melt crystallization. Magmatic melts whose volatile content amounts to several percent display the maximal difference in density as compared to holocrystalline rocks. Based on the mean density of ultrabasic, basic, and intermediate rocks (Birch et al. 1942), the average density of magmatic melts which gave rise to these rocks may be approximately estimated at 3.0, 2.7, and 2.5 g cm^{-3}, respectively.

When volatiles amount to above 10%, magmatic melts begin to acquire physical properties intrinsic in fluids with attendant much lower density. According to Kostrovitsky (1976) the kimberlite fluid is a residual liquid of essentially carbonate-magnesian composition rich in H_2O and CO_2. The density of the fluid system — liquid-solid suspension — was determined from the content of olivine phenocrysts amounting to 30 vol.% of the system. The value obtained for a fluid with suspended crystals at a gas phase of 10–75%, at a pressure of 50 MPa, and temperature of 200–600° varies within 0.67 and 2.48 g cm^{-3}.

CHAPTER 18

Dynamics and Mechanism of Pipe Formation

Velocity of Ascending Flow of Mantle Material

When magma rises from the subcrustal depth to the Earth's surface, the paths of its radial movement are subjected to important alterations. This also holds true for thermodynamic conditions, and as a result the rate of endogenic matter movement also changes greatly. It should be emphasized that the major trend of the radial movement of deep-seated matter, which is to be regarded as the most regular feature of these processes, shows an inversely proportional relationship between the depth at which the process is operating at each instant of time and the rate of movement.

For example, at maximal subcrustal depths where generation and initial evolution of magmatic melts take place, the rate of radial movement is often governed only by that of the convection flows of the mantle material, estimated by Pikeris (1935) and Latynina (1958) at about 1 cm year^{-1}. Even in cases when magma chambers being generated in unsealed substratum blocks are involved in the convection flow, the evolution of the blocks follows the mechanism of zone melting and the ascent rate remains of the same order of magnitude because the rate of the melt movement by zone melting, according to Magnitsky (1965), is about $0.1-1.0$ cm year^{-1}.

Similar rates were reported for the subhorizontal spreading of the apical portions of mantle material convection currents reaching the base of the Earth's crust. In the course of such (mushroom-like) mass movement, the highest rate is attained in the axial zone of the vertical portion of their path. In the case of subhorizontal spreading of the mantle material, the rate of its movement decreases as it recedes from the site where the trajectory changes. There are estimates for the "velocity" of the kimberlite volcanism within the Central Siberian province as a whole and also for the area from the central to the transitional and from the latter to the peripheral zones of this province (Table 28). Similar values ($0.3-0.4$ cm year^{-1}) were reported for the South African province where the intrusion of the diamond-subfacies kimberlites seems to have taken place in the Late Triassic-Early Jurassic (Jagersfontein, Kimberley and other fields), the pyrope-subfacies kimberlites (Sutherland field) being dated at 58 m.y. (Milashev 1972a).

Fedotov (1976) believes that high rates are inherent in columns of basaltic magmas ascending from the focal layer through the asthenosphere to volcanic centers due to hydrostatic forces. For the volcanoes of Kamchatka the rate of the intrusion of rounded cylindrical magma columns with an optimal radius of $350-1000$ m in the asthenosphere was estimated at 1 m year^{-1}.

Table 28. "Rate" of kimberlite volcanic activity in different areas of the Central Siberian province. (Milashev 1972a)

Areas of kimberlite occurrence	Age of kimberlites	Time interval m.y.	Distance between centers of areas km	"Rate" of kimberlite volcanic activity cm year^{-1}
Daldyn-Alakit Nizhnii Olenek	$D_3 - C_1$ $\}$ K	240	630	0.26
Daldyn-Alakit Srednii Olenek	$D_3 - C_1$ $\}$ $T_3 - J_1$	150	450	0.30
Srednii Olenek Nizhnii Olenek	$T_3 - J_1$ $\}$ K	90	180	0.20

Unlike the processes operative at depth, the eruptions of volcanoes are accompanied by the ejection of gases and other products at a rate of several hundreds of meters per second. It is natural to assume that the rate of ascending flows of the mantle material when it is carried from the subcrustal depth to the Earth's surface falls within the above values (some centimeters per year — hundreds of meters per second) and its change is inversely proportional to the depth of magma occurrence at every instant.

The rate of movement of magmatic melts differing in composition near the base of the crust cannot be estimated for lack of data. However, as a first approximation, it may be considered to be close to the rate of the basaltic magma flow estimated from geophysical data for the Kilauea and Mauna Loa volcanoes at 1 cm s^{-1} at a depth of 40 km (Sheinmann 1968). Seismological observations on earthquakes predating the 1975—1976 Bolshoi Tolbachik eruption suggest that the viscous, highly magnesian basaltic magma rose from the base of the crust (from a depth of about 30 km) at a rate of 3—4 cm s^{-1} (Fedotov 1976).

The strong dependence of the viscosity of silicate melt on their composition (see Chap. 17) does not permit identification of the ascent rates obtained by observations on the recent volcanoes with rates, for instance, of kimberlite magma flows. Therefore, to estimate probable rates of ascending flows of the deep-seated material and those of flows participating in the formation of diatremes filled with rocks which have no equivalents among recent volcanites, indirect evidence and contentions should be used, some of which may be applied to magmatic melts which took part only in the formation of diamondiferous kimberlites.

Ukhanov and Malysheva (1973), using the Mössbauer effect, have found the distribution of Fe^{2+} between M1 and M2 octahedrons in orthopyroxene from a lherzolite inclusion in kimberlite of the Obnazhennaya pipe and in the same mineral heated to 621 and 1000°C. The Mössbauer spectra of orthopyroxene contain two quadripole doublets corresponding to ions of ferrous iron in two types of coordination octahedrons, M1 and M2, the doublet areas being proportional to the Fe^{2+} content in each of the two structures. Fe^{2+} ions show a stronger tendency to fill distorted octahedrons M2 than Mg^{2+} ions, hence the ordered structure of orthopyroxene. On heating, the ordering decreases because Fe^{2+}, filling the space of Mg^{2+}, passes into the position of M1. The constant of this exchange equilibrium is a function of temperature and when the iron content of orthopyroxene is below 60%, it complies with an ideal solution model.

Using experimental data available on the Fe^{2+} distribution between M1 and M2 octahedrons and a simple thermodynamic relationship, one can determine an equilibrium temperature recorded in the orthopyroxene crystal. However, the importance of the temperature in terms of geology remains uncertain. The equilibrium distribution of Fe and Mg ions in orthopyroxenes sets is very fast: at $1000°C$ in 11 h and at $500°C$ in 50 h (Virgo and Hafner 1970). Thus, the orthopyroxene thermometer can provide the most valuable information about rocks subjected to drastic cooling. Because solidification of metamorphites and magmatites (except thin lava flows in small dykes) is a very slow process in the Earth's crust, the orthopyroxene thermometer shows not the temperature of the mineral formation but only a limit ($480°C$) below which the diffusion of ions cannot take place.

Records of orthopyroxenes disordered more strongly than those heated up to $600°C$ first appeared only in 1973 and this temperature was reported only for pyroxenes from chilled volcanites (Virgo and Hafner 1970). In 1973 Ukhanov and Malysheva, in the above-mentioned paper, stated that they found a zonal change in orthopyroxene ordering in lherzolite from kimberlite of the Obnazhennaya pipe, viz., pyroxene is highly ordered in the center of the nodule and disordered in 1.5-cm-wide outer zone.

The authors have not ventured to construct a temperature section for the lherzolite nodule and have not used the published data to determine the temperature because high magnesian orthopyroxene has not been studied experimentally. Nevertheless, they think "there are some grounds to suggest that the temperature in the core (of the nodule) was below $600°C$, while on the surface it was above $1000°C$ and may be even $1200°C$" (Ukhanov and Malysheva 1973, p. 1470). Ukhanov and Malysheva believe that the deep-seated inclusions coming in the kimberlite magma could not be heated to the full depth and were then subjected to a momentary quenching. "Therefore, as a result of fast cooling, such as quenching, the heated outer zone and the core which could not be heated above the temperature of lherzolite engulfed somewhere at a depth by the kimberlite melt and remained 'canned' in the outer homogenous nodule for tens of millions years" (p. 1470).

Further, the authors conclude that "the time during which the xenolith was carried by the kimberlite magma from the place of its original occurrence at a depth to the present level, as a first approximation, may be expressed by the time required for its heating. The latter can be determined for bodies of simple geometry provided the temperature distribution within them and the thermal conductivity are known" (p. 1471). Approximating the nodule as a ball and taking the temperature to be equal to $1200°C$ and $600°C$ respectively on its surface and in the center, they found that it took the sample only 1 min to be heated. Hence, the rate required for the kimberlite melt to rise to the Earth's surface should be above 500 m s^{-1} (!).

Such a rapid movement of magma through the crust (at the speed of a bullet in a gun barrel) does not seem feasible; the authors consider, therefore, that the figure was overestimated because they had not taken into account the wear of the xenolith by its upward movement as a result of which the outer heated layer was scraped off, and no allowance was made for the rate of orthopyroxene disordering on heating. Therefore, had the wear been allowed for, then the residence time of the xenolith in the melt should be "twice or three times as long or may be not that large" (p. 1471). The authors also mention that the rate of disordering is lower than that of heating, which means that the "orthopyroxene thermometer" shows the temperature with some delay and, therefore, the xenolith remained in the melt apparently longer than 1 min.

There is no doubt that the inclusion of garnet lherzolite remained in the kimberlite melt much longer than 1 min, even without allowing for the time adequate for the rise from the subcrustal depth to the Earth's surface (the authors tried to estimate this) because the kimberlite cement of the eruption breccia in the Obnazhennaya pipe had all the properties pertinent to the crystallization from the magmatic melt just within the diatreme. The melt had intruded into the channel developed and heated by emanations coming before the magma column, and for a long time it retained fluidity, as evidenced by the flow traces of the kimberlite groundmass and primary flow structures of the rock as a whole (see Chap. 2). These factors, along with a considerable depth of occurrence (several hundreds of meters from the paleosurface) of the sample studied and a large volume of the melt (hundreds of thousands or millions of cubic meters), are difficult to reconcile with the "instant" crystallization of the melt, its "quenching" and, therefore, lend support to the view of a moderately rapid (days-weeks-months) solidification of the mass.

A great body of data obtained by geologists imply that it took a long time for the fairly large volumes of silicate melts to solidify. According to their estimates, a complete solidification of rapidly intruded (country rocks have not been heated by earlier batches of melt) basalt dykes, say 20 m thick, will take place approximately in 2–2.6 years (Fedotov 1976), whereas small hypabyssal intrusions can solidify in about 6 m.y. (Masurenkov 1979). Fairly long periods of both total and vigorous activity of recent volcanoes, in particular the alkali-ultrabasic and carbonate volcanism of the Oldoinyo Lengai, imply that temperature remains high for long time intervals (see Chap. 12). Four phases (and several stages) of powerful activity, each lasting from several months to 3 years, were recorded for Oldoinyo Lengai during the period between 1913 and 1960 inclusively (there were no observations from January 1955 to early 1960).

In this context we should mention that the garnet lherzolite inclusion (reported by Ukhanov and Malysheva) from the kimberlite of Obnazhennaya pipe had enough time to be heated in its whole "depth" to the ambient temperature ($1000-1200°C$) and for the following gradual cooling when orthopyroxene was able to restore its high ordering, as in the case of orthopyroxenes from remaining inclusions in holocrystalline ultrabasic rocks from the same diatreme (mentioned on p. 1471 of the cited paper).

Although Ukhanov and Malysheva attribute the presence of a narrow surface zone of orthopyroxene disordering in one inclusion of garnet lherzolite to the nodule heating from the outside, their own data show that it is not such an easy problem to solve. They reported the Mössbauer spectra for orthopyroxene from different portions of the nodule as well as the control weights of the mineral preheated at 1000 and $621°C$ for 22 and 36 h, respectively. The pyroxene heated up to $1000°C$ displays not only higher disordering but widening of spectral lines and stronger chemical shift as compared to pyroxene heated at $621°C$. Orthopyroxene from the outer zone of the lherzolite inclusion, apart from higher disordering, exhibits narrowing of spectral lines and smaller chemical shift as compared to the species from the inner parts of the nodule.

The experimental coincidence of one of three characters does not allow us to ascribe the disordering of orthopyroxene in the narrow surface zone of the nodule to its subsequent "quenching" in the kimberlite magma and, hence, all the inferences made by Ukhanov and Malysheva become groundless, including that of the ascent of kimberlite melts at a rate of 500 m s^{-1}.

Very high rates for the ascent of kimberlite magma in the Earth's crust were reported by Galimov (1973), who postulated cavitation within a melt as an important factor of diamond formation. Using the equation describing the hydraulic losses in the magma channel, he found that cavitation may take place in melts moving at an average rate of 300–500 m s^{-1} and up to 1200 m s^{-1} through narrow sites of the channel at permissible physical parameters of magma and in case of its radial movement. These parameters are: the pressure drop Δp = 2000 MPa along the channel length of 10 km, the pressure of gas complete dissolution in a melt of 1100 MPa, density of a melt-fluid of 2.5 g cm^{-3}, its viscosity of 10^4 Pa s^{-1}, Reynolds number 1000–4000, and cavitation number 0.5.

The parameters taken by Galimov are valid in principle, although we cannot accept some of the values. For example, the length of the channel during the melt ascent from the upper mantle should be much greater, while within the diatreme it should be much smaller than 10 km. The ultrabasic melt-fluid rich in volatiles at a density of 2.5 g cm^{-3} should have much lower (by several orders of magnitude) viscosity.

The refinement of the parameters will entail a change in the estimated rate of kimberlite magma ascent, which may prove to be smaller than that obtained by Galimov. In this context, it should be stressed that even a rate of 300–500 m s^{-1} seems highly improbable for the magma movement within the crust so that higher values make estimates even more unreal. A rate for kimberlite magma movement of less than 300–500 m s^{-1} is insufficient for the generation of the cavitation regime. In this context all the discussed reconstructions make no sense, because cavitation is invoked not to support the high rate achieved by the melt, but is taken a priori as the mechanism of synthesis of natural diamonds, which according to Galimov, obviates all difficulties inherent in the hypotheses of diamond formation in kimberlites.

It is noteworthy that while eliminating some of the difficulties in other hypotheses of diamond genesis, the cavitation hypothesis has its own disadvantages. Apart from extremely high rates of magma movement achievable only in a free space, diamond crystallization necessitates the formation of cavitation bubbles, whose collapse exerts very high pressure and temperature. Galimov calculated that a diamond of 9 carats can crystallize from a bubble of diameter of no less than 40 cm. This means that diamonds of hundreds of carats can form from bubbles of several meters (!) in diameter. An additional requirement: the bubbles have to retain their spherical stability. The bubbles can form only in narrow portions of the channel and collapse as they reach wide portions. If narrow portions of the channel where the rate of the melt movement reached 1200 m s^{-1} accommodated bubbles up to several meters in diameter without disturbing the sphericity, then at ratio allowed by Galimov of diameter narrowing to average diameter of the channel of $0.1 \div 0.5$, the width of the channel has to be many tens of meters.

It also seems doubtful that crustal chasms or subisometric channels some tens of kilometers deep (from the upper mantle to the Earth's surface) and many tens of meters wide could exist even during a short period. If the channel were not a chasm, and magma pierced its way to the earth surface, its movement even through tectonically weakened zones at an average rate of $300-500$ m s^{-1} would have necessitated the concentration of huge additional amount of energy, in the form of an extremely powerful beam. Unfortunately, Galimov does not discuss these problems.

If the rate of hundreds of meters per second for the ascent of ultrabasic, basic, and intermediate magmas to the surface seems unrealistically high, a very slow rate of their movement compared to that of radial movement of magma in the upper mantle (cm y^{-1}) should be considered unlikely, at least for the melts from which diamondiferous kimberlites were derived.

Many researchers have made thermodynamic calculations of the stability of two major carbon modifications, diamond and graphite, at different PT conditions. The calculations were made to a different degree of approximation, but all of them, despite some variations in the position of the equilibrium curve, imply that diamonds are extremely barophilic. The positive isobaric potential of the graphite-diamond reaction under standard conditions and the fact that they are slightly temperature-dependent suggest that such a transition cannot take place at surface and under hypabyssal conditions. Knowing that the discussed polymorphic transformation is accompanied by decrease in volume, the increase of pressure at a constant temperature tends to diminish the isobaric potential. As a result, the diamond stability field lies within a high pressure area. This conclusion was confirmed experimentally.

As pressure drops at high temperature, the barophilic diamond becomes a metastable phase and turns into graphite, which is stable at similar PT conditions. Because the transformation of diamond into graphite requires an important alteration in atomic structure, such a transformation is relatively slow even in near-surface conditions and takes place at fairly high temperatures. At a temperature below $1200°$ or below $1000°C$, the process ceases and diamond retains its metastability phase for a period of uncertain length (over 1600 m.y. in the Premier Pipe).

The temperature of ultrabasic melts from which diamondiferous kimberlites are formed at plutonic and hypabyssal phases of the magmatic stage, that is by the ascent from the base of the crust to a depth of diatreme emplacement, varies from 1600 to $1200°C$ and pressure drops from 2000 to $300-400$ MPa (Milashev 1972a). It was shown experimentally that rise of temperature accelerates while increase of pressure (within a metastable field) retards diamond transformation into graphite. In kimberlite melts, diamond graphitization is a catalytic process and its rate, all things being equal, depends on the chemical composition of magma and, primarily, on the concentration of $[Ti^{4+}O_3]^{2-}$ anions of metatitanium acid (Milashev 1965).

Experiments on diamond heating free of oxygen (in nitrogen) were run at atmospheric pressure. Graphitization which started from crystal tops and edges, gradually incorporated crystal facets, and continued into the crystal until a complete transformation of diamond into carbon modification stable under such conditions. An important graphitization of medium-sized diamond on heating to $1500°C$ lasted for $20-40$ h.

Graphitic jackets and additions occurring on diamond crystals when kimberlite melt rises along deep fractures are usually not retained due to high magma viscosity and the stripping off the minerals from abundant phenocrysts. In the inner parts of the spongy diamond growths, protected from mechanical action, graphite amounts to 50%.

When kimberlite melts rise from the base of the crust to a depth of diatreme emplacement, some factors (high temperatures and anions of metatitanium acid) speed up or slow down (high pressure) graphitization of diamond, therefore, as a first approximation, the time of intense diamond graphitization in kimberlites is estimated at about 30 h. In most areas of kimberlite volcanism, the Moho discontinuity lies at a depth of about 35 km (Milashev 1974a; Milashev and Rozenberg 1974). Identifying the time to active diamond graphitization in kimberlite melts with that of the melt movement within the Earth's crust up to their intrusion in a diatreme, where a drastic temperature decrease below $1000°C$ (see Chap. 18) tends to "can" diamonds within the metastable field, we obtain an average rate for kimberlite magma ascent equal to about 0.33 m s^{-1}.

If the contention of the inverse relationship between the depth at which these processes are operating and the rate of material ascent is valid, then one should assume that the rate of melt ascent at the base of the crust is much lower than that at the middle levels and, hence, much lower than that of magma movement through the upper horizons of the crust. Taking the rate of melt movement at a depth of 40 km to be equal to 1 cm year^{-1} (Sheinmann 1968) and an averaged rate of their movement within the crust of about 0.33 m s^{-1} at a crustal thickness of 35 km, we can obtain tentative average rates for kimberlite magma movement along subsequent portions on their way to the Earth's surface, viz., 0.1, 0.3, 1, and 3 m s^{-1} respectively for depths of 35–30, 30–17, 17–5, and 5–0 km.

Attempts were also made to estimate rates of endogenic matter erupting from the subcrustal depths to the Earth's surface. For example, McGetchin and coworkers (McGetchin and Ulrich 1973; McGetchin et al. 1973) made their estimates proceeding from the statement that kimberlite magma is a residual, essentially carbonate, melt containing inclusions of silicate minerals. At a depth of about 3 km such magma "boils" and as a result of adiabatic expansion turns into a mixture of strongly cooled gas and hot fragments of rocks and minerals. Based on these models, the rate of the material flow is about 20 m s^{-1}, and it increases near the surface to 380 m s^{-1}. It should be stressed that, although overestimated, these values are still acceptable, whereas the geologo-petrological statements involved seem to be dubious.

The results obtained by Mercier (1979) are more reliable; he estimated an average rate for kimberlite magma ascent using the size of platy olivine crystals from inclusions of porphyroclastic peridotites. Considering these fine crystals as neoblasts formed after the capture of peridotite fragments by the kimberlite magma, Mercier found the residence time for the fragments in magma, with allowance made for depth of capture and also a probable rate of magma ascent to the surface. The growth rate of platy olivine crystals in peridotite nodules was inferred from experimental data; it was shown that the value is strongly temperature-dependent. The initial temperature and depth (pressure) at the time when peridotite fragments sunk into magma were estimated on the basis of calcium and aluminum contents in pyroxenes (Mercier 1976). Thus, the ascent rate of magmas from which the Southern African kimberlites of the Premier and Thaba Putsoa pipes were derived range from 11 to 20 m s^{-1} at initial temperatures of 1625 and 1725°C, respectively.

McCalister et al. (1979) reported very interesting observations on the structure decay in clinopyroxene of megacrysts and in clinopyroxene from lherzolite nodules from the Thaba Putsoa kimberlite pipe. Similar microstructures from exsolution of clinopyroxene of similar composition (tiny flakes of pigeonite in diopside) were found in experiments run at a regulated rate of cooling, which permitted the authors to estimate the time of cooling for natural samples at 15 h. The exsolution reaction proceeds over a fairly narrow temperature range stretching from 1300 to 1000°C (McCalister and Yund 1977). To estimate a probable ascent rate of the magma with engulfed peridotite fragments, the authors, using the published geotherms for the area of the pipe studied, take an initial depth of 190 km and so obtain an average rate of 3.5 m s^{-1}.

However, the kimberlite magma had a temperature of 1300°C most probably not at the intratelluric phase but at the close of the plutonic or onset of the hypabyssal phases (Milashev 1972a, 1974a), i.e., at a depth of 15–20 km. Moreover, it is not solidified rock but silicate melt, which, despite a high proportion of the solid phase (phenocrysts, xenoliths, and the like) undoubtedly retains plastic flow for some time (hours?), i.e., has a temperature of 1000°C, and intrudes the already formed diatreme. If the pyroxene decomposition reaction proceeds within a gradually cooling kimberlite over even half of the period obtained by McCalister, then an average rate for the melt ascent over the last 13–20 km may be estimated at 0.5–0.7 m s^{-1}.

The rate of the upward movement of the endogenic matter within the already formed diatremes was determined with due regard for the transporting ability of flows withdrawing rock fragments from the lower horizons of the sedimentary cover, from the platform crystalline basement, and from the upper mantle. The estimates are naturally strongly dependent on an assumed density and viscosity of the endogenic material. Kostrovitsky (1976) obtained estimates for kimberlite fluids containing 10, 50, and 70% of the gaseous phase and having a density of 2.4, 1.4, and 0.8 g cm^{-3}, respectively. The presence of garnet peridotite and eclogite nodules in kimberlitic rocks filling explosion pipes allowed him to use the formulae of applied hydraulics to determine the velocity for isolated flows of kimberlite fluids within the pipe vent as equal to 4 m s^{-1}.

It should be stressed that there is an alternative hypothesis explaining the presence of nodules of holocrystalline ultrabasic rocks in diatremes and, hence, different estimates for the rate of the endogenic matter flow within the pipe vent. The kimberlite tuffs are known to be comparatively rare (several percent of the total volume of kimberlite bodies) and most diatremes are composed of eruption breccias of kimberlite and massive varieties of these rocks, which bear strong evidence of their crystallization from magmatic melts. The homogeneous but not amygdaloidal structure of the rocks implies that these melts had never contained any noticeable amount of gas pockets.

In this context, not rejecting the importance of fluids in the formation of diatremes, we arrive at the conclusion that nodules of garnet peridotites and eclogites had been withdrawn from the subcrustal depth and brought into pipe vents by silicate magmas whose density and viscosity are much higher than those used by Kostrovitsky in his estimates for the rate of flow through the pipe. The rate of sinking of inclusions of the mantle rocks in such a medium is much lower than 4 m s^{-1}; however, it does not mean that the flow rates in diatremes should be lower as well. It became obvious that this calculation technique is invalid, because the presence of mantle rock nodules in diatremes suggests only that the rate of magma movement within the pipe vent was higher than that of nodule sinking, but in this case we cannot determine the upper limit of the flow rate, which may well be much higher than the rate of nodule sinking (in fact, infinitely higher).

In terms of the two-stage formation of explosion pipes (see Chap. 16), we should distinguish between the rate of gas flows at the time of diatreme formation and that of magmatic melt flow by infilling of the diatreme. The rate of a gas flow at the time of diatreme emplacement is not constant and continuously increases as the flow approaches the Earth's surface, and with the widening of the channel, it reaches several hundreds of meters per second near the channel mouth, as is evidenced by eruptions from recent volcanoes. On the contrary, the movement of magmatic melts as they rise within the pipe vents becomes slower, because owing to their funnel shape their cross-sectional area steadily increases toward the Earth's surface, whereas the size of the feeder in the root parts of diatremes actually remains the same. An average rate of melt ascent within diatremes most probably has the same order of magnitude as that in vents of recent volcanoes, i.e., not higher than several meters per second.

Thermomechanical Abrasion of Intruded Strata and the Formation of Diatreme Vents

Comprehensive geological data imply that it is expeditious to recognize two major phases in diatreme formation (see Chap. 16). The first phase is the formation of a pipe vent, most probably by means of a flow of highly compressed and very hot gases. The importance of gases in diatreme formation and transformation of an initially fissured vent into a pipe vent has been stated by many workers, who laid special emphasis on the stronger abrasion effect of gas flows by the agency of material particles contained in them (Daly 1933; Cloos 1941; Harris et al. 1970; MacDonald 1972; Lorenz 1975; Woolsey et al. 1975; Novikov and Slobodskoi 1978; Strakhov 1978).

The spatial distribution of kimberlite and other diatremes is undoubtedly governed by fractures which are not easily deciphered by means of geological observations or inferred from geophysical data. However, the genetic relation between explosion pipe localization and linear dislocations of country rocks stems from the fact that in areas of recent volcanism, scoria cones often display a linear arrangement: on deeply eroded sites necks composed of eruption breccias are usually confined to dykes of appropriate magmatic rocks. Kimberlite fields are marked not only by the linear arrangement of pipes and dykes but contain dykes and long (in plan) axes of diatremes running concordant with a trend of "chains", there we find also the transition of the lower parts of pipes into dykes.

The accounts discussed suggest that isometric and oval-elongate in plan diatremes formed as a result of a successive transformation of vents, which at earlier phases of eruption were strongly elongate, that is vents of fissured type emplaced in linear zone of the highly permeable crust.

Most statements on the mechanism and major phases in the formation of tubular channels are, in fact, fairly similar: compressed gas, regardless of its origin, migrates toward the lower pressure area or here toward the Earth's surface. According to Lorenz (1975) and some other workers, the formation of pipe vents takes place over the whole length of the initially fissured channel from the apical part of the magmatic column to the surface; however, most researchers believe that this process starts at the surface and goes deep down (Cloos 1941; Novikov and Slobodskoi 1978; Strakhov 1978).

Among the hypotheses of the second group we can recognize two subgroups on the basis of the supposed manner of gas eruption to the surface. Some workers believe that this is an explosion of gases which then descends along the fissure "like an explosion of a column charge of weak disruptive explosive"; with attendant shattering of the fissure walls and extrusion of clastic material this also stimulates the extrusion of pyroclasts due to boiling of pyromagma (Strakhov 1978).

Others do not consider gas release to the surface as an explosive outburst, and suggest that at considerable stresses due to high pressure and flow rate it is a fairly extended process (several hours – days – weeks) (Cloos 1941; Novikov and Slobodskoi 1978). The observations in areas of recent volcanism lend support to the second viewpoint. For example, the formation of volcanic vents crowned by maars at the 1957 phreatic eruptions on Iwo Jima Volcano Islands lasted for 65 min (Corwin and Foster 1959) and the 1973 eruption on the northern flank of the Tyatya Volcano continued for 3–4 h (Pavlov and Slobodskoi 1976).

A gradual transformation of fissure vents into pipe vents takes place also in magmatic eruptions when large quantities of gases are ejected. Fedotov, describing the creation of a scoria of cones in the 1975–1976 fissure eruption at Bolshoi Tolbachik, says: "Following the earthquake swarm a fissure opened up to 400–500 m long. Near the eruption site shocks were felt the day before and one could hear 'metallic' underground knocks. When the fissure opened, gas and ash were ejected along it, the fountaining of lava began. With the extension of the fissure the number of fountaining vents reached 10–30; however in a few hours the eruption had concentrated in one or two vents and a scoria cone began to grow" (Fedotov 1976, p. 19). In this case eruption from the first, the second, and third cones continued for 35, 38, and 8 days, respectively. The transformation of fissure vents into circular craters by the 1973 eruption on Heimay Island (Iceland) continued for about 24 h (Self et al. 1974) and in the 1970 eruption of the Beerenberg volcano on Jan Mayen Island (Birkenmajer 1972) terminated only after 5 days.

Fissure vents turn into pipe vents due to the shattering of walls and gradual widening of fissures by the dynamic and thermal agency of gas flow, and due to the impact-abrasion effect of particles engulfed by the flow. The differing initial width of the fissure and variable physicomechanical properties of rocks at different sites of the fissure lead to irregular widening already at the first phase of eruption. The enlargements formed at the intersection of fissures or in sites of lower-strength rocks suffer the most intense shattering, because the abrasive dynamic action of the gas flow with engulfed particles is directly proportional to the rate squared, which is in turn directly proportional to the diameter squared of a subisometric (or width square of elongate) channel.

An intense widening of initially favorable portions of fissures is accompanied by an increase in flow rate, followed by the shattering of surrounding strata, as a result of which the specific discharge of gas drastically increases. The narrow portions of the fissure gradually "die off" (clogged), adjacent enlargements merge, and a fissure vent isometric in plan is steadily transformed into a pipe channel. The splitting of a "flat" gas flow into a number of jets causes the uneven heating of enclosing strata at individual portions with attendant increase of stresses, stronger fissuring and collapse of rocks.

The formation of a pipe-shaped channel from a fissured vent in the case of the eruption discussed may take place only from above: starting from the Earth's surface and subsequently spreading deep into the depth. In this context, the volcanic process will lead to a gradual decrease in the degree of isometry (i.e., the degree of development) of a channel with depth, and orientation of the long axis of its cross-section will follow the trend of the initial fissure. Both phenomena were in fact geologically confirmed for all diatremes, and we may state that inferred gas emissions from fissured vents comply with the observations.

The above contentions on the formation of diatreme vents explain also features such as the typical conic pattern of these bodies, that is the decrease of their cross-sectional area with depth. Novikov and Slobodskoi (1978) have provided a comprehensive account. The formation of the vent from above is responsible for a long-term effect of destructive agents within the upper parts of a diatreme, and hence for their greater widening. By the formation of conic vents, very important apart from the time factor are: the increase of the rate ("momentum") of gas flow and that of engulfed debris in their movement from the roots to the mouth of a diatreme, the increase in amount and degree of comminution of debris whose abrasive ability increases as it approaches the surface. It should be stressed that when clasts attain a critical size, the lifting force of gas jets becomes smaller than the gravity affecting them, and then blocks of country rocks chipped off the diatreme walls move only downward.

Novikov and Slobodskoi (1978) attribute the generation of a funnel-shaped enlargement at the diatreme mouth to the broadening of a gas jet at the outlet from the vertical channel. The horizontal component of velocity vector of the gaseous flow and transported particles strongly affects the mouth projection formed at the intersection of the channel and the present exposure surface. As a result, the walls of the mouth portion are smoothed and the outlines of profiles line up with the velocity vector direction in each point of the near-wall part of the flow.

Dynamics of Diatreme Infilling and Formation of Major Genetic Rock Groups

After reaching its climax, the volcanic activity dies away. Mild volcanic activity marks a transition to the second phase in diatreme formation, viz., the infilling of pipe vents generated in the course of the preceding eruptions. With allowance made for a considerable length required for pipe-shaped channel formation and the observed pulsing eruptions of recent volcanoes, it should be stressed that intermittent decline in volcanic activity also takes place at the first stage of diatreme formation. However, at the stage when the diatreme vent is formed, paroxysmal eruptions not only help to clear the channel of debris collapsed during relative quiescence, but greatly widen it. At the second stage, eruptions remain pulsing, but owing to a general decline of volcanic activity the vent is filled both with material fallen from the walls and endogenic material.

The leading role of gases in channel formation and the gradual decrease in their quantity among eruption products is reflected in the fact that xenotuffisite and tuffisite breccias are the first to be formed and accumulated at the initial stage of diatreme infilling. At the end of volcanic activity, they give way to eruption breccias and massive varieties of magmatites, which solidify, usually not reaching the apical parts of diatremes.

The presence of powerful upward gaseous flows is very important not only for diatreme formation but for grading of clastic material which leads to the generation of banded "stratified" tuffs. The latter was reported by Geikie (1897) and described in a comprehensive account by Kostrovitsky (1976). Based on well-known laws of air flow in clastic rocks widely used in the study of separation and pneumatic transport of loose material, Kostrovitsky examines the hydrodynamic inter-relation of particles in a moving flow, the character of motion and distribution of clastic material, as well as the mechanism of banding formation in the vent-facies volcanites.

The change of a bulk body from steady to unsteady state, when the gas flow passing through it attains an adequate rate, incorporates several successive states of a layer, viz., (a) steady, (b) transient from steady to unsteady (pseudoliquified), (c) boiling, and (d) suspended converting into entrainment. In state (a) all parameters of clastic material (porosity — the ratio expressed as a percentage of the volume of the pore space to the total volume of a bulk body material; the relative positions of particles and the like) remain unaltered at a relatively low rate of gas flow. State (b) (transitional layer) sets in as the flow rate increases which leads to higher porosity as if the whole mass "swells up" and material is being graded and fine particles occur on top.

The further increase in flow rate governs the transition to state (c) ("boiling layer") characterized by an intense agitation of clastic material. The entire mass resembles a boiling fluid, the contacts between particles being continuously broken; porosity and volume (height in the vertical plane) of the mass increase. The previous sorting of the material is disturbed. The increase in the flow rate leads to the complete failure of the bulk material, and as a result its constituents are converted into suspension when the entraining force of the flow balances the gravity forces of the fragments. Such a rate is called a "rate of free soaring". If the rate of the upward movement of gas is higher than the rate of "free soaring", particles will be carried beyond the dynamic system.

It is quite evident that the above sequence of qualitatively different states of bulk material is retained at a higher rate of piercing gas flows only if fragments composing this material are fairly similar in size. Actually, the sizes of country rock fragments vary from fractions of a millimeter to many tens and in some cases to several hundreds of meters at the time of diatreme emplacement. From the available equations (Strakhovich 1934) the "rate of free soaring" is estimated at 15 m s^{-1}, 100 m s^{-1}, and 500 m s^{-1} respectively for particles of 1 mm radius with a density of 2.5 g cm^{-3} and at a density of gas of 0.003 m cm^{-3}, for similar fragments 10 cm in diameter and for blocks 2 m in diameter of the same rock.

Such a high velocity can be observed apparently only at the state of diatreme formation. This rate of gas emission (500 m s^{-1}) was recorded by Perret at the 1906 Vesuvius eruption (Rittman 1960). A comparatively small proportion of large blocks in kimberlite and other diatremes may be attributed to the extrusion rather than to the shattering of older erratics at the time the diatremes formed. Fragments derived from country rocks mainly vary in size from a few to about 20 cm; the rate of upward gas flows at the stage of diatreme emplacement may therefore be estimated at about 100 m s^{-1}. Under these conditions small particles whose rate of free soaring does not exceed several tens of meters per second are carried beyond the diatremes, whereas large blocks of country rocks sink in an upward gas flow at a rate which is directly proportional to the mass of blocks and inversely proportional to their cross-section area.

Abrasion and shattering of sinking blocks, "soaring" fragments and diatreme walls "bombarded" by fine particles engulfed by the flow involve a decrease in the proportion of large fragments and as a result the amount of fine clastic material increases. As volcanic activity declines, the intensity of eruptions drops, which is reflected in smaller quantities and lower rates of gas emissions, and as a result the transporting ability of upward flows decreases, and small fragments can now accumulate in a diatreme.

Diverse tuffogenic products including accreted volcanic bombs, lapilli, autoliths, banded, "stratified", tuffs, and tuffisite breccias appear to be formed in the early stage of diatreme infilling. These components and varieties of tuffs attract special attention and the possible mechanisms of their formation are widely discussed.

Kharkiv considers ball autoliths several millimeters to several centimeters in diameter composing some varieties of kimberlitic rocks as fragments derived from older kimberlites and formed ". . . by the solidification of the apical and near-contact parts of a magmatic chamber at depth during a temporal hault of kimberlite magma as it ascends to the surface" (1967a, p. 90). He also stresses the concentric arrangement of porphyric phenocrysts of olivine and pyrope and the sporadic occurrence of altered rock fragments in the center of ball autoliths. The section of the Legkaya pipe presented in the paper cited shows that kimberlitic rocks of small-ball structure fill the upper part of the diatreme, whereas the large-ball rocks occur below and are underlain in turn by porphyric massive kimberlites. According to Kharkiv the occurrence of ball-textured rocks in the upper part of the diatreme should be attributed to their withdrawal or extrusion from depth. This contention does not seem very persuasive and in this context the vertical "stratification" of rocks having small- and large-ball structure remains absolutely inexplicable.

Novikov and Slobodskoi (1978) provide a more substantiated explanation for the mechanism of formation of the concentrically zoned bombs and lapilli known to occur in ancient diatremes and in ejectamenta of recent volcanoes. Large fragments are transported by the flow at a rate slower than that of particles in a dispersed melt. Colliding, these particles stick to the central core — xenolith — gradually forming a mantle composed of magmatic rock. Flattened gas pockets and elongate grains of minerals observed in the mantle owe their orientation parallel to the outer surface of a bomb to the flattening of drops on collision. Round outlines of bombs and lapilli imply they were in free suspension at the time of their formation and reflect collision over the entire surface of the body.

Banded structures of rocks filling diatremes are fairly common. These structures, known from eruption breccias and from some (commonly endocontact) varieties of igneous rocks, are primary structures of unambiguous genesis. The same holds true for the genesis of stratified deposits crowning sections of weakly eroded diatremes (Mwadui, Orapa, and other kimberlite pipes, trap diatremes of the Angara-Ilim province). There are a number of cases showing a regular change in vertical and lateral distribution of grain size as reflected in the occurrence of slump breccias and pebble-stones at the base of the succession and at diatreme walls, whereas silty-clay rocks appear higher in the section and away from pipe walls. Freshwater Mesozoic and other fossils found in these deposits not only provided means for the establishment of the upper age limit of volcanism, but imply their affinity to crater lake deposits (which is unrelated to the problems discussed in the present book).

It is different with banded formations commonly filling the upper horizons of diatremes and consisting of variable amounts of magmatic (as a rule, predominant) and enclosing sedimentary and/or metamorphic rocks but nonfossiliferous and not displaying regular change either in grain size distribution or in lithology. As to the environments in which these rock varities were formed, opinions differ, which may be partially due to contradictory views on the reasons and mechanism of the generation of identical or even the same bodies. In most cases inadequate statements about the origin of banded rocks of the group discussed are probably due to actual differences in the genesis of these outwardly similar formations.

Steeply dipping to vertical bodies of banded kimberlite breccias filling the Aeromagnitnaya and Aikhal pipes were recorded by Zolnikov and Egorov (1970). The rocks owe their structure to the alternation of thin (0.5–3.0 cm) bands differing in the proportion of kimberlite cement and fine (1–5 mm) autoliths of earlier generations of kimberlite and fragments of country rocks set in the cement. Microstructure analysis shows a linear and plano-parallel orientation of primary minerals and xenoliths lined up with microstriation. With due regard for the geological setting of rocks, we may suggest a possible mechanism of their formation: kimberlite material, most probably in the plastic state, was squeezed from the central not yet solidified part of a pipe along fissures formed by solidification and cracking of apical and near-contact parts of diatremes.

Zolnikov and Egorov do not explain why only small fragments are present in breccias and how the thin-banded structure was formed, so that their opponents and, in particular, Kostrovitsky took it as impossible to explain these factors in terms of the hypothesis presented. However, we cannot agree with this. The thin-banded structure of breccias, the small size, and a certain orientation of fragments of rocks and minerals composing them may well be attributed to an intermittent regime of fissure opening. Discrete "layers" of banded breccias could well be formed by multiple intrusions of almost solidified melt into pulsatory opening fissures, with an amplitude of a single opening of several centimeters.

Kostrovitsky (1976) stresses the importance of the material sorting in banded rocks discussed by Zolnikov and Egorov, which he classes as tuffogenic formations. He parallels the conditions of their formation to a state of a "transitional layer" of bulk bodies as responsible for a grain size distribution of clastic material. However, he "forgets" a steep to vertical dipping of rocks and the presence of a primary structure, which cannot be explained for the case of tuffogenic formations even in terms of a "transitional layer".

Without rejecting in principle the possibility of the formation of banded structure in volcanites filling diatremes either by gaseous sorting when ejected fragments fell back (Geikie 1897) or by a "transitional layer" mechanism (Kostrovitsky 1976), allowance should be made that their proportion is very small even in structurally similar rocks because these structures can be formed and preserved under very peculiar conditions if they exist. For example, sorting of tuffogenic material following the "transitional layer" mechanism can take place only within a narrow range of gas flow rates (somewhere between the pseudoliquefaction and boiling of gases). Even a short-term increase in the flow rate completely eliminates the existing sorting of material. However, the Legkaya pipe, in which small-ball kimberlitic rock lies above a large-ball variety, provides a good example of the formation and preservation of volcanite sorting which most probably followed the "transitional layer" mechanism. A relatively short duration, moderate strain, and fairly smooth "dying off" of eruptions from such small (less than 0.2 ha) vents may be listed among the favorable factors.

Apart from rocks with vertical banding, varieties with horizontal banding were found in the Aikhal kimberlite pipe (Kharkiv and Prokopchuk 1973). There is a lens of these rocks about 40 m long and 10–12 m thick in the north-eastern part of the diatreme. The thickness decreases to 2–3 m in the contact part and can then be traced in the country rocks. In the middle part of this strongly elongate in plan diatreme, at its south-eastern contact with country rocks, banded rocks fill trough-like depressions about 60 m long, which seem to be the relics of a larger lens removed by erosion. The first lens displays an increase in grain size from psammitic to psephitic as it approaches the diatreme contact. The material of the second lens is characterized by distinct bedding persistent in thickness along the strike. The rock consists of rounded and oval, rarely angular fragments of limestone and clay limestone with long axes running along the bedding. Kharkiv and Prokopchuk classify rocks accumulated in karst caves among strongly carbonatized kimberlitic rocks as sedimentary.

Banded rocks filling separate isometric and elongate in plan parts of the upper horizons of the largest kimberlite diatreme in Yakutia were formed according to Zolnikov et al. (1979) by injection of kimberlite melts into the space between the layers of sedimentary rocks, the large xenoliths and "floating reefs" in kimberlite breccia. It is noteworthy that in some portions we see sharp contacts between bed-by-bed injections of kimberlite material and sedimentary rocks, whereas others are marked by gradual transitions from medium-clastic kimberlite breccia to fine-clastic aphyric varieties of kimberlitic rocks and then to altered sedimentary rocks. Here the above-listed varieties of kimberlitic rocks form bands oriented parallel with layers and joints of sedimentary rocks. A transition zone from one variety of kimberlitic rocks to the other is several tens of centimeters thick with a macroscopically regular distribution of all constituents in each rock variety.

The above statements imply the polygenic nature of the banded rocks filling the diatremes. Only special studies will provide an opportunity to determine conditions under which these outwardly fairly similar rocks were generated in each particular case.

Clastic material and pyroclasts are engulfed and forced toward the Earth's surface by the supply and ascent of magmatic melts within already formed diatremes. As a result, we have a typical vertical zoning to explosion pipes, viz., their parts are filled with tuffogenic rocks, below which come eruption breccias, replaced at a greater depth by massive magmatites.

CHAPTER 19

Energetics and Geology of Processes Responsible for Pipe Formation

Geothermal Energy Consumption by Diatreme Formation

If we take into account the large size of the diatremes within which rocks are shattered and extruded, the huge amount of energy consumed by the formation of these bodies becomes evident. It is also quite obvious how difficult it is to estimate quantitatively the amount of energy consumed, because such estimates can be based only on the data available, whereas volcanic edifices, being a kind of heat engine, have an efficiency of less than 100%. However, even an approximate estimate of the amount of energy required for the formation of a medium-sized diatreme is very important to understand many problems related to the genesis of pipes, and, in particular, the probable volume of compressed and overheated gases involved, sources from which they originated, size, depth of occurrence of magmatic chambers, and the like. The energetics of the formation of the most widespread — kimberlite and trap diatremes has been already discussed. For example, Strakhov (1971, 1972) had estimated the energy consumed by the formation of pipe vents of volcanic edifices of the Angara-Ilim iron-ore province, together with the amount of gases required. The estimates were based on the assumption of a complete evacuation of comminuted material from a pipe vent by the emission of volcanic gases and pyromagma, whose pressure was higher than resistance of roof rocks (Strakhov 1978).

The Krasnoyarskaya pipe, measuring 800 m by 1200 m in plan, has been investigated, based on its morphology; its vertical section is subdivided into three parts, of which each was estimated separately. The total volume of the pipe vent was estimated at 12.4×10^9 m^3. The energy required to overcome gravity was allowed for the upward movement of the gravity centers of each part of the pipe to a height of 200 m above a probable Triassic surface, i.e., at 500, 1500, and 2500 m, respectively. The resultant energy was 9×10^{16} J.

The extrusion of rocks from a pipe had to be accompanied by rock shattering, so that the energy consumed to overcome gravity is only a part of the total energy used by the diatreme formation. In this context Strakhov had estimated the minimal energy consumed by the formation of the Krasnoyarskaya pipe at $n \times 10^{17}$ J, an estimate confirmed by the assessment of the total energy consumed by the formation of the upper part of the pipe (down to a depth of 1200 m from the probable Triassic surface) using the empirical equation of Pokrovsky (1967) for vents formed by explosions at great depths. The resultant value was 2×10^{17} J. The volume of the lower part of the pipe is 1.5 times smaller, but its depth of occurrence is twice as much and, hence a large amount of energy is consumed. Therefore, the total consumption of energy required for the pipe formation could not be less than n $\times$ 10^{17} J.

Similar results were reported for recent volcanoes. If the energy required to overcome gravity by the formation of the vent of the Bezymyanny volcano was estimated at 3.2 $\times$ 10^{15} J on the basis of ejecta (Strakhov 1978) and that determined from empirical equations was 1.3 $\times$ 10^{16} (Strakhov 1978), then the total energy of the explosion inferred from the shock wave amounted to 1.2 $\times$ 10^{17} J (Gorshkov and Bogoyavlenskaya 1965).

The estimates obtained by Strakhov for the amount of gas required for pipe emplacement at a depth of 2–5 km suggest that an appropriate energy of the processes will be released by adiabatic expansion of gases of n $\times$ $(10^7 - 10^9)$ at an initial temperature of 1100°C and a pressure of 200–400 MPa (ie., PT conditions which can exist in an intermittent vent before the explosion). Based on observations of recent volcanoes, where the amount of gases emitted by explosive eruptions accounts for 3% of that of pyroclasts and lavas, Strakhov (1978) concludes that the volume of pyromagma responsible for trap pipe formation at a depth of 2–5 km amounts to $10^8 - 10^{10}$ m^3.

Energy estimates, as applied to various mechanisms of the kimberlite pipe formation, have been reported elsewhere (Kostrovitsky 1976). An empiric relation between the energy and depths of the charge occurrence, on the one hand, and radii of the vent formed and a column of shattered rocks formed by the underground nuclear explosions, on the other, shows that the minimal energy of a single explosion at a depth of 2 km whose destruction action reaches the Earth's surface amounts to 4.6 $\times$ 10^{16} J. In this case the radius of the vent formed by the explosion in the root part of the pipe was estimated at 185 m. Because of the absence of such enlargements in the root parts of well-known diatremes, Kostrovitsky and Vladimirov (1971) do not share the view about the formation of pipes by explosion from an intermittent vent. The thickness of feeding dykes is in fact several meters to several tens of meters, and the authors used it along with averaged dimensions of diatremes: a depth of 2 km, diameters on the surface and at a depth of 2 km are 300 m and 20 m, respectively.

Taking the volume of the pipe as 2.4 $\times$ 10^7 m^3 and the empirically found energy of shattering of this volume of rocks as 4.7 $\times$ 10^{11} J, and based on the assumption (see Pokrovsky 1967) that the value obtained amounts to 0.19% of the total energy consumption, Kostrovitsky and Vladimirov (1971) estimated the amount of energy at 2.4 $\times$ 10^{14} J. The estimate of energy released by chemical explosion within the confines of a feeding fissure was presented for comparison. Taking the volume of the fissure to be equal to $10^4 - 10^5$ m^3 and the amount of gases in it by PT conditions at a depth of 2 km to be 4.54 $\times$ 10^6 n mole, the energy of one mole of the gas phase (67% of inert components) to be 4.18 $\times$ 10^4 J, Kostrovitsky and Vladimirov estimated the energy of a single explosion in the fissure at $10^{12} - 10^{13}$ J. This amount of energy is insufficient for the formation of a medium-sized pipe and, therefore, they classify such action as a confined explosion.

Kostrovitsky and Vladimirov, in an attempt to confirm diatreme emplacement as a result of repeated confined explosions whose front moves toward the land surface, classify funnels in the upper parts of tubular channels as typical vents formed when explosion chambers reach near-surface horizons whose energy is sufficient not only to shatter the rocks but for their extrusion as well. For this purpose they estimated the maximal depth of a vent by the explosion with an energy equal to 10^{13} J, i.e., the value identical to the energy which they obtained for chemical explosions within a fissure. Using the equation of Pokrovsky and Chernigovsky (Pokrovsky 1967) for the energy from powerful explosions at great depths, they estimated the maximal depth of the vent at 110 m. They have stressed a slight difference in maximal calculated and observed depths of a vent, the latter being seldom 150 m deep. They name two main reasons leading to the discrepancy: first, the inconsistency between the effect of volcanic gases and that of explosives for which the formula of Pokrovsky and Chernigovsky appears valid; second, a possible slightly lower value of the maximal energy from a single fissure explosion. An energy of 4×10^{13} J, i.e., four times that of the calculated one, is required for a vent with a depth of 150 m to be formed.

We consider these statements of Kostrovitsky and Vladimirov to be very speculative and not substantiated geologically. Indicating identical or somewhat higher values of energy released by near-surface explosion as a result of which a vent was formed as compared to the amount of energy released from the explosion at a depth of 2 m, the authors forget that when they calculated a possible amount of energy in a feeding fissure they took its volume to be equal to $(10^4 - 10^5 \text{ m}^3)$ and estimate the amount of gases which can fill this volume under appropriate PT conditions at a depth of 2 km (50 MPa, 1100°C). However, the same 4.54×10^6 n mole of gases approaching the Earth's surface because of much lower PT parameters (especially pressure), and whose explosion could release an adequate amount of energy, will fill a volume of 10^7 m^3, i.e., almost the entire volume of a diatreme which has not yet formed. The assumption that the volume of the explosion chamber in fact does not change with the radial distribution of confined explosion front, naturally presumes a less powerful explosion and hence a vent 100 m deep cannot be formed.

Kostrovitsky and Vladimirov (1971) present a section of the Kimberley and De Beers Pipes crowned by funnel-shaped enlargements with sites of the inferred origination of explosion; these plots occur at the base of the funnel some 130 and 120 m below the present exposure surface. Thus we obtain an impression that this section supports the estimates by which the maximal depths of vent occurrence is 110 m, and reaches 150 m in extreme cases. However, if we take into consideration that since the time of their formation (K) these pipes had suffered denudation to a minimum depth of several tens of meters, then the section will rule out rather than confirm their statement. The least eroded Mwadui kimberlite pipe, having a funnel-shaped widening to a depth of 350 m (see Fig. 9), is direct evidence for the existence of funnels up to several hundreds of meters deep.

Evaluation of the work by adiabatic expansion of gases liberated from magmatic melts, made by Kostrovitsky and Vladimirov, was based on the estimates of Graton (1945) for the relative extent of the melt-gas system as a function of the magma ascent to the surface. Taking the volatile content as 4% and the volume of magma involved in the explosion as several times larger than the volume of a medium-sized pipe, first they calculated the work of discrete quantities of gases emitting from the magma at depths of 2, 1, 0.5, 0.25, 1.1, 0.02, and 0 km from the surface. The resultant energy released by adiabatic expansion of gases and consumed by the formation of a medium-sized kimberlite pipe equalled 4.2×10^{15} n J; the number n being of several unities, the total energy consumption may be estimated at 10^{16} J.

In fact, the same value, 3.4×10^{15} n, was obtained by Kostrovitsky and Vladimirov for a chemical explosion of gases released from magma whose volume was assumed to be equal to several volumes of a medium-sized diatreme. If the emission of self-exploding gaseous mixtures from magma seems questionable, then the amount of energy released by chemical explosion cannot be equal to the amount released by adiabatic expansion of gases emitted from magma. Because according to the above assumptions the magma volume (2.4×10^7 m^3) and gas content (4%) are the same in both cases, then both by the emission and adiabatic expansion of gases the same work has to be done, while by the generation of explosive mixtures it will increase owing to the energy of the chemical transformations. Therefore, the value obtained by Kostrovitsky and Vladimirov should be doubled to comply with the above version.

Apart from these errors, another inaccuracy affected all their energy estimates. Using a truncated cone as a very approximate model of the most common diatremes, they erroneously underestimated the volume of a medium-sized diatreme. With average dimensions used by the authors (h = 2000 m, R = 150 m, r = 10 m), the volume of a diatreme was taken to be 5.05×10^7 m^3 and not 2.4×10^7 m^3, which is easy to find by substituting the appropriate values into the well-known equation $V_{tr.cone} = (1/3) \pi h (R^2 + r^2 + Rr)$.

The amount of energy of 7.1×10^{14} J is required to overcome gravity by the extrusion of material to the surface when a diatreme of the volume given above is being formed in rocks with a mean density of 2.6 g cm^{-3}. The extrusion of material by diatreme formation is so far accompanied by shattering of rocks. It is difficult to determine energy consumption by the disintegration of faulted strata. The empirical equation expressing the relation between the amount of energy required for rock disintegration and modulus of elasticity and breaking strength, which is used in mining engineering in the calculation of explosive charges (Pokrovsky 1967), has also been applied to estimates of energy balance by the diatreme formation (Kostrovitsky 1976 and other works). However, the geological and methodological validity of this approach seems dubious. Comprehensive geological and petrological data argue not only against diatreme formation as a result of a single powerful explosion, but also against an explosive mechanism of their formation including multiple confined explosions whose front gradually moves toward the surface (see Chaps. 15, 16, 18).

If, in diatreme formation, as in the eruption of many modern central type volcanoes, highly compressed and heated gases play a leading role, then as the process procedes it will, apart from shattering of monolithic blocks due to collision, also involve a dense network of tectonic fissures emplaced long before the volcanic activity began. In fact no energy is required for the disintegration of rocks along the fissures: blocks bounded by cracks easily fall out of vent walls and these blocks, even on a slight collision or being engulfed by the impetuous upward flow of incandescent gases, spread along cutting fissures as small fragments. Fragments of rock whose size is below critical are extruded from the diatreme, whereas larger fragments sink at a rate proportional to their mass. Being continuously "bombarded" by small particles engulfed by the flow, the fragments inevitably become smaller, and on reaching critical size they are also ejected from the pipe vent.

The above mechanism of diatreme formation presumes very high energy losses, most of it actually "flies" into the atmosphere with the flow of hot gases; the efficiency of such a mechanism of diatreme formation is probably much lower than that of explosives used for excavation and rock shattering. Although it is difficult to estimate energy losses, we can assume the efficiency to be $1-10\%$. The vent at the Bezymyanny volcano provides indirect evidence for the validity of this assumption. Energy consumed for its formation was calculated from a mass of extruded rocks and from the empirical equations and proved to be respectively two and one order of magnitude lower than the amount of energy determined on the basis if the shock wave from the volcanic explosion (Gorshkov and Bogoyavlenskaya 1965; Strakhov 1978).

With due regard for these statements, the total amount of energy consumed by the formation of a medium-sized kimberlite pipe may be estimated at $n \times (10^{15} - 10^{16})$ J.

Sources and Probable Forms of Energy Conversion at the Time of Diatreme Formation

The existence of diatremes itself implies the concentration of adequate amounts of energy in the appropriate parts of the Earth's crust. To choose the least contradictory hypothesis of pipe formation, apart from the quantitative aspect of the energetics of the process, one should also take into account the source of energy and the "working body" contributing to the formation of pipe vents.

All the hypotheses of formation of diatremes filled mainly with igneous rocks are subdivided into two groups according to the probable source of energy. Most hypotheses of the first group envisage that the source of energy is the material ascending from the Earth's interior; this high pressure and high temperature material fills a pipe vent formed in the near-surface horizons of the crust. The hypotheses of the second group are based on the assumption of the presence of some exterior (with respect to deep-seated material) energy source.

The hypotheses relating diatreme formation to the internal energy sources of the acending deep-seated material from depth are in turn subdivided according to the kinds of energy and "working bodies" playing the leading role in these processes. The following main subgroups are classified on the basis of the following reasons: (a) mechanical energy head of a magma or highly concentrated fluid generated in subcrustal conditions without contribution from alkali-ultrabasic melts; (b) mechanical energy (head) of gases liberated from magma and, partly, the heat energy of the melt smoothing over drastic differences in gas temperature inevitable on the adiabatic espansion of gases; (c) heat energy of magma converted into mechanical energy of water vapor when fiery-liquid melts come into contact with groundwater; (d) mechanical energy of gases emitted from magma and heat energy of magma being converted into mechanical energy of water vapor when melt comes into contact with underground waters (i.e., combination of the previous two).

Those who ascribe diatreme formation to the effect of external sources of energy refer to: (a) hydrocarbons "sucked" by magma from the country rocks and transformed into self-detonating explosive mixtures when they come into contact with magma; (b) tectonic stresses (strains) giving rise to the plastic flow and diapirism of fairly cold primary-magmatic masses or to the generation of breccia-filled fissures whose intersection forms a path for the ascent of magma, which forces out and also partly catches and cements clastic material.

A critical account of the hypotheses (see Chap. 15), and the analysis of the physical state of the ejected matter, dynamics and mechanism of the formation and infilling of pipe vents (see Chaps. 17 and 18) show the importance of the gas phase in these processes. Estimates based on a hydrostatic diatreme model, especially for the case when diatremes are filled with ultrabasic rocks, lend support for the statement. A volcano hydrostatic model, according to Masurenkov (1979), using a basic equation of hydrostatics, establishes a relation between a height, L_1, of a column of magmatic melt as a function of its density, d_1, at a given density, d, and a thickness, L, of the crystalline shells of the Earth:

$$dL = d_1 L_1. \tag{19}$$

For volcanoes having onland edifices $L_1 = L + b$, i.e., the entire length of the channel is the depth (L) to its floor from the Earth's surface plus the height (b) of the volcanic cone.

For magmatic columns of ultrabasic composition which sink below the Moho discontinuity, Eq. (10) will be written as

$$d_2 L + d_2 l = dL + d_1 l, \tag{20}$$

where L and l are lengths of a magmatic column respectively below and above the Moho discontinuity; d, d_1, and d_2 are densities of the mantle and crustal rocks, and the magmatic melt.

$$1 - [(d - d_2)/(d_2 - d_1)] L. \tag{21}$$

At an average density of 3.4, 2.75 (Belousov 1966) and 3.3 g cm^{-3} respectively for the mantle and crustal rocks, and ultrabasic melts, we obtain l = 0.18 L. Hence, in the platform areas with a crustal thickness of 35–40 km, ultrabasic melts can reach the surface due to hydrostatic forces only when the length (depth) of the magmatic column is no less than 200–220 km. The generation of ultrabasic magmas at these or close depths is consistent with the present petrological and geophysical concepts; however, even the fairly short-term existence of a magmatic column extending to a depth of 160–180 km seems dubious. Knowing that extrusion can take place at very high excessive pressure (see Chap. 16), we should exclude the assumption of diatreme formation due to the hydrostatic head of ultrabasic magmas as not being substantiated. The assumption of excessive pressure of a melt due to tectonic stresses is also quite unrealistic because most workers agree that the generation of magma takes place at low strain, extension, and the like.

It should be stressed that because all magmas contain some volatiles liberated at low PT conditions, and because the circulation of groundwater takes place at a depth of several kilometers from the surface and the water coming in contact with melts starts boiling, we classify the statements about diatreme formation due to pressure exerted only by magmatic melt as speculative.

To accept a leading role of gases means also to solve the problem of a mechanism responsible for the excessive pressures required for the formation of pipe vents due to internal resources of even the "heaviest" ultrabasic magmas. Experimental data on the variability of pressure in a liquid column due to the floating up of gas bubbles in it provide strong evidence for the pressure increase in the system when part of the gases solved in magma are liberated by low PT parameters.

Each gas bubble in the lower part of the liquid column is under a pressure equal to the product of liquid density by column height. Coming to the upper part of the chamber, a gas bubble is no longer subjected to pressure and tries to expand; however, the volume cannot increase in a closed vessel, so that the pressure in the upper part of the chamber tends to be equal to the initial pressure in the lower part of the chamber, and as a result the pressure in the lower part is doubled (Chekalyuk 1970). A drastic increase of pressure also takes place when magma enters the crustal horizons saturated with underground waters.

The amount of gases sufficient for the formation of trap diatremes 2–5 km deep was determined by Strakhov (1978). He estimated that the required effect can be achieved by adiabatic expansion of volcanic vapors heated to $1100°C$ and compressed up to 200–400 MPa (PT conditions inferred for the chamber before the explosion) and with mass of 10^7 t and 10^9 t, respectively, at the depth of a pipe of 2 and 5 km. With a content of volatiles of 3% of magma and their complete separation from the melt, the volume of pyromagma required for the formation of a pipe 2–5 km deep was estimated at $10^8 – 10^{10}$ m^3. Knowing that basaltic magmas are poor in volatiles and that trap sills are not large enough to provide appropriate accumulations of gases in the apical parts of sills, Strakhov concluded that explosion pipes cannot form above the basic sills.

The explosive outburst of water vapors and gases appearing by the contact of trap magma and gas-water mixtures contained in the cracks and pores of rocks is also ruled out. The denial of such a mechanism of diatreme formation is motivated by the insufficient amount of such mixtures in a zone of direct contact with a melt and by the gradual warming up of distant portions of sedimentary strata, and hence the impossibility of a fairly strong explosion adequate for pipe formation.

Strakhov says: "For volcanic explosion to take place, a fairly large amount of volcanic gases should expand very fast, almost instantly. There are three possibilities: a drastic opening of a wide fissure up which gases will come to the surface, and the ability of infinite enlargement; the outburst of the chamber roof rocks with attendant collapse and extrusion of fragments through a newly formed pipe or vent; the combination of the above two, i.e., the opening of fissure above the chamber, filling the fissure with gases and pyromagma, and attendant outburst of the upper not fissured layer, and as a result the formation of a cone- or funnel-shaped vent" (1978, p. 77).

Without casting doubt on the correctness of the energy estimates obtained by Strakhov, we should stress the internal inconsistency of his statements on the genetic relationships of volcanic processes responsible for diatreme formation, and primarily his inability to discriminate between the reasons and consequences of the phenomena discussed. For example, if in the first case the role of the "trigger" is attributed to the external reasons of tectonic stresses or processes operating at depth, then the second case, in the sense of genetic relations, is a closed circle: the explosion ("almost instantaneous expansion of a fairly large amount of gases") presupposes an outstripping outburst of the roof with the extrusion of fragments and the attendant formation of a tubular channel, but the outburst and diatreme formation cannot take place without explosion!

Kostrovitsky and Vladimirov (1971) suggest that when the emanation of gases amounts to 4% of the kimberlite melt within a pipe then in terms of energy it is enough for shattering and comminution of an appropriate volume of the country rocks. In this context the explosion activity of gases can be manifested in two ways: (1) an abrupt adiabatic expansion of gases due to pressure release in the melt-gas system; (2) the gas phase is responsible for the formation of binary systems whose chemical interaction is characterized by a chain mechanism of explosion.

In the first case the mechanism is discussed but not the cause of energy liberation (we do not know what gave rise to an "abrupt release of pressure in the melt-gas system"). Moreover, one of the authors proposing the hypothesis of pipe formation due to repeated confined explosions with the front moving normally to the surface, has in fact ruled out the adiabatic expansion of gases as a driving mechanism of the diatreme formation (Kostrovitsky 1976).

Earlier several statements were reported in which volcanic explosions were attributed to chain chemical reactions proceeding in mixtures of volcanic gases (hydrogen, chlorine, and others) (Gushchenko 1962). However, observations on modern volcanoes carried out over many years provide no grounds to ascribe a leading role, if any, to these reactions as responsible for the mechanism and energetics of volcanic eruptions. If the above assumption were valid, the explosion volcanic clouds should contain a considerable amount of chloric hydrogen, but this simply does not exist. Even in the catastrophic 1956 eruption of the Bezymyanny volcano, the content of chlorine in the atmosphere amounted to only 0.106 mg l^{-1}, while that of carbonic acid and sulfur dioxide was larger by a factor of 2–3 than that of chloric hydrogen (Gorshkov 1956). Neither is there any evidence of high concentration of explosive components in gaseous ejectamenta of ancient volcanoes.

Therefore, despite the importance of gases in diatreme formations, the statement on adiabatic expansion and chemical explosion of volatiles liberated from magma as a leading mechanism have not been substantiated either theoretically or by observations. An alternative hypothesis is that of a phreatomagmatic origin of explosion pipes, which like any hypothesis has its advocates (Lorenz 1975, 1979) and opponents (Strakhov 1978).

We shall not dwell upon the well-known statements of the phreatomagmatic hypothesis and shall consider only the energy aspects; if the data obtained are consistent with the estimates of energy consumption on diatreme formation, we shall discuss the geological conditions optimal for the appropriate conversions of energy. It is reasonable to make estimates for diatremes composed of ultrabasic and alkali-ultrabasic melts because their smaller size, as compared to that of pipes filled with more acid rocks, is sometimes attributed to a more difficult ascent and outburst of magma. As a first approximation, ultrabasic and alkali-ultrabasic melts in composition and thermal and physical parameters may be identified with pyrolite with a melting heat of 503 J g^{-1}, a heat capacity of 1.26 J g^{-1} °C, a solidus temperature of 1300°C, and a liquidus temperature of 1500°C. The density of a melt containing volatiles is taken to be 3.2 g cm^{-3}.

The estimates show that when 1 cm^3 of the melt with the above parameters is being cooled from 1500 to 360°C, the amount of heat liberated totals 6285 J. At a depth of 2 km, corresponding to the depth of occurrence of a medium-sized diatreme, with a geothermal gradient of 33 m/°C, the temperature will be equal to 60°C, at a hydrostatic pressure of 20 MPa, and the boiling temperature of water of about 360°C. Under these conditions an energy of about 3520 J is adequate to convert 1 g of water into vapor. Hence, 1 m^3 of the ultrabasic melt, when cooled from 1500 to 360°C by the attendant crystallization, will liberate heat energy sufficient for the conversion of 1.76 t into vapor under the PT conditions apparently prevailing in the platform areas at a depth of 2 km from the Earth's surface. At a pressure of 20 MPa 1.76 t of water vapor occupies a volume of 5.4 m^3. A threefold increase of the volume accompanying the conversion of water from the liquid to the gas state leads to a drastic increase of pressure, and as a result vapor is forced from below into the overlying rocks of the sedimentary cover.

The temperature of gas drops steadily when it moves within strata whose temperature does not exceed several degrees centigrade, and pressure also steadily decreases as it approaches the surface. As a result of the change in pressure from 20 to 0.1 MPa in an ideal gas, the initial temperature of 633 K would fall to 3 K. However, water vapor exhibits a different temperature regime: heat losses due to warming up of surrounding rocks and vapor expansion lead to its partial condensation accompanied by heat release, owing to which the temperature of the remaining vapor remains at the boiling point of water at the appropriate pressure. Because of the great heat of evaporation-condensation (2258 J at 100°C), on the condensation of 1 g of water the amount of heat released is sufficient to raise 1100 g of the water vapor by 1°C.

We may thus conclude as a first approximation that only 1.32 t of 176 t of vapor reaches the surface when 1 m^3 of the ultrabasic melt comes into contact with underground waters at a depth of 2 km. This vapor, liberated into the atmosphere at a temperature of 100°C, carries away 35.2×10^8 J, of which 33.1×10^8 J have come from magma (at an initial water temperature of 60°C). The amount of energy required for the adiabatic expansion is about 19.3×10^8 J. A similar value (4.4×10^8 J) was obtained from the formula

$$A = \frac{\gamma}{\gamma - 1} \, p_1 \, V_1 \, (1 - \frac{p}{p_1} \times \frac{\gamma - 1}{\gamma}), \tag{22}$$

where p_1 and V_1, are the pressure and volume of the gas phase at the onset of the process; p is the pressure at the termination of the process; γ denotes polytropes for three-atomic gases (hence, for water) $\gamma = 1.33$.

A considerable amount of heat which cannot be reliably estimated is consumed for the warming-up of the rocks when water vapor rises to the surface. It is absorbed from the first fractions of vapor and is accompanied by condensation; as a result a fairly small volume of vapor is ejected into the atmosphere, heat is also taken from the juvenile volatiles which are always present in magmatic melts and liberated on their cooling.

According to the values presented at the beginning of the chapter, the amount of energy required for the formation of a medium-sized kimberlite pipe, even with allowances made for large losses (90–99%), was estimated at n × $(10^{15} - 10^{16})$ J.

This work is done by the adiabatic expansion of $n \times (10^6 - 10^7)$ t of water vapor rising from a depth of 2 km to the surface when it comes into contact with $n \times (10^6 - 10^7)$ m^3 of the ultrabasic melt. The resultant volume of the magmatic melt which can release enough energy is by an order of magnitude smaller or similar to that of a medium-sized kimberlite pipe $(5.05 \times 10^7$ $m^3)$; in this context the estimates are beyond doubt. Now we shall discuss energy conversions in terms of geological setting, and, in particular, the possibility of large masses of underground waters being involved in the process.

Geological Setting Responsible for Energy Conversion and Regular Features of Diatreme Distribution

Comprehensive study of the occurrence, localization, mineral composition, and many other geological features of explosion pipes suggests their formation as due to gas effect in the weakened parts of the crust within shatter zones located in blocks marked by isometrically oriented and densely spaced fissures. To elucidate a source supplying gases in a volume sufficient for such a result, we should rule out the assumption of their predominant juvenile nature, otherwise we should be suggesting that gases amount to 80–90% of the magma, which cannot be reconciled with the known structure and texture of volcanites. An alternative hypothesis envisages the conversion of heat energy of magma into the work of water vapor formed when magmatic melts come into contact with underground waters. The limited content of gas-water mixtures in fissured rocks made some workers doubt the efficiency of such a mechanism of diatreme formation (Strakhov 1978). This important problem should be discussed in detail.

We agree with the view about a small amount of pore and cavern water in sedimentary rocks in a zone of intense thermal effect of magma. These conditions seem to be favorable for the emplacement of small discrete dykes and stock-like bodies not confined to diatremes; the rise of the bodies up the syngenetic fractures fairly close to the surface is not accompanied by explosive activity. However, the geological setting and structural control suggest the formation of the bodies in a different environment.

In this context the peculiar geological setting of explosion pipes may be exemplified by the kimberlite and picrite fields containing numerous and highly concentrated diatremes, making the recognition of statistical rules possible. The main feature of the spatial distribution of explosion pipes within the confines of a given field, apart from their localization in the form of linear and isometric groups, is their occurrence in the areas of the present and ancient channels. In this case we should exclude fields containing thick intrusives with intrinsic contraction jointing differing greatly in orientation and spacing from fissures of the country rocks. The attempt to establish a relationship between the spatial distribution of explosion pipes and a drainage network using such areas as examples may result in a false correlation, because the developments of channels are strongly affected there by the high mechanical strength of intrusive rocks, and contraction jointing of magmatites is not directly related to the geological history of the region.

In those cases when kimberlites and picrites are sought by sampling of heavy concentrates based on trace minerals, the above tendency in diatreme distribution may be artificially intensified because this technique makes the discovery of pipes on a valley bottom and walls easier than in watershed areas. This tendency is evidenced also by the well-known fields where different prospecting techniques have been used, including those independent of topography or thickness of sediments overlying the pipes.

In terms of geology, the genetic relation of the discussed phenomena seems quite natural: the drainage system is known to be emplaced along fissured rocks, in zones of weakness and other faulting of the Earth's crust (Golbraikh et al. 1968). Recently it has been found that these dislocations control the distribution of diatremes as well (Milashev 1979). In the portions of heavily fissured and shattered rocks, with the rare exceptions of the accumulators of groundwater, the intrusion of magmatic melts rising from the subcrustal depths inevitably "involves" a continuous-discontinuous process of vapor generation. An explosive type of vapor ejection toward a minimum pressure up the fissure zone to the surface are intermittent and partly overlapping at the time of an influx of groundwater from the adjacent parts of a fault containing magma.

Intense incorporation of groundwater into explosive and hydrothermal activity is observed in areas of present volcanism. The small recent Ebeko volcano (Paramushir Island) provides an example of the extent of these phenomena. Twenty two years after a mild eruption of the volcano, fumarolic and thermal activity still continues, its flow totaling above 2 t s^{-1} or about 64 m t year^{-1}, with minimal thermal capacity estimated at 10^8 W (Zelenov 1959; Nekhoroshev 1959). The estimates show that almost the same amount of water vapor is required for the formation of a medium-sized kimberlite pipe by adiabatic expansion. At the onset and in the middle of volcanic activity the volume of water involved in the process per unit time was much (one to two orders of magnitude) greater and hence the required amount of vapor could be generated from water in a fairly short time interval (several days or weeks).

It should be stressed that the time factor as a requirement of a transient process is very important under the assumption of an explosive (especially one-act) mechanism of diatreme formation, but it is less important when a pipe vent is formed by the liberation of highly compressed and heated gases over a more extended period (days-weeks). With allowance made for the limited yield of even the most water-saturated fracture in the Earth's crust and the relatively small thickness (meters, rarely several meters) of rocks filling the fissures along which magma moves upward to the root parts as, for example, in well-known pipes, the generation of huge quantities ($10^6 - 10^7$ t) of vapor cannot take place very rapidly in several seconds or even minutes.

An appropriate amount of volatiles also cannot liberate from magma over a commesurable time interval. Volatiles are liberated from the melt mainly due to the pressure drop forcing the magma from below into the upper layers of the crust. Graton (1945) estimated that at the depth of pipe emplacement (2–2.5 km) about half of the total amount of volatiles is liberated, viz., at an initial content of 3–4% it accounts for 1–2% of the melt. At permissible dimensions (10×500 m) of the feeder for a medium-sized kimberlite diatreme and with a depth of efficient gas liberation of 500 m, the gas phase may amount to $10^4 - 10^5$ t, which is approximately two orders of magnitude smaller than the amount required for the formation of the pipe vent through any explosion mechanism.

It is noteworthy that statements of Chekalyuk (1970) about the increase of pressure due to the liberation and floating up of gas bubbles in a melt seem to be valid only for fairly small P-gradients. Should the first portions of gas bubbles floating up to the upper part of the magmatic column give rise to considerable pressure increase, the degasation of magma would cease until for some reason (gas leakage, ascent of the melt close to the surface) the pressure dropped again to adequate values.

The statement about the long duration (days – weeks) of the processes responsible for the generation of vapor in quantities sufficient for diatreme formation when groundwater come into contact with magmatic melts is quite consistent with the data available for the creation of recent volcanoes and with estimates for ancient volcanic edifices (see Chaps. 11 and 18). A steady intensification of eruptions observed during the first half of the explosive activity may probably be attributed to the extrusion of large quantitites of magma through apical parts of chimneys widened in the course of volcanic explosions, and also when large volumes of groundwater, in particularly from intruded (and hence drained by vents) aquifirs, become incorporated into the process.

The conversion of heat energy of magmatic melts into *mechanical work* of water vapor by the contact of a magma column with groundwater is accompanied by solidification, shattering, capture, and withdrawal due to upward vapor flows of some part of the endogenic material as well as by saturation with water vapors, lowering of crystallization temperature, and total cooling of the remaining melt.

The above statements explain many factors otherwise hardly explicable. We shall now deal with the most important of them. A peculiar feature of even adjacent pipes is the varying depth of the transition from pipe to dyke. Provided the pipes are formed by the violent liberation of volatiles from the magma at a depth where overlying strata have optimal pressure, then all the bodies of similar age in a region occur at almost the same depth from the paleosurface. However, the well-known closely spaced (0.1–0.3 km) diatremes whose depth of transition into dykes differs by many hundreds of meters (Mir – Sputnik Pipes, Kimberley – Saint Augustine, and others) are not only at variance with the above statement, but in fact eliminate its single geological premise. Alternatively, the depth of parts providing a maximal yield of groundwater and confined to fissures, karst caves, and other structures even within a small field differs and, according to the phreatomagmatic hypothesis of emplacement of any diatreme (or the transition into dykes), takes place at different depth.

A low-temperature effect on rocks surrounding a diatreme and on the country rock xenoliths captured by volcanic rocks is commonly attributed to the cooling of a melt by the adiabatic expansion of gases liberated from the melt. Two important points are commonly not taken into consideration. In the melt-gas system the melt is naturally a heat carrier and, therefore, 3–4% of volatiles from liberated magma cannot considerably decrease the amount of heat energy accumulated in the magma. In fact, the temperature of the melt should not drop, because the expansion of volcanic gases (when the regime to a greater or lesser extent approaches the adiabatic one) takes place only on their way to the surface, i.e., at some distance from the parent melt, where gases cannot affect its temperature. No discrepancies arise if we allow for the conversion of the greater amount of heat energy of magma into the mechanical work of water vapor when it comes into contact with groundwater, which leads also to the decrease of the total temperature of the system and that of the solidification of the magmatic melt by its saturation with water vapor.

Strong serpentinization typical of kimberlites and comagmatic porphyric ultra-basic and alkali-ultrabasic rocks is consistent with a large content of water in the magma and at the same time is at variance with the contention of the liberation of most volatiles from magma, and hence dewatering of the latter by diatreme formation. The phreatomagmatic hypothesis, on the contrary, not only indicates a "working body" as source responsible for the formation a pipe vent but also provides arguments for the indispensible saturation of a melt with water vapor.

Of note also is that among the hypotheses on the diatreme origin (see Chap. 15) only the phreatomagmatic hypothesis postulates and explains the genetic relation between the localization of explosion and water-saturated rocks filling the fractures in the Earth's crust. This means that the observed confinement of most diatremes to present and paleo-drainage systems and to fissure zones "not used" by surface channels but accumulating underground waters was substantiated theoretically and, therefore, may now be classed not as an empirical but as a scientifically valid state-ment. The conclusion as to the recognition of ancient and paleovalleys as probable sites for prospecting for diamondiferous diatremes within both already known and new kimberlite fields is of practical importance.

Conclusions

This book is a synthesis of data on the geostructural setting, morphology, internal structure, mineral composition, and other features of tubular bodies. It deals with many problems relevant to the mechanism and conditions of diatreme formation. Special emphasis is given to the analysis of thermodynamic and physical parameters of the material involved in pipe formation and to the dynamics, energetics, and mechanism of emplacement and infilling of diatremes. The geological environments favorable for their localization are also considered. The following major conclusions were reached:

1. Pipes filled with porphyric ultrabasic, alkali-ultrabasic, basic, intermediate, acid rocks, carbonatites, eruption, and tuffisite-breccias do not exhibit any fundamental differences in setting, morphology, and internal structure, which allow them to be placed in a single group of geological bodies, viz., diatremes.
2. The eruption products of recent volcanoes being in composition and textural and structural features identical to rocks filling most diatremes (except for kimberlite diatremes), all diatremes may be classed as erosionally exposed feeders or vents of ancient volcanoes of central type.
3. Despite a certain variability in diatreme shape, both horizontally and vertically, the general morphology of these bodies narrows downward. The least eroded pipes are crowned by funnels surrounded by remnants of erupted material at the surface. This material seems to have composed the body of volcanoes now destroyed.
4. The size of a pipe, including the horizontal cross-section and depth of emplacement, has no direct correlation with the composition of the infilling rocks. However, the fact that trap diatremes are generally larger than pipes filled with other rocks is not coincidental and may be related to the greater volumes of basic magmas generated at subcrustal depths and reaching the surface.
5. In general, diatremes occur in the weakened zones of the crust characterized by high permeability and therefore serve as paths for the ascent of deep-seated matter to the surface. These are mainly in rocks with dense fissure systems of isometric plan and in particular occur at sites of intersection of several fissures.
6. In areas where pipes are abundant, they commonly form linear groups and more or less isometric "clusters". Diatremes of different sizes do not exhibit any regular pattern in distribution, either in respect of the center or boundaries of the area, or within the local groups.

7. Among volcanic rocks filling pipes and formed in the course of eruption of magmatic melts of various (ultrabasic to acid) composition, the following structural-genetic groups can be recognized: porphyric massive and amygdaloidal magmatic rocks, their eruption breccia and tuffisite breccia and end varieties of tuffisite breccia, xenotuffisite with subordinate or negligible amount of eruption material. Tuffisite and xenotuffisite breccias occur mainly in the upper horizons of weakly eroded diatremes.

8. The period of diatreme formation consists of two stages, closely related and partially overlapping in time: the first stage is the formation of a pipe vent and the second its infilling.

9. The duration of volcanic eruption associated with the formation of the most structurally simple diatreme is several days to many weeks or months. The concept of almost instantaneous (several minutes) solidification of deep-seated intrusive matter in the pipe channels near the surface is not confirmed by the geological structure of ancient diatremes, and is at variance with the observed length of time required for the formation of the present-day volcanic vents.

10. The explosions themselves, i.e., the phenomena caused by a drastic increase and subsequent drop of pressure during the formation of diatremes, are probably not of particular importance. The driving mechanism of the process most probably is the emission of highly compressed and heated gases leading to thermomechanical abrasion of intruded strata and gradual transformation of a fissured channel into a cone-shaped one. Hence, the widely used expression "explosion pipes" is not quite appropriate, despite its figurativeness and deep historical roots.

11. Diatremes owe their formation to energy released by magmatic melts ascending from subcrustal depths, but the role of a catalyst ("working body") is played by gases among which vapor produced by the contact of magma and groundwater are the most important. The phreatomagmatic hypothesis of diatreme formation provides an explanation for well-known but poorly understood factors such as the low temperature character of the contact metamorphism of rocks and xenoliths surrounding pipes; the abundance of hydroxyl-bearing secondary minerals in volcanic rocks despite losses of gases by diatreme formation; the varying depth of emplacement of even adjacent pipes; and the confinement of most diatremes to recent and ancient valleys.

References

Agafonov LV, Pinus GV, Lesnov FP (1975) Deep-seated inclusions in alkali basaltoids of the Shavaryn Tsaram Pipe (Mongolia). Akad Nauk SSSR 224:1163–1165 (Russian)

Alekseev VV, Diakov AG (1961) Diamond-bearing formations of the Siberian platform and some regular features of the distribution of diamondiferous deposits. In: Kobelyatsky IA et al (eds) Materials concerning the geology and mineral resources of Yakutian ASSR. Yakut Knizhn Izd, Yakutsk VI:5–16 (Russian)

Antipov GI, Ivashchenko MA, Korabelnikova VV, Kosygin MK, Kuznetsov PM, Pekarin PM, Roslyakov GV, Strakhov LG (1960) Angara-Ilim iron deposits. Gosgeoltekhizdat, Moscow, 371 pp (Russian)

Arseniev AA (1961) The laws of the distribution of kimberlites in the eastern part of the Siberian platform. Dokl Akad Nauk SSSR Earth Sci Sect 137:1170–1173 (Russian)

Atkinson WJ, Hughes FE, Smith CB (1984) A review of the kimberlitic rocks of Western Australia. Proc 3rd Int Conf Kimb, Elsevier 1:195–224

Atlasov IP (1960) Tectonics of the north-eastern part of the Siberian platform. In: Saks VN (ed) Tectonics of the north-eastern part of the Siberian platform and Taimyr Trough. Gosgeoltekhizdat, Leningrad, pp 3–169 (Russian)

Babayan GD, Molchanov YuD, Savrasov DI (1976) Reflection of fault tectonics of the Malaya Botuobiya district in gravity field. In: Kovalsky VV (ed) Application of geophysical techniques in prospecting for kimberlite bodies in Yakutian province. Yakut Knizhn Izd, Yakutsk, pp 97–100 (Russian)

Bailley DK (1963–1964) Temperature and vapor composition in carbonatite and kimberlite. Ann Rept Dir Geophys Lab Carnegie Inst, Washington, pp 79–81

Barashkov YuP, Nikolaev NS, Marshintsev VK, Migalkin KN (1976) Distribution, peculiar features of morphology and composition of olivine of kimberlitic rocks of the Udachnaya-Zapadnaya and Udachnaya-Vostochnaya Pipes. In: Kovalsky VV, Oleinikov BV (eds) Geology, petrography and geochemistry of magmatic rocks in the north-east of the Siberian platform. Izd YaF Akad Nauk SSSR, Yakutsk, pp 131–147 (Russian)

Baratov RB, Kukhtikov MM, Mushkin IV (1970) Volcanic explosion pipes and some features of abyssal structure of the South Gissar. Donizh, Dushanbe, 114 pp (Russian)

Bardet MG (1964) Contrôle géotectonique de la répartition des venues diamantifères dans le monde. Chron Mines Rech Minière 328-9:67–89

Bazilevsky AG (1966a) On the temperature of emplacement of ultrabasic intrusions. Geokhimiya 4:404–409 (Russian)

Bazilevsky AG (1966b) Calculation of the temperature of a dyke of micaceous perioditite at the time of intrusion. Geokhimiya 8:1004–1005 (Russian)

Belousov VV (1962) Main problems of geotectonics. Izd Akad Nauk SSSR, Moscow, 608 pp (Russian)

Belousov VV (1966) The Earth's crust and upper mantle of continents. Nauka, Moscow, 123 pp (Russian)

Bergman SC (1984) Lamproites and other potassium-rich igneous rocks: a review of their occurrence, mineralogy and geochemistry. Proc Conf Alkaline Igneous Rocks, Edinbourgh, pp 1–78

Bessolitsyn EP (1970) On the origin of ore-enclosing forms of iron deposits of the Angara-Ilim district. In: Baryshev AS et al. (eds) Geology and mineral resources of the southern Siberian platform. Nedra, Moscow, pp 161–168 (Russian)

Bilibina TV, Dashkova AD, Donakov VI, Titov VK, Shchukin SI (1967) Petrology of alkaline vol-
canogenic-intrusive complex of the Aldan shield. Nedra, Moscow, 264 pp (Russian)

Birch Fr, Schairer JF, Spicer HC (1942) Handbook of physical constants. Geol Soc Am, Spec Pap
N 36, 289 pp

Birkenmajer K (1972) Geotectonic aspects of the Beerenberg volcano eruption 1970, Jan Mayen
Island. Acta Geol Pol 22:1–15

Blagulkina VA, Gubanov VA, Umanetz VN, Futergendler SI (1975) Ilmenite microcrysts from
kimberlite of the Luchakan district. In: Tatarinov PM et al (eds) Minerals and mineral parageneses of endogenic deposits. Nauka, Leningrad, pp 11–18 (Russian)

Blanchet PH (1957) Development of fracture analysis as exploration method. BAAPG 41-8:1748–
1759

Bobrievich AP, Ilupin IP, Kozlov IT, Lebedev LI, Pankratov AA, Smirnov GI, Kharkiv AD (1964)
Petrography and mineralogy of the kimberlitic rocks of Yakutia. Nedra, Moscow, 191 pp
(Russian)

Bogatykh IYa (1976) New data on dynamic influence of traps on kimberlite bodies. Dokl Akad
Nauk SSSR 226:166–167 (Russian)

Borley GD (1967) Potassium-rich volcanic rocks from southern Spain. Miner Mag 36:364–379

Borodin LS, Lapin AV, Pyatenko IK (1976) Petrology and geochemistry of dykes of alkali ultrabasic rocks and kimerlites. Nauka, Moscow, 244 pp (Russian)

Botkunov AI (1964) Some regulr pattern in the distribution of diamond in the Mir Pipe. Zap Vses
Miner Obshch 4:424–435 (Russian)

Brakhfogel FF, Petrov EK, Frolov VI, Shamshina EA (1975) On the Middle Carboniferous –
Lower Triassic deposits on the southern slope of the Anabar Dome. In: Kovalsky VV, Oleini-
kov BV (eds) Magmatic rocks of the north-east of the Siberian platform. YaF Akad Nauk SSSR,
Yakutsk, pp 3–18 (Russian)

Brakhfogel FF, Kovalsky VV, Korzilov AN, Lashkevich IV, Petrova EK, Shamshina EA (1979)
Age and erosion surface of one of kimberlite pipes in the Alakit field. In: Kovalsky VV, Oleini-
kov BV (eds) Mineralogy and geochemistry of kimberlite and trap rocks. Izd YaF Akad Nauk
SSSR, Yakutsk, pp 40–51 (Russian)

Bretshnaider SI (1966) Properties of gases and liquids. Technical methods of calculation. Khimiya,
Moscow, 535 pp (Russian)

Chekalyuk EV (1970) Retrograde phenomena in the process of energy accumulation in the Guten-
berg layer in the upper mantle. In: Magnitsky VA et al (eds) Problems of the structure of the
Easrth's crust and upper mantle. Nauka, Moscow, pp 286–289 (Russian)

Clement CR, Skinner EMW, Scott BH (1977) Kimberlite redefined. 2nd Int Conf Kimb, Santa Fe,
Ext Abst (unpaged)

Cloos H (1941) Bau und Tätigkeit von Tuffschloten. Untersuchungen an den Schwäbischen Vul-
kanen. Geol Rundsch 32:709–800

Construction of non-polynomial surfaces in trend surface analysis (1976) In: Oleinikov AN,
Romanovsky SI (eds) Programs for mixed computer systems. All-Union Geological Institute,
Leningrad, 13-24:23–24 (Russian)

Cornelissen AK, Verwoerd WJ (1975) The Bushmanland kimberlites and related rocks. Phys Chem
Earth 9:71–80

Corwin G, Foster HL (1959) The 1957 explosive eruption on Iwo Jima, Volcano Island. Am Sci J
257:161–171

Cross W (1897) The igneous rocks of the Leucite Hills and Pilot Butte, Wyoming. Am J Sci 4:
115–141

Daly RA (1933) Igneous rocks and depth of the Earth. Haffner Publ, New York, 598 pp

Davies KA (1952) The building of Mount Elgon (East Africa). Mem Geol Surv Uganda 7

Davis GL, Sobolev NV, Kharkiv AD (1980) New U-Pb datings of Yakutian kimberlites on zircon.
Dokl Akad Nauk SSSR 254:175–179 (Russian)

Dawson JB (1962a) The geology of Oldoinyo Lengai. Bull Volcanol, Napoli 24:349–387

Dawson JB (1962b) Sodium carbonate lavas from Oldoinyo Lengai, Tanganyika. Nature (Lond)
4846:1075–1076

Dawson JB (1964) Carbonate tuff cones in northern Tanganyika. Geol Mag 101:129–137

Dawson JB (1971) Advances in kimberlitic geology. Earth Sci Rev 7:187–214

Dawson JB, Hawthorne JB (1973) Magmatic sedimentation and carbonatic differentiation in kimberlite sills at Benfontein, South Africa. J Geol Soc Lond 129:61–85

Dawson JB, Powell DG, Reid AM (1970) Ultrabasic lavas and xenoliths from the Lashaine Volcano, Tanzania. J Petrol 11:519–548

Dukhanin SF, Lopatin BG (1961) On the effect of Sinian volcanism in the north-east of the Siberian platform. Inf Bull Nauchno Issled Inst Geol Arktiki, Leningrad 25:45–48 (Russian)

Du Toit AL (1906) The diamondiferous and allied pipes and fissures. In: Eleventh Annual Rept Geol Comm of the Cape of the Good Hope

Eckermann H von (1948) The alkaline district of Alnö Island. Sver Geol Unders, Stockholm, s Ca, no 36, 176 pp

Edwards CB, Howkins JB (1966) Kimberlites in Tanganyika with special reference to the Mwadui occurrence. Econ Geol 61:537–554

Egorov LS (1969) Melilitic rocks of the Maimecha-Kotui province. Nedra, Leningrad, 247 pp (Russian)

Erlikh EN (1958) On the tectonics of the central part of the Sukhan trough and regular pattern of distribution of kimberlite bodies in the Olenek River basin. Inf Bull Nauchno Issled Inst Geol Arktiki 12:16–25 (Russian)

Fedotov SA (1976) On the rise of basic magmas in the Earth's crust and the mechanism of fissure basalt eruptions. Izv Akad Nauk SSSR Earth Sci Sect 10:5–23 (Russian)

Feoktistov GD (1972) Contact metamorphism of sand-clayey rocks. Nauka, Moscow, 99 pp (Russian)

Filippov AG, Lelyukh MI (1980) Deep, mantled and armoured karst of the upper Alakit River (Yakutia). Dokl Akad Nauk SSSR 253:942–944

Filippov LV, Lipovsky YuO, Kapitonova TA (1976) Potassic basaltoids of Central Mongolia and some problems of abyssal magma formation. Geokhimiya 4:475–489 (Russian)

Frantsesson EV (1960) Contact alteration of country rocks related to the Egientei kimberlite vein. Nauchn Soobshch YaF Akad Nauk SSSR, Yakutsk 4:12–16

Frantsesson EV (1962) Composition and structure of the Mir kimberlite pipe. In: Menyailov AA (ed) Petrography and mineralogy of primary diamond deposits. Izv Akad Nauk SSSR, Moscow, pp 19–38 (Russian)

Frantsesson EV (1976) On peculiar features of the structure of kimberlite fields. Geol Rudn Mest 4:66–69 (Russian)

Frantsesson EV, Ilupin IP, Vaganov VI (1976) Petrological and structural indications of diamond content of kimberlite fields. In: Materials for V All-Union Petrographic Meeting. Nauka, Alma-Ata, pp 69–71 (Russian)

Galimov EM (1973) Cavitation as a mechanism of synthesis of natural diamonds. Izv Akad Nauk SSSR Earth Sci Sect 2:22–37 (Russian)

Galimov EM, Prokhorov DV, Fedoseyev DV, Varnin VP (1973) Heterogeneous carbon isotope effects in synthesis of diamond and graphite from gas. Geokhimiya 3:416–424 (Russian)

Geikie A (1897) The ancient volcanoes of Great Britain. Macmillan, London, I, II

Gerling EK, Morozova IM, Nikitin YuV (1972) Radiological interpretation of anomalous age values of terrestrial and lunar rocks. In: Tugarinov AI (ed) Essays of modern geochemistry and analytical chemistry. Nauka, Moscow, pp 429–440 (Russian)

Golbraikh IG, Zabaluev VV, Lastochkin AN, Mirkin GR, Reinin IV (1968) Morphostructural method of study of the tectonics of closed platform oil- and gas-bearing regions. Nedra, Leningrad, 152 pp (Russian)

Gorshkov GS (1956) Eruption of the Bezymyanny volcano. Bull Volc stations 26:19–72 (Russian)

Gorshkov GS, Bogoyavlenskaya GE (1965) Bezymyanny volcano and peculiarities of its last eruption. Nauka, Moscow, 121 pp (Russian)

Graton LC (1945) Conjectures regarding volcanic heat. Am J Sci 243A:135

Green DH, Hibberson W (1970) Experimental duplication of conditions of precipitation of high-pressure phenocrysts in a basaltic magma. Phys Earth Planet Int 3:247–254

Guest NJ (1956) The volcanic activity of Oldoinyo Lengai, 1954. Rec Geol Surv Tanganyika 4: 56–59

Gushchenko II (1962) On the mechanism of formation of pyroclasts. Tr Lab Vulkan Akad Nauk SSSR 22:131–142 (Russian)

Harger HS (1906) The diamond pipes and fissures of South Africa. Trans Geol Soc S Afr 8:110–134

Harker RJ, Tuttle OF (1956) The lower limit of stability of akermanite ($Ca_2 MgSi_2 O_7$). Am J Sci 254:468–478

Harris PG, Kennedy WQ, Scarfe CM (1970) Volcanism versus plutonism – the effect of chemical composition. In: Newall G, Rast N (eds) Mechanism of igneous intrusion. Gallery Press, Liverpool, pp 187–200

Hawthorne JB (1968) Kimberlite sills. Trans Geol Soc S Afr 71:291–311

Hawthorne JB (1975) Model of a kimberlite pipe. Phys Chem Earth 9:1–15

Hearn BC (1968) Diatremes with kimberlitic affinities in north-central Montana. Science 159:522–625

Hobley CS (1918) A volcanic eruption of East Africa. J E Afr Uganda Nat Hist Soc 111:339–342

Hodgson RA (1965) Genetic and geometric relations between structures in basement and overlying sedimentary rocks, with examples from Colorado Plateau and Wyoming. BAAPG 49-7:935–949

Holmes A (1950) Petrogenesis of katungite and its associates. Am Mineral 35:772–792

Ilupin IP, Kaminsky FV, Frantsesson EV (1978) Geochemistry of kimberlites. Nedra, Moscow, 352 pp (Russian)

Ivankin PF, Musatov DM (1973) Some aspects of the tectonics, magmatism and metallogeny of the Siberian platform. In: Mezhelovsky NV (ed) Problems of the geology of ancient platforms. Krasnoyarskoe o-vo Gornoe, Krasnoyarsk, pp 47–64 (Russian)

Izarov VT, Kharkiv AD, Chernyi ED (1963) On the age of kimberlite bodies of the Daldyn-Alakit district. Geol Geofiz 9:102–117 (Russian)

Jaques AL, Lewis JD, Smith CB, Gregory GP, Ferguson J, Chappell BW, McCulloch MT (1984) The diamond-bearing ultrapotassic (lamproitic) rocks of the West Kimberley Region, Western Australia. Proc 3rd Int Kimb Conf, Elsevier 1:225–254

Kalmykov NT (1963) On the volcanic pipes of the Minusinsk intermontane trough. Izv Akad Nauk SSSR Earth Sci Sect 2:80–88 (Russian)

Kaminsky FV (1969) Peculiar features of kimberlite rocks of the Aldan Shield. Sov Geol 4:161–165 (Russian)

Kaminsky FV (1976) Alkali basaltoid breccia of Onoga Peninsula. Izv Akad Nauk SSSR Earth Sci Sect 7:50–59 (Russian)

Kaminsky FV (1980) Garnetiferous alkali basaltoids of the Shavaryn Tsaram district, Mongolia, and conditions of their formation. Geol Geofiz 3:23–35 (Russian)

Kaminsky FV, Potapov SV (1968) Kimberlite bodies of the Ingili district. Geol Geofiz 11:30–36 (Russian)

Kaminsky FV, Potapov SV (1969) Petrographic and mineralogic characteristics of kimberlitic rocks of the Ingili district. Geol Geofiz 1:50–55 (Russian)

Kaminsky FV, Klyuev YuA, Konstantinovsky AA, Piotrovsky SV, Sochneva EG, Yuzhakov VM (1975) Diamond indications of alkali basaltoids of the north of Russian platform. Dokl Akad Nauk SSSR 222:939–941 (Russian)

Kaminsky FV, Lavrova LD, Sandomirsky SM (1979) On the deep-seated inclusions in alkali basaltoids of Mongolia. Geol Geofiz 5:53–65 (Russian)

Kepezhinskas VV, Devyatkin EV, Dashdava Z (1975) Cenozoic basaltoids of the Tapyat depression (Mongolia). Geol Geofiz 4:3–14 (Russian)

Khanukaev AN (1958) On the physical essence of rock breakdown by explosion. In: Meknikov NV (ed) Problems of the theory of rock breakdown by the action of explosion. Uzd Akad Nauk SSSR, pp 7–44 (Russian)

Kharkiv AD (1967a) Early generation kimberlite balls in eruption kimberlite breccia. Izv Akad Nauk SSSR Earth Sci Sect 1:87–91 (Russian)

Kharkiv AD (1967b) New data on the age of kimberlite pipes of Daldyn-Alakit region. Geol Geofiz 4:124–128 (Russian)

Kharkiv AD (1967c) Instances of high temperature metamorphism connected with kimberlite magmas. Geol Geofiz 6:124–162 (Russian)

Kharkiv AD (1978) Mineralogical principles of search for diamond deposits. Nedra, Moscow, 136 pp (Russian)

Kharkiv AD, Melnik YuP (1970) Old weathering crust on kimberlitic rocks of the Imeni XXIII S'ezda KPSS Pipe (Malaya Botuobiya diamond district). Nauka, Moscow, pp 230–246 (Russian)

Kharkiv AD, Prokopchuk BI (1973) On the question of the genesis of laminated rocks in the Aikhal kimberlite pipe. Izv Akad Nauk SSSR Earth Sci Sect 7:11–15 (Russian)

Kharkiv AD, Boris EI, Ivanov IN, Shchukin VN (1972) On the characteristics of explosion pipes of the Malaya Botuobiya district. Sov Geol 8:51–65 (Russian)

King BC, Sutherland DL (1960) Alkaline rocks of eastern and southern Africa. Sci Progr XLVIII: 298–321, 709–720

Komarov AN, Ilupin IP (1978) New data on the age of kimberlites of Yakutia obtained by the fission-track method. Geokhimiya 7:1004–1014 (Russian)

Kopecky L, Sattran V (1966) Buried occurrences of pyrope-peridotite and the structure of the crystalline basement in the extreme SW of the Ceske Stredhori Mts. Krystalinikum 4:65–86 (Czeck)

Kopecky L, Pisova J, Pokorny L (1967) Pyrope-bearing diatremes of the Ceske Stredhori Mountains. Sb Geol Ved 12:81–130 (Czeck)

Kostrovitsky SI (1972) Peculiar features of the formation of kimberlite explosion pipes. Geol Geofiz 11:35–43 (Russian)

Kostrovitsky SI (1976) Physical conditions, hydraulics and kinematics of filling of kimberlite pipes. Nauka, Novosibirsk, 96 pp (Russian)

Kostrovitsky SI (1980) Fluid brecciation as a driving mechanism of the formation of kimberlite pipes. In: Ovcharenko FD (ed) Physico-chemical mechanics and lyophility of dispersal systems. Naukova Dumka, Kiev, p 13 (Russian)

Kostrovitsky SI, Vladimirov BM (1971) Energetical calculations for the mechanism of the formation of kimberlite pipes. Geol Geofiz 6:31–38 (Russian)

Kovalsky VV (1963) The kimberlitic rocks of Yakutia and main principles of their petrogenetic classification. Izd Akad Nauk SSSR, Moscow, 184 pp (Russian)

Kovalsky VV, Brakhfogel FF, Nikishov KN (1973) Cambrian fauna in xenoliths from kimberlite pipes of the eastern slope of the Anabar Arch. Dokl Akad Nauk SSSR 211:1161–1164 (Russian)

Kryukov AV (1962) On the explosion pipes of the North Minusinsk depression. Tr YaF Akad Nauk SSSR Earth Sci Sect 6:301–304 (Russian)

Kryukov AV (1964) Geology of the Kongara explosion pipe in the North Minusinsk depression. Nedra, Moscow, Tr SNIIGGIMS 35:190–202 (Russian)

Kushiro J, Yoder HS Jr (1964) Breakdown of monticellite and akermanite at high pressures. Ann Rep Dir Geophys Lab Carn Inst, Year Book, pp 63

Kutolin VA, Frolova VM, Mushkin IV, Zhukova EN (1973) Perology of ultrabasic inclusions in the Tuvish explosion pipe (Southern Gissar). Geol Geofiz 6:37–44 (Russian)

Lacroix A (1933) Les roches éruptives potassiques leucitiques ou non du Tonkin accidental. CR Acad Sci, Paris, Ser D 197:625–627

Lakatos S, Miller DS (1973) Problems of dating mica by the fission-track method. Can J Earth Sci 10-3:403–407

Latynina LA (1958) Thermal convection in the Earth's shell. Izv Akad Nauk SSSR 9:1085–1098 (Russian)

Lebedev AA (1960) On the serpentinization of kimberlites. Zap Vost Sib Otd Vses Miner Obshch 2 (Russian)

Lewis HC (1887) On a diamondiferous peridotite and genesis of the diamond. Geol Mag New Ser 4:22–24

Locke A (1926) The formation of certain ore bodies by mineralization stoping. Econ Geol 21: 431–453

Lodochnikov VN (1936) Serpentines and serpentinites (Ilchir and others). Izd TsNIGRI, Moscow 38:850 pp (Russian)

Lorenz V (1975) Formation of phreatomagmatic maar-diatreme volcanoes and its relevance to kimberlite diatreme. Phys Chem Earth, Pergamon Press 9:17–28

Lorenz V (1979) Phreatomagmatic origin of the olivine melilitite diatremes of the Swabian Alb, Germany. In: Boyd FR, Meyer HOA (eds) Kimberlites, diatremes and diamonds, their geology, petrology and geochemistry. AGU, Washington, pp 354–363

Luchitsky IV (1957) On young basalts of the Minusinsk intermontane trough. Izv Akad Nauk SSSR Earth Sci Sect 10:94–97 (Russian)

Luchitsky IV (1971) Fundamentals of paleovolcanology. Nauka, Moscow 1-2:383 pp (Russian)

MacDonald GA (1972) Volcanoes. Prentice Hall, Inc., Englewood Cliffs, New Jersey, 430 pp

Magnitsky VA (1965) The internal structure and physics of the Earth. Nedra, Moscow, 379 pp (Russian)

Malkov BA, Gustomesov VA (1975) Finding of Jurassic belemnite in the Obnazhennaya kimberlite pipe on the Olenek uplift (Northern Yakutia). Izv Akad Nauk SSSR Earth Sci Sect 11:137–140 (Russian)

Marshintsev VK (1974) Carbonatitic rocks of the eastern slope of the Anabar Arch. Yakut Knizhn Izd, Yakutsk, 120 pp (Russian)

Marshintsev VK, Balakshin GD (1969) On the nature of carbonatitic rocks on the eastern slope of the Anabar Arch. Dokl Akad Nauk SSSR 188:645–648 (Russian)

Marshintsev VK, Tursky VE, Migalkin KN, Nikolaev NS (1975a) Peculiarities of xenolith distribution in kimberlite rocks of the Udachnaya Pipe. In: Kovalsky VV, Oleinikov BV (eds) Magmatic rocks of the north-east of the Siberian platform. Izd YaF Akad Nauk SSSR, Yakutsk, pp 132–146 (Russian)

Marshintsev VK, Kovalsky VV, Migalkin KN, Nikolaev NS, Tursky VE (1975b) Peculiarities of the composition, structure and mechanism of formation of kimberlitic rocks of the Udachnaya-Vostochnaya Pipe. In: Kovalsky VV, Oleinikov BV (eds) Magmatic rocks of the north-east of the Siberian platform. Izd YaF Akad Nauk SSSR, pp 112–132 (Russian)

Masurenkov YuP (1979) Volcanoes over intrusions. Nauka, Moscow, 220 pp (Russian)

McCallister RH, Yund RA (1977) Coherent exsolution in Fe-free pyroxenes. Am Miner 62:721–726

McCallister RH, Meyer HOA, Aragon R (1979) Partial thermal history of two exsolved clinopyroxenes from the Thaba Putsoa kimberlite pipe, Lesotho. In: Boyd FR, Meyer HOA (eds) The mantle sample: inclusions in kimberlites and other colvanics. AGU, Washington, pp 244–248

McCallum ME, Eggler DH, Burns LK (1975) Kimberlitic diatremes in northern Colorado and southern Wyoming. Phys Chem Earth 9:149–162

McGetchin TR, Ullrich GW (1973) Xenoliths in maars and diatremes with inferrence for the Moon, Mars and Venus. J Geophys Res 78:1833–1853

McGetchin TR, Nihan YS, Chodos AA (1973) Carbonatite-kimberlite relations in the Cane Valley diatreme, San Juan County, Utah. J Gephys Res 78:1854–1869

McIver JR, Ferguson J (1979) Kimberlitic, melilitic, trachytic and carbonatite eruptives at Saltpetre Kop, Sutherland, South Africa. In: Boyd FR, Meyer HOA (eds) Kimberlite, diatremes and diamonds: their geology, petrology and geochemistry. AGU, Washington, pp 111–128

Menyailov AA (1962) Tuffs and kimberlites of the Siberian platform, their origin. Izd Akad Nauk SSSR, Moscow, 228 pp (Russian)

Mercier JCC (1976) Single-pyroxene geothermometry and geobarometry. Am Miner 61:601–615

Mercier JCC (1979) Peridotite xenoliths and the dynamics of kimberlite intrusion. In: Boyd FR, Meyer HOA (eds) The mantle sample: inclusions in kimberlites and other volcanics. AGU, Washington, pp 197–212

Meyer de Stadelhofen (1963) Les brèches kimberlitiques du territoire de Bakwanga (Congo). Arch Sci 16:87–143

Mikheenko VI (1972) Mechanism of the formation of kimberlite pipes. Dokl Akad Nauk SSSR 205:428–430 (Russian)

Mikheenko VI, Nenashev NI (1961) Absolute age of the formation and relative age of the emplacement of the kimberlites of Yakutia. Izd Akad Nauk SSSR, Moscow, pp 146–164 (Russian)

Milashev VA (1960) The application of structural analysis to kimberlite bodies. Geol Geofiz 6: 49–59 (Russian)

Milashev VA (1962) Secondary alterations of kimberlites. Tr Nauchno Issled Inst Geol Arktiki 121:165–185 (Russian)

Milashev VA (1963a) The term "kimberlite" and classification of kimberlitic rocks. Geol Geofiz 4:42–52 (Russian)

Milashev VA (1963b) Paragenetic associations of secondary rock-forming minerals in kimberlitic rocks. Geokhimiya 6:557–564 (Russian)

Milashev VA (1964) Petrochemical characteristics of kimberlitic rocks. Geol Geofiz 3:138–142 (Russian)

Milashev VA (1965) Petrochemistry of the kimberlites of Yakutia and characteristics of their diamond mineralization. Nedra, Leningrad, 160 pp (Russian) English translation by M Constable)

Milashev VA (1968a) Kimberlites and allied rocks of the Kuonamka district. Tr Nauchno Issled Inst Geol Arktiki 13:5–31 (Russian)

Milashev VA (1968b) Problems of kimberlite formation and the upper mantle of the Earth. Tr Nauchno Issled Inst Geol Arktiki 12:158–170 (Russian)

Milashev VA (1968c) The causes of the discrepancy between the results of absolute age determination of kimberlites and geological data on the time of their formation. Tr Nauchno Issled Inst Geol Arktiki 12:181–188 (Russian)

Milashev VA (1969) Petrochemistry of kimberlites and allied rocks of the Kuonamka district. Tr Nauchno Issled Inst Geol Arktiki 16:30–44 (Russian)

Milashev VA (1972) Factors of kimberlite localization. In: Rabkin MI et al (eds) Kimberlite volcanism and potentials of primary diamond content in the north-east of the Siberian platform. Izd Nauchno Issled Inst Geol Arktiki, Leningrad, pp 48–56 (Russian)

Milashev VA (1972a) Physico-chemical conditions of kimberlite formation. Nedra, Leningrad, 176 pp (Russian)

Milashev VA (1972b) Main principles and optimal schemes of zoning of kimberlite provinces. Sov Geol 1:125–128 (Russian)

Milashev VA (1973) The dependence of the thermodynamic regime and the scale of kimberlite volcanism on Earth crust permeability. In: Puminov AP, Zhukov VV (eds) Placer diamond content of the central part of Siberia. Izd Nauchno Issled Inst Geol Arktiki, Leningrad, pp 39–43 (Russian)

Milashev VA (1974a) Kimberlite provinces. Nedra, Leningrad, 238 pp (Russian)

Milashev VA (1974b) Main tendencies of distribution and factors of localization of kimberlites. In: Milashev VA (ed) Mineralogy, geochemistry and prediction of diamond deposits. Izd Nauchno Issled Inst Geol Arktiki, Leningrad, pp 78–88 (Russian)

Milashev VA (1974c) Major principles and criteria of prediction of primary diamond deposits. Izd Nauchno Issled Inst Geol Arktiki, Leningrad, pp 89–100 (Russian)

Milashev VA (1979) Structures of kimberlite fields. Nedra, Leningrad, 183 pp (Russian)

Milashev VA, Rozenberg VI (1974) Structure of the crust and distribution of kimberlites of the Siberian platform. Geol Geofiz 1:61–73 (Russian)

Milashev VA, Shulgina NI (1959) New data on the age of kimberlite of the Siberian platform. Dokl Akad Nauk SSSR 126:1320–1322 (Russian)

Milashev VA, Sokolova VP (1984) Major crustal fracturing and structural boundaries of kimberlite fields. Geol Geofiz 10:133–140 (Russian)

Milashev VA, Krutoyarsky MA, Rabkin MI, Erlikh EN (1963) Kimberlitic rocks and picritic porphyries of the north-east of the Siberian platform. Gosgeoltekhizdat, 216 pp (Russian)

Milashev VA, Tabunov SM, Tomanovskaya YuI (1971) Kimberlite fields of the north-east of the Siberian platform. In: Rabkin MI et al (eds) Kimberlite volcanism and potentials of primary diamond content in the north-east of the Siberian platform. Izd Nauchno Issled Inst Geol Arktiki, Leningrad, pp 5–42 (Russian)

Mitchell RH (1979) Kimberlite and related rocks – a critical reappraisal. J Geol 78:686–706

Nasedkin VV, Saltykovsky AYa, Genshaft YuS (1969) Peculiar features of evolution of silicate matter according to composition, pressure and temperature (from high PT experimental data). In: Markhinin EK et al (eds) Volcanism, thermal springs, and Earth's depth. Dalnevost knizhn Izd, Petropavlovsk-Kamchatsky, pp 126–131 (Russian)

Nekhoroshev AS (1959) Heat flow of the Ebeko volcano on Paramushir Island. In: Magakyan IG (ed) Problems of volcanism. Izd Akad Nauk Arm SSR, Erevan, pp 75–80 (Russian)

Nekrasov IYa, Gorbachev NS (1978) On the possible mechanism of kimberlite formation. Dokl Akad Nauk SSSR 240:181–184 (Russian)

Newton AR, Gurney JJ (1975) Discussion of "A plate tectonics origin for diamond-bearing kimberlites". Earth Planet Sci Lett 27:356–358

Niggli P (1923) Gesteins- und Mineralprovinzen. Bd 1. Gebrüder Borntraeger, Berlin, 602 pp

Nikishov KN, Altukhova ZA (1978) Structural features of kimberlite breccia in complex-structure pipes (Daldyn field). Geol Geofiz 7:32–43 (Russian)

Nikishov KN, Bogatykh IYa, Bogatykh MM (1975) Intrusive kimberlite body in the Alakit field. In: Kovalsky VV, Oleinikov BV (eds) Magmatic rocks of the north-east of the Siberian platform. YaF Akad Nauk SSSR, Yakutsk, pp 158–168 (Russian)

Nikitin BM (1980) Deformation of enclosing rocks during the formation of kimberlite pipes. Izv Akad Nauk SSSR Earth Sci Sect 11:41–49 (Russian)

Nikolaev VA, Dolivo-Dobrovolsky VV (1961) Fundamentals of the theory of the processes of magmatism and metamorphism. Gosgeoltekhizdat, Moscow, 338 pp (Russian)

Novikov LA, Slobodskoi RM (1978) Mechanism of diatreme formation. Sov Geol 8:3–14 (Russian)

Odintsov MM (1958) Data on the geology and diamond content of the Siberian platform. Tr Vost Sib F Akad Nauk SSSR Earth Sci Sect 14:37–75 (Russian)

Osann A (1906) Über einige Alkaligesteine aus Spanien. Featsch Rosenbusch, Stuttgart

Pavlinov VN (1971) Kimberlite bodies in the scheme of morphological classification of intrusive bodies. Izv Vusov Geol Razv 9:9–16 Russian)

Pavlov AL, Slobodskoi RM (1976) Mechanism of eruption of the Tyatya volcano, 1973. Geol Geofiz 6:46–53 (Russian)

Pikeris CL (1935) Thermal convection in the interior of the Earth. Monthly Notes Roy Astron Soc, Geophys Suppl 3:343–367

Pokrovsky GI (1967) Explosion. Nedra, Moscow 174 pp (Russian)

Pokrovsky GI, Chernigovsky AA (1960) Calculation of charges for mass excavating explosions. Gosgeoltekhizdat, Moscow, 46 pp (Russian)

Poturoev AA (1976) Structural relationships of kimberlite pipes and enclosing rocks. In: Kovalsky VV, Oleinikov BV (eds) Geology, petrography and geochemistry of magmatic rocks in the north-east of the Siberian platform. YaF Akad Nauk SSSR, Yakutsk, pp 12–22 (Russian)

Prider RT (1960) The leucite lamproites of the Fitzroy Basin, Western Australia. J Geol Soc Aust 6:71–118

Radkevich EA (1960) On the relation between major and minor fractures. In: Belousov VV, Gzovsky MV (eds) Problems of tectonophysics. Gosgeoltekhizdat, Moscow, pp 168–174 (Russian)

Reck H (1914) Oldoinyo Lengai, ein tätiger Vulkan im Gebiet der Deutsch-Ostafrikanischen Bruchstufe. Branca Festschrift VII:373–409

Reverdatto VV (1968) The temperature of hyperbasite formation relative to tectonic conditions. In: Shtreis NA et al (eds) Volcanism and tectogenesis. Nauka, Mscow, pp 235–236 (Russian)

Richard JJ (1942) Oldoinyo Lengai. The 1940–41 eruption. Volcanological observations in East Africa. J E Afr Uganda Nat Hist Soc XVI:89–108

Rittmann A (1951) Magmatic character and tectonic position of the Indonesian volcanoes. Nomenclature of volcanic rocks. Bull Vocanol Ser II tome 12:75–109

Rittmann A (1960) Vulkane und ihre Tätigkeit. Enke, Stuttgart, 333 pp

Roslyakov GV (1960) Rudnaya Gora deposit. In: Antipov GI et al. (eds) Angara-Ilim iron deposit. Gosgeoltekhizdat, Moscow, pp 104–155 (Russian)

Rozhkov IS, Melnik YuM, Kharkiv AD (1969) Ancient weathering crust of kimberlites of the Imeni XXIII S'ezda KPSS Pipe (Yakutia). Dokl Akad Nauk SSSR 183:1130–1133 (Russian)

Rust GW (1937) Preliminary notes on explosive volcanism in south-eastern Missouri. J Geol 1:48–75

Saether E (1957) The alkaline rock province of the Fen area in southern Norway. Nor Vidensk Selsk Skr 1

Sahama ThG (1974) Potassium-rich alkaline rocks. In: Sorensen H (ed) The alkaline rocks. Wiley, London, pp 96–109

Sarsadskikh NN (1968) Structural factor of the distribution of kimberlites on the Siberian platform and prediction of primary diamond content. In: Kavardin GI et al (eds) Prediction and methods of search for nickel, tin and diamond in the Soviet Arctic. Izd Nauchno Issled Inst Geol Arktiki, Leningrad, pp 72–77 (Russian)

Savrasov DI (1963) On the application of the paleomagnetic method to determine the age of kimberlite and traps. Tr YaF Akad Nauk SSSR Earth Sci Sect 9:17–19 (Russian)

Sawkins FJ (1969) Chemical brecciation, an unrecognized mechanism for breccia formation? Econ Geol 64:613–617

Self S, Sparks SJ, Booth B, Walker GPL (1974) The 1973 Heimay strombolian scoria deposit, Iceland. Geol Mag 111-6:539–548

Selivanov BP, Shpeizman VM (1937) Viscosity of scoria of the system O_2–FeO–Ca. Metallurg 10:17–22 (Russian)

Serenko VP, Kharkiv AD (1975) New data on the thermal action of kimberlite magma on xenoliths of enclosing rocks. Geol Geofiz 2:94–101 (Russian)

Shamshina EA (1979) Weathering crusts of kimberlitic rocks of Yakutia. Nauka, Novosibirsk, 151 pp (Russian)

Sharp WE (1974) A plate tectonic origin for diamond-bearing kimberlites. Earth Planet Sci Lett 21:351–354

Shaw HR (1968) The viscosity of basaltic magma on analysis of field measurements in Makaopuhi lava lake, Hawaii. Am J Sci 266-4:225–264

Shcherbina VV (1964) On the geochemistry of silicate melt. Tr Vses Miner Obshch 5:537–544 (Russian)

Shcherbina VV (1972) New data on the physics and chemistry of silicate melts. In: Tatarinov PM (ed) Problems of mineralogy and petrology. Nauka, Leningrad, pp 30–35 (Russian)

Shchukin VN, Kharkiv AD, Boris EI (1967) On the discovery of a new diamondiferous pipe in the Malaya Botuobiya district. Dokl Akad Nauk SSSR 177:193–196 (Russian)

Shchukin VN, Minorin VE, Kharkiv AD (1972) Tectonic control, stages in the formation and diamond content of kimberlite of the Malaya Botuobiya district. In: Ivensen YuP (ed) Ore formation and its relation with magmatism. Nauka, Moscow, pp 171–177 (Russian)

Sheimann YuM (1968) Essays of deep-seated geology. Nedra, Moscow, 231 pp (Russian)

Sillitoe RN, Sawkins FJ (1971) Geologic, mineralogic and fluid inclusion studies relating to the origin of copper-bearing tourmaline breccia pipes, Chile. Econ Geol 66:1028–1041

Sinitsyn NM (1957) Scheme of the tectonics of the Tien Shan. Vestn Leningr Gos Univ 12:5–25 (Russian)

Sobolev NV (1974) Deep-seated inclusions in kimberlites and the problem of the composition of the upper mantle. Nauka, Novosibirsk, 262 pp (Russian) (English translation: Am Geophys Un, Washington, 1977)

Sobolev VS (1960) Conditions of the formation of diamond deposits. Geol Geofiz 1:7–22 (Russian)

Sobolev VS (1964) Physico-chemical conditions of mineral formation in the earth's crust and mantle. Geol Geofiz 1:7–22 (Russian)

Spizharsky TN (1960) Geologic zoning of the Siberian platform and main regularities of distribution of mineral resources over its territory. In: Bardin IP et al (eds) Development of production forces of the Eastern Siberia. Izd Akad Nauk SSSR, Moscow, pp 29–43 (Russian)

Stankovsky AF, Danilov MA, Grib VP, Sinitsyn AV (1973) Explosion pipes of the Onega Peninsula. Sov Geol 8:69–79 (Russian)

Stankovsky AF, Verichev EM, Grib VP (1979) The new type of magmatism in the Vendian of the north of the Russian platform. Dokl Akad Nauk SSSR 247:1456–1460 (Russian)

Strakhov LG (1964) On the time of formation of volcanic pipes and iron deposits in the south of the Siberian platform. Geol Geofiz 7:60–73 (Russian)

Strakhov LG (1971) On the genesis of explosion pipes. In: Volcanism and Earth's depth. Nauka, Moscow, pp 52–55 (Russian)

Strakhov LG (1972) Energy of the formation of explosion pipes. Dokl Akad Nauk SSSR 203: 664–666 (Russian)

Strakhov LG (1978) Ore-bearing volcanoes in the south of the Siberian platform. Nauka, Novosibirsk, 118 pp (Russian)

Strakhov LG, Shiryaev PM (1965) On the cluster structure of iron deposits of the Angara-Ilim type. Geol Geofiz 7:120–124 (Russian)

Strakhovich KN (1934) Fundamentals of theory and calculation of pneumatic transport installations. ONTI, Leningrad-Moscow, 112 pp (Russian)

Strauss CA, Truter FC (1951) The alkali complex at Spitzkop, Sekukuniland, Eastern Transvaal. Trans Geol Soc S Afr 53:169–190

Strekeisen A (1967) Classification and nomenclature of igneous rocks – final report of an enquiry. Neues Jahrb Miner Abh 107:144–214

Stutzer O (1935) Die Lagerstätten der Edelsteine und Schmucksteine. Berlin, 200 S

Tabunov SM (1971a) On the question of the relationship between kimberlite volcanism and fractures in the Earth's crust. In: Rabkin MI et al (eds) Kimberlite volcanism and potentials of primary diamond content of the north-east of the Siberian platform. Izd Nauchno Issled Inst Geol Arktiki, Leningrad, pp 62–64 (Russian)

Tabunov SM (1971b) Sizes and diamond content of kimberlite diatremes. In: Rabkin MI et al (eds) Kimberlite volcanism and potentials of primary diamond content of the north-east of the Siberian platform. Izd Nauchno Issled Inst Geol Arktiki, Leningrad, pp 144–147 (Russian)

Taljaard MS (1936) South African melilite basalts and their relations. Trans Geol Soc S Afr 39: 281–316

Trofimov VC (1967) On the peculiarity of distribution and formation of diamond deposits. Nedra, Moscow, 300 pp (Russian)

Troger WE (1935) Spezielle Petrographie der Eruptivgesteine. Dtsch Mineral Ges, Berlin

Tursky AA (1969) Some structural features of the Daldyn-Alakit kimberlite field. Zap Leningr Gorn Inst 58-2:85–90 (Russian)

Ukhanov AV, Malysheva TV (1973) Warming of ultrabasic xenoliths in kimberlite magma (from data of Fe^{57} Mössbauer spectrum in orthopyroxenes). Geokhimiya 10:1467–1472 (Russian)

Vaganov VI, Konstantinovsky AA (1978) Ultrapotassic trachytes on the eastern slope of the Anabar massif. Dokl Akad Nauk SSSR 241:691–694 (Russian)

Valkenburg A, Rynders GF (1958) Synthetic cuspidine. Am Min 43:11–12

Vasiliev VG, Kovalsky VV, Chersky NV (1961) The problem of the genesis of the diamond. Yakut Knizhn Izd, Yakutsk, 150 pp (Russian)

Virgo D, Hafner S (1970) Fe^{2+}, Mg order-disorder in natural orthopyroxenes. Am Min 55:1–2

Vladimirov BM, Tverdokhlebov VA, Kolesnikova TP (1971) Geology and petrography of igneous rocks of the south-western part of the Guinea shield. Nauka, Moscow, 242 pp (Russian)

Volcanic Centre (1980) Structure, dynamics, composition (Karym structure). Nauka, Moscow, 300 pp (Russian)

Vulkalovich MP, Altunin VV (1965) Thermophysical properties of carbon dioxide. Atomizdat, Moscow, 455 pp (Russian)

Vulkalovich MP, Rivkin SL, Aleksandrov AA (1969) Tables of thermophysical properties of water and water vapor. Izd standartov, Moscow, 408 pp (Russian)

Wade A, Prider RT (1940) The leucite-bearing rocks of the West Kimberley area, Western Australia. Q J Geol Soc, London 96:39–98

Wagner PA (1909) Die diamantführenden Gesteine Südafrikas, ihr Abbau und ihre Aufbereitung. Berlin, 132 S

Williams AF (1932) The genesis of the diamond. Ernest Benn, London, 636 pp

Woolsey TS, McCallum ME, Schumm SA (1975) Modelling of diatreme emplacement by fluidization. Phys Chem Earth 9:29–42

Zelenov KK (1959) Thermal springs of active volcanic zones as sources of accumulation of iron
and aluminium in sea basins as exemplified by the Ebeko volcano. In: Problems of volcanism.
Izd Akad Nauk Arm SSR, Erevan, pp 97–99 (Russian)

Zolnikov GV, Egorov OS (1970) Eutaxitic structure of kimberlitic rocks from some pipes in
Yakutia. Geol Geofiz 5:64–71 (Russian)

Zolnikov GV, Filippov ND (1976) Kimberlitic varieties of the Mir Pipe and the succession of their
formation. In: Kovalsky VV, Oleinikov BV (eds) Magmatic rocks of the north-east of the
Siberian platform. Izd YaF Akad Nauk SSSR, Yakutsk, pp 99–111 (Russian)

Zolnikov GV, Kovalsky VV (1976) The relationship between diamond content and content of
heavy fraction minerals in the kimberlitic rocks of the Mir Pipe. In: Kovalsky VV, Oleinikov
BV (eds) Geology, petrography and geochemistry of magmatic rocks in the north-east of the
Siberian platform. Izd YaF Akad Nauk SSSR, Yakutsk, pp 164–172 (Russian)

Zolnikov GV, Kovalsky VV, Zimin LA, Kornilova VP, Korzilov AN, Nikishov KN, Bogatykh IYa,
Kryuchkov AI, Lashkevich IV, Shamshina EA, Gamyanina VV (1979) Peculiarities of rocks
and conditions of formation of one of kimberlite pipes of the Alakit field. In: Kovalsky VV,
Oleinikov BV (eds) Mineralogy and geochemistry of kimberlite and trap rocks. Izd YaF Akad
Nauk SSSR, Yakutsk, pp 5–39 (Russian)

Zuenko VV, Makovskaya NS, Kharkiv AD, Chernetskaya NA (1973) Diamond distribution in
deeper levels of one of kimberlite pipes in Yakutia. Geol Geofiz 4:7–14 (Russian)

Zuev PP (1973a) Kimberlite-like bodies of the Central Aldan district and their geological and structural features. Geol Geofiz 5:40–48 (Russian)

Zuev PP (1973b) On the petrology of kimberlite-like rocks of the Central Aldan district (Aldan
shield). Dokl Akad Nauk SSSR 212:205–208 (Russian)

Zveder LN, Shchukin VN (1960) On the nature of fractures in the Daldyn kimberlite field. Geol
Geofiz 6:132–134 (Russian)

Subject Index